BASIC STATISTICS

BASIC STATISTICS

Understanding Conventional Methods and Modern Insights

Rand R. Wilcox

2009

OXFORD
UNIVERSITY PRESS

Oxford University Press, Inc., publishes works that further
Oxford University's objective of excellence
in research, scholarship, and education.

Oxford New York
Auckland Cape Town Dar es Salaam Hong Kong Karachi
Kuala Lumpur Madrid Melbourne Mexico City Nairobi
New Delhi Shanghai Taipei Toronto

With offices in
Argentina Austria Brazil Chile Czech Republic France Greece
Guatemala Hungary Italy Japan Poland Portugal Singapore
South Korea Switzerland Thailand Turkey Ukraine Vietnam

Published by Oxford University Press, Inc.
198 Madison Avenue, New York, New York 10016

www.oup.com

Library of Congress Cataloging-in-Publication Data
Wilcox, Rand R.
Basic statistics : understanding conventional methods and
modern insights / Rand R. Wilcox.
p. cm.
Includes bibliographical references and index.
ISBN 978-0-19-531510-3
1. Mathematical statistics—Textbooks. I. Title.
QA276.12.W553 2009
519.5—dc22 2009007360

9 8 7 6 5 4 3 2 1

Printed in the United States of America
on acid-free paper

Preface

There are two main goals in this book. The first is to describe and illustrate basic statistical principles and concepts, typically covered in a one-semester course, in a simple and relatively concise manner. Technical and mathematical details are kept to a minimum. Throughout, examples from a wide range of situations are used to describe, motivate, and illustrate basic techniques. Various conceptual issues are discussed at length with the goal of providing a foundation for understanding not only what statistical methods tell us, but also what they do not tell us. That is, the goal is to provide a foundation for avoiding conclusions that are unreasonable based on the analysis that was done.

The second general goal is to explain basic principles and techniques in a manner that takes into account three major insights that have occurred during the last half-century. Currently, the standard approach to an introductory course is to ignore these insights and focus on methods that were developed prior to the year 1960. However, these insights have tremendous implications regarding basic principles and techniques, and so a simple description and explanation seems warranted. Put simply, when comparing groups of individuals, methods routinely taught in an introductory course appear to perform well over a fairly broad range of situations when the groups under study do not differ in any manner. But when groups differ, there are general conditions where they are highly unsatisfactory in terms of both detecting and describing any differences that might exist. In a similar manner, when studying how two or more variables are related, routinely taught methods perform well when no association exists. When there is an association, they might continue to perform well, but under general conditions, this is not the case. Currently, the typical introductory text ignores these insights or does not explain them sufficiently for the reader to understand and appreciate their practical significance. There are many modern methods aimed at correcting practical problems associated with classic techniques, most of which go well beyond the scope of this book. But a few of the simpler methods are covered with the goal of fostering modern technology. Although most modern methods cannot be covered here, this book takes the view that it is important to provide a foundation for understanding common misconceptions and weaknesses, associated with routinely used methods, which have been pointed out in literally hundreds of journal articles during the last half-century, but which are currently relatively unknown among most non-statisticians. Put another way, a major goal is to provide the student with a foundation for understanding and appreciating what modern technology has to offer.

The following helps illustrate the motivation for this book. Conventional wisdom has long held that with a sample of 40 or more observations, it can be assumed that observations are sampled from what is called a normal distribution. Most introductory books still make this claim, this view is consistent with studies done many years ago, and in fairness, there are conditions where adhering to this view is innocuous. But numerous journal articles make it clear that when working with means, under very general conditions, this view is not remotely true, a result that is related to the three major insights previously mentioned. Where did this erroneous view come from and what can be done about correcting any practical problems? Simple explanations are provided and each chapter ends with a section outlining where more advanced techniques can be found.

Also, there are many new advances beyond the three major insights that are important in an introductory course. Generally these advances have to do with the relative merits of methods designed to address commonly encountered problems. For example, many books suggest that histograms are useful in terms of detecting outliers, which are values that are unusually large or small relative to the bulk of the observations available. It is known, however, that histograms can be highly unsatisfactory relative to other techniques that might be used. Examples that illustrate this point are provided. As another example, a common and seemingly natural strategy is to test assumptions underlying standard methods in an attempt to justify their use. But many papers illustrate that this approach can be highly inadequate. Currently, all indications are that a better strategy is to replace classic techniques with methods that continue to perform well when standard assumptions are violated. Despite any advantages modern methods have, this is not to suggest that methods routinely taught and used have no practical value. Rather, the suggestion is that understanding the relative merits of methods is important given the goal of getting the most useful information possible from data.

When introducing students to basic statistical techniques, currently there is an unwritten rule that any major advances relevant to basic principles should not be discussed. One argument for this view, often heard by the author, is that students with little mathematical training are generally incapable of understanding modern insights and their relevance. For many years, I have covered the three major insights whenever I teach the undergraduate statistics course. I find that explaining these insights is no more difficult than any of the other topics routinely taught. What is difficult is explaining to students why modern advances and insights are not well known. Fortunately, there is a growing awareness that many methods developed prior to the year 1960 have serious practical problems under fairly general conditions. The hope is that this book will introduce basic principles in a manner that helps bridge the gap between routinely used methods and modern techniques.

Rand R. Wilcox
Los Angeles, California

Contents

Partial List of Symbols

α alpha: Probability of a Type I error

β beta: Probability of a Type II error

β_1 Slope of a regression line

β_0 Intercept of a regression line

δ delta: A measure of effect size

ϵ epsilon: The residual or error term in ANOVA and regression

θ theta: The population median or the odds ratio

μ mu: The population mean

μ_t The population trimmed mean

ν nu: Degrees of freedom

Ω omega: The odds ratio

ρ rho: The population correlation coefficient

σ sigma: The population standard deviation

ϕ phi: A measure of association

χ chi: χ^2 is a type of distribution

Δ delta: A measure of effect size

$\sum$ Summation

τ tau: Kendall's tau

BASIC STATISTICS

1

Introduction

At its simplest level, statistics involves the description and summary of events. How many home runs did Babe Ruth hit? What is the average rainfall in Seattle? But from a scientific point of view, it has come to mean much more. Broadly defined, it is the science, technology and art of extracting information from observational data, with an emphasis on solving real world problems. As Stigler (1986, p. 1) has so eloquently put it:

> Modern statistics provides a quantitative technology for empirical science; it is a logic and methodology for the measurement of uncertainty and for examination of the consequences of that uncertainty in the planning and interpretation of experimentation and observation.

The logic and associated technology behind modern statistical methods pervades all of the sciences, from astronomy and physics to psychology, business, manufacturing, sociology, economics, agriculture, education, and medicine—it affects your life.

To help elucidate the types of problems addressed in this book, consider an experiment aimed at investigating the effects of ozone on weight gain in rats (Doksum and Sievers, 1976). The experimental group consisted of 22 seventy-day-old rats kept in an ozone environment for 7 days. A control group of 23 rats, of the same age, was kept in an ozone-free environment. The results of this experiment are shown in table 1.1.

What, if anything, can we conclude from this experiment? A natural reaction is to compute the average weight gain for both groups. The averages turn out to be 11 for the ozone group and 22.4 for the control group. The average is higher for the control group suggesting that for the typical rat, weight gain will be less in an ozone environment. However, serious concerns come to mind upon a moment's reflection. Only 22 rats were kept in the ozone environment. What if 100 rats had been used or 1,000, or even a million? Would the average weight gain among a million rats differ substantially from 11, the average obtained in the experiment? Suppose ozone has no effect on weight gain. By chance, the average weight gain among rats in an ozone environment might differ from the average for rats in an ozone-free environment. How large of a difference between the means do we need before we can be reasonably certain that ozone affects weight gain? How do we judge whether the difference is large from a clinical point of view?

Table 1.1 Weight gain of rats in ozone experiment

Control:	41.0	38.4	24.4	25.9	21.9	18.3	13.1	27.3	28.5	−16.9
Ozone:	10.1	6.1	20.4	7.3	14.3	15.5	−9.9	6.8	28.2	17.9
Control:	26.0	17.4	21.8	15.4	27.4	19.2	22.4	17.7	26.0	29.4
Ozone:	−9.0	−12.9	14.0	6.6	12.1	15.7	39.9	−15.9	54.6	−14.7
Control:	21.4	26.6	22.7							
Ozone:	44.1	−9.0								

What about using the average to reflect the weight gain for the typical rat? Are there other methods for summarizing the data that might have practical value when characterizing the differences between the groups? The answers to these problems are nontrivial. The purpose of this book is to introduce the basic tools for answering these questions.

The mathematical foundations of the statistical methods described in this book were developed about two hundred years ago. Of particular importance was the work of Pierre-Simon Laplace (1749–1827) and Carl Friedrich Gauss (1777–1855). Approximately a century ago, major advances began to appear that dominate how researchers analyze data today. Especially important was the work of Karl Pearson (1857–1936) Jerzy Neyman (1894–1981), Egon Pearson (1895–1980), and Sir Ronald Fisher (1890–1962). During the 1950s, there was some evidence that the methods routinely used today serve us quite well in our attempts to understand data, but in the 1960s it became evident that serious practical problems needed attention. Indeed, since 1960, three major insights revealed conditions where methods routinely used today can be highly unsatisfactory. Although the many new tools for dealing with known problems go beyond the scope of this book, it is essential that a foundation be laid for appreciating modern advances and insights, and so one motivation for this book is to accomplish this goal.

This book does *not* describe the mathematical underpinnings of routinely used statistical techniques, but rather the concepts and principles that are used. Generally, the essence of statistical reasoning can be understood with little training in mathematics beyond basic high-school algebra. However, if you put enough simple pieces together, the picture can seem rather fuzzy and complex, and it is easy to lose track of where we are going when the individual pieces are being explained. Accordingly, it might help to provide a brief overview of what is covered in this book.

1.1 Samples versus populations

One key idea behind most statistical methods is the distinction between a sample of participants or objects versus a population. A *population* of participants or objects consists of all those participants or objects that are relevant in a particular study. In the weight-gain experiment with rats, there are millions of rats we could use if only we had the resources. To be concrete, suppose there are a billion rats and we want to know the average weight gain if all one billion were exposed to ozone. Then these one billion rats compose the population of rats we wish to study. The average gain for these rats is called the population mean. In a similar manner, there is an average weight gain for all the rats if they are raised in an ozone-free environment instead. This is the population mean for rats raised in an ozone-free environment. The obvious problem is that it is impractical to measure all one billion rats. In the experiment, only 22 rats were exposed to ozone. These 22 rats are an example of what is called a *sample*.

Definition A *sample* is any subset of the population of individuals or things under study.

Example 1. Trial of the Pyx

Shortly after the Norman Conquest, around the year 1100, there was already a need for methods that tell us how well a sample reflects a population of objects. The population of objects in this case consisted of coins produced on any given day. It was desired that the weight of each coin be close to some specified amount. As a check on the manufacturing process, a selection of each day's coins was reserved in a box ('the Pyx') for inspection. In modern terminology, the coins selected for inspection are an example of a sample, and the goal is to generalize to the population of coins, which in this case is all the coins produced on that day.

Three fundamental components of statistics

Statistical techniques consist of a wide range of goals, techniques and strategies. Three fundamental components worth stressing are:

1. *Design*, meaning the planning and carrying out of a study.
2. *Description*, which refers to methods for summarizing data.
3. *Inference*, which refers to making predictions or generalizations about a population of individuals or things based on a sample of observations available to us.

Design is a vast subject and only the most basic issues are discussed here. Imagine you want to study the effect of jogging on cholesterol levels. One possibility is to assign some participants to the experimental condition and another sample of participants to a control group. Another possibility is to measure the cholesterol levels of the participants available to you, have them run a mile every day for two weeks, then measure their cholesterol level again. In the first example, different participants are being compared under different circumstances, while in the other, the same participants are measured at different times. Which study is best in terms of determining how jogging affects cholesterol levels? This is a design issue.

The main focus of this book is not experimental design, but it is worthwhile mentioning the difference between the issues covered in this book versus a course on design. As a simple illustration, imagine you are interested in factors that affect health. In North America, where fat accounts for a third of the calories consumed, the death rate from heart disease is 20 times higher than in rural China where the typical diet is closer to 10% fat. What are we to make of this? Should we eliminate as much fat from our diet as possible? Are all fats bad? Could it be that some are beneficial? This purely descriptive study does not address these issues in an adequate manner. This is not to say that descriptive studies have no merit, only that resolving important issues can be difficult or impossible without good experimental design. For example, heart disease is relatively rare in Mediterranean countries where fat intake can approach 40% of calories. One distinguishing feature between the American diet and the Mediterranean diet is the type of fat consumed. So one possibility is that the amount of fat in a diet, without regard to the type of fat, might be a poor gauge of nutritional quality. Note, however, that in the observational study just described, nothing has been done to control other factors that might influence heart disease.

Sorting out what does and does not contribute to heart disease requires good experimental design. In the ozone experiment, attempts are made to control for factors that are related to weight gain (the age of the rats compared) and then manipulate the single factor that is of interest, namely the amount of ozone in the air. Here the goal is not so much to explain how best to design an experiment but rather to provide a description of methods used to summarize a population of individuals, as well as a sample of individuals, plus the methods used to generalize from the sample to the population. When describing and summarizing the typical American diet, we sample some Americans, determine how much fat they consume, and then use this to generalize to the population of all Americans. That is, we make inferences about all Americans based on the sample we examined. We then do the same for individuals who have a Mediterranean diet, and we make inferences about how the typical American diet compares to the typical Mediterranean diet.

Description refers to ways of summarizing data that provide useful information about the phenomenon under study. It includes methods for describing both the sample available to us and the entire population of participants if only they could be measured. The average is one of the most common ways of summarizing data. In the jogging experiment, you might be interested in how cholesterol is affected as the time spent running every day is increased. How should the association, if any, be described?

Inference includes methods for generalizing from the sample to the population. The average for all the participants in a study is called the *population mean* and typically represented by the Greek letter mu, μ. The average based on a sample of participants is called a *sample mean*. The hope is that the sample mean provides a good reflection of the population mean. In the ozone experiment, one issue is how well the sample mean estimates the population mean, the average weight-gain for all rats if they could be included in the experiment. That is, the goal is to make inferences about the population mean based on the sample mean.

1.2 Comments on teaching and learning statistics

It might help to comment on the goals of this book versus the general goal of teaching statistics. An obvious goal in an introductory course is to convey basic concepts and methods. A much broader goal is to make the student a master of statistical techniques. A single introductory book cannot achieve this latter goal, but it can provide the foundation for understanding the relative merits of frequently used techniques. There is now a vast array of statistical methods one might use to examine problems that are commonly encountered. To get the most out of data requires a good understanding of not only what a particular method tells us, but what it does not tell us as well. Perhaps the most common problem associated with the use of modern statistical methods is making interpretations that are not justified based on the technique used. Examples are given throughout this book.

Another fundamental goal in this book is to provide a glimpse of the many advances and insights that have occurred in recent years. For many years, most introductory statistics books have given the impression that all major advances ceased circa 1955. This is not remotely true. Indeed, major improvements have emerged, some of which are briefly indicated here.

1.3 Comments on software

As is probably evident, a key component to getting the most accurate and useful information from data is software. There are now several popular computer programs for analyzing data. Perhaps the most important thing to keep in mind is that the choice of software can be crucial, particularly when the goal is to apply new and improved methods developed during the last half century. Presumably no software package is best, based on all of the criteria that might be used to judge them, but the following comments might help.

Excellent software

The software R is one of the two best software packages available. Moreover, it is free and available at http://cran.R-project.org. All modern methods developed in recent years, as well as all classic techniques, are easily applied. One feature that makes R highly valuable from a research perspective is that a group of academics do an excellent job of constantly adding and updating routines aimed at applying modern techniques. A wide range of modern methods can be applied using the basic package. And many specialized methods are available via packages available at the R web site. A library of R functions especially designed for applying the newest methods for comparing groups and studying associations is available at www-rcf.usc.edu/~rwilcox/.[1] Although not the focus here, occasionally the name of some of these functions will be mentioned when illustrating some of the important features of modern methods. (Unless stated otherwise, whenever the name of an R function is supplied, it is a function that belongs to the two files Rallfunv1-v7 and Rallfunv2-v7, which can be downloaded from the site just mentioned.)

S-PLUS is another excellent software package. It is nearly identical to R and the basic commands are the same. One of the main differences is cost: S-PLUS can be very expensive. There are a few differences from R, but generally they are minor and of little importance when applying the methods covered in this book. (The R functions mentioned in this book are available as S-PLUS functions, which are stored in the files allfunv1-v7 and allfunv2-v7 and which can be downloaded in the same manner as the files Rallfunv1-v7 and Rallfunv2-v7.)

Very good software

SAS is another software package that provides power and excellent flexibility. Many modern methods can be applied, but a large number of the most recently developed techniques are not yet available via SAS. SAS code could be easily written by anyone reasonably familiar with SAS, and the company is fairly diligent about upgrading the

1. Details and illustrations of how this software is used can be found in Wilcox (2003, 2005).

routines in their package, but this has not been done as yet for some of the methods to be described.

Good software

Minitab is fairly simple to use and provides a reasonable degree of flexibility when analyzing data. All of the standard methods developed prior to the year 1960 are readily available. Many modern methods could be run in Minitab, but doing so is not straightforward. Like SAS, special Minitab code is needed and writing this code would take some effort. Moreover, certain modern methods that are readily applied with R cannot be easily done in Minitab even if an investigator was willing to write the appropriate code.

Unsatisfactory software

SPSS is certainly one of the most popular and frequently used software packages. Part of its appeal is ease of use. When handling complex data sets, it is one of the best packages available and it contains all of the classic methods for analyzing data. But in terms of providing access to the many new and improved methods for comparing groups and studying associations, which have appeared during the last half-century, it must be given a poor rating. An additional concern is that it has less flexibility than R and S-PLUS. That is, it is a relatively simple matter for statisticians to create specialized R and S-PLUS code that provides non-statisticians with easy access to modern methods. Some modern methods can be applied with SPSS, but often this task is difficult. However, SPSS 16 has added the ability to access R, which might increase its flexibility considerably. Also, zumastat.com has software that provides access to a large number of R functions aimed at applying the modern methods mentioned in this book plus many other methods covered in more advanced courses. (On the zumastat web page, click on robust statistics to get more information.)

The software EXCEL is relatively easy to use, it provides some flexibility, but generally modern methods are not readily applied. A recent review by McCullough and Wilson (2005) concludes that this software package is not maintained in an adequate manner. (For a more detailed description of some problems with this software, see Heiser, 2006.) Even if EXCEL functions were available for all modern methods that might be used, features noted by McCullough and Wilson suggest that EXCEL should not be used.

2

Numerical Summaries of Data

To help motivate this chapter, imagine a study done on the effects of a drug designed to lower cholesterol levels. The study begins by measuring the cholesterol level of 171 participants and then measuring each participant's cholesterol level after one month on the drug. Table 2.1 shows the change between the two measurements. The first entry is -23 indicating that the cholesterol level of this particular individual decreased by 23 units. Further imagine that a placebo is given to 176 participants resulting in the changes in cholesterol shown in table 2.2. Although we have information on the effect of the drug, there is the practical problem of conveying this information in a useful manner. Simply looking at the values, it is difficult determining how the experimental drug compares to the placebo. In general, how might we summarize the data in a manner that helps us judge the difference between the two drugs?

A basic strategy for dealing with the problem just described is to develop numerical quantities intended to provide useful information about the nature of the data. These numerical summaries of data are called *descriptive measures* or *descriptive statistics*, many of which have been proposed. Here the focus is on commonly used measures, and at the end of this chapter, a few alternative measures are described that have been found to have practical value in recent years. There are two types that play a particularly important role when trying to understand data: measures of location and measures of dispersion. Measures of location, also called *measures of central tendency*, are traditionally thought of as attempts to find a single numerical quantity that reflects the 'typical' observed value. But from a modern perspective, this description can be misleading and is too narrow in a sense that will be made clear later in this chapter. (A clarification of this point can found in section 2.2.) Roughly, measures of dispersion reflect how spread out the data happen to be. That is, they reflect the variability among the observed values.

2.1 Summation notation

Before continuing, some basic notation should be introduced. Arithmetic operations associated with statistical techniques can get quite involved and so a mathematical shorthand is typically used to make sure that there is no ambiguity about how the computations are to be performed. Generally, some letter is used is to represent whatever

Table 2.1 Changes in cholesterol level after one month on an experimental drug

−23 −11 −7 −13 4 −32 −20 −18 1 7 −32 −14 −18 6 10 −4 −15 −7
−21 −10 10 −20 −15 −10 −11 −10 −5 0 −13 −14 −6 9 −19 −10 −19 −11
5 −6 −17 −6 −15 6 −8 −17 −8 −16 2 −6 −14 −22 −11 −23 −6 −5
−12 −12 0 0 −3 −14 −34 −8 −19 −30 −17 −17 −1 −30 −31 −17 −16 −5
8 −23 −12 9 −33 4 −18 −34 −2 −28 −10 −8 −20 −8 19 −12 −11 0
−19 −12 −10 −20 −11 −2 −17 −24 −18 −18 −13 25 4 −13 −1 −7 −2
−22 −25 −19 −8 −17 −10 −27 −1 −6 −19 4 −16 −29 4 −8 −16
−16 1 −7 −31 −9 0 −4 −16 −5 −6 −14 −3 0 31 −10 −23
−14 −24 −11 −2 20 −5 −21 −1 −2 −3 −21 −5 −10 −12 0 −5
10 −26 −9 −10 16 −15 −26 1 −18 −19 −16 10 0 4 −9 −4

is being measured; the letter X is the most common choice. So in tables 2.1 and 2.2, X represents the change in cholesterol levels, but it could just as easily be used to represent how much weight is lost using a particular diet, how much money is earned using a particular investment strategy, or how often a particular surgical procedure is successful. The notation X_1 is used to indicate the first observation. In table 2.1, the first observed value is −21 and this is written as $X_1 = -23$. The next observation is −11, which is written as $X_2 = -11$, and the last observation is $X_{171} = -4$. In a similar manner, in table 2.2, $X_1 = 8$, $X_6 = 26$, and the last observation is $X_{177} = -19$. More generally, n is typically used to represent the total number of observations, and the observations themselves are represented by

$$X_1, X_2, \ldots, X_n.$$

So in table 2.1, $n = 171$ and in table 2.2, $n = 177$.

Summation notation is simply way of saying that a collection of numbers is to be added. In symbols, adding the numbers $X_1, X_2, \ldots, X_n$ is denoted by

$$\sum_{i=1}^{n} X_i = X_1 + X_2 + \cdots + X_n,$$

where $\sum$ is an upper case Greek sigma. The subscript i is the *index of summation* and the 1 and n that appear respectively below and above the symbol $\sum$ designate the range of the summation. So if X represents the changes in cholesterol levels in table 2.2,

$$\sum_{i=1}^{n} X_i = 8 + 7 + 2 \cdots = 22.$$

Table 2.2 Changes in cholesterol level after one month of taking a placebo

8 7 2 5 11 26 2 0 −10 6 −28 3 −14 2 −27 1 12 0
17 68 14 −16 10 10 30 −27 −35 6 −1 22 2 1 0 −11 −5 −36
10 4 7 15 −6 10 −8 −4 6 −2 −2 −1 10 34 39 4 15 −4
−7 1 −8 −4 −7 −3 −12 0 −17 −1 −17 7 −16 −1 15 20 1 −9
1 −3 −14 0 2 1 7 2 −17 −25 −7 −16 3 −1 −2 9 11 0
13 8 −20 0 −3 10 −1 −4 −9 −7 9 −7 9 −43 10 −17 −10 −18
11 −11 −22 0 11 11 10 6 −5 8 71 −11 −9 −1 12 0 −6 −1
−21 11 5 −3 24 −11 −36 −1 4 18 −8 −8 1 −1 3 0 6 3
−5 8 0 −4 −7 11 0 16 −1 −3 −11 −16 −14 −12 6 −5 21 −16
−11 6 −10 3 13 −5 13 5 −1 −1 −8 5 −9 18 −19

In most situations, the sum extends over all n observations, in which case it is customary to omit the index of summation. That is, simply use the notation

$$\sum X_i = X_1 + X_2 + \cdots + X_n.$$

Example 1

Imagine you work for a software company and you want to know, when customers call for help, how long it takes them to reach the appropriate department. To keep the illustration simple, imagine that you have data on five individuals and that their times (in minutes) are:

$$1.2, 2.2, 6.4, 3.8, 0.9.$$

Then

$$\sum_{i=2}^{4} X_i = 2.2 + 6.4 + 3.8 = 12.4$$

and

$$\sum X_i = 1.2 + 2.2 + 6.4 + 3.8 + 0.9 = 14.5.$$

Another common arithmetic operation consists of squaring each observed value and summing the results. This is written as

$$\sum X_i^2 = X_1^2 + X_2^2 + \cdots + X_n^2.$$

Note that this is not necessarily the same as adding all the values and squaring the results. This latter operation is denoted by

$$\left(\sum X_i\right)^2.$$

Example 2

For the data in example 1,

$$\sum X_i^2 = 1.2^2 + 2.2^2 + 6.4^2 + 3.8^2 + 0.9^2 = 62.49$$

and

$$\left(\sum X_i\right)^2 = (1.2 + 2.2 + 6.4 + 3.8 + 0.9)^2 = 14.5^2 = 210.25.$$

Let c be any constant. In some situations it helps to note that multiplying each value by c and adding the results is the same as first computing the sum and then multiplying by c. In symbols,

$$\sum cX_i = c \sum X_i.$$

Example 3

Consider again the data in example 1 and suppose we convert the observed values to seconds by multiplying each value by 60. Then the sum, using times in seconds, is

$$\sum 60X_i = 60 \sum X_i = 60 \times 14.5 = 870.$$

Another common operation is to subtract a constant from each observed value, square each difference, and add the results. In summation notation, this is written as

$$\sum(X_i - c)^2.$$

Example 4

For the data in example 1, suppose we want to subtract 2.9 from each value, square each of the results, and then sum these squared differences. So $c = 2.9$, and

$$\sum(X_i - c)^2 = (1.2 - 2.9)^2 + (2.2 - 2.9)^2 + \cdots + (0.9 - 2.9)^2 = 20.44.$$

One more summation rule should be noted. If we sum a constant c n times, we get nc. This is written as

$$\sum c = c + \cdots + c = nc.$$

Problems

1. Given that
   $X_1 = 1$ $\quad X_2 = 3$ $\quad X_3 = 0$
   $X_4 = -2$ $\quad X_5 = 4$ $\quad X_6 = -1$
   $X_7 = 5$ $\quad X_8 = 2$ $\quad X_9 = 10$
   Find
   (a) $\sum X_i$, (b) $\sum_{i=3}^{5} X_i$, (c) $\sum_{i=1}^{4} X_i^3$, (d) $(\sum X_i)^2$, (e) $\sum 3$, (f) $\sum(X_i - 7)$
   (g) $3\sum_{i=1}^{5} X_i - \sum_{i=6}^{9} X_i$, (h) $\sum 10X_i$, (i) $\sum_{i=2}^{6} iX_i$, (j) $\sum 6$
2. Express the following in summation notation. (a) $X_1 + \frac{X_2}{2} + \frac{X_3}{3} + \frac{X_4}{4}$,
   (b) $U_1 + U_2^2 + U_3^3 + U_4^4$, (c) $(Y_1 + Y_2 + Y_3)^4$
3. Show by numerical example that $\sum X_i^2$ is not necessarily equal to $(\sum X_i)^2$.

2.2 Measures of location

As previously noted, measures of location are often described as attempts to find a single numerical quantity that reflects the typical observed value. Literally hundreds of such measures have been proposed and studied. Two, called the sample mean and median, are easily computed and routinely used. But a good understanding of their relative merits will take some time to achieve.

The sample mean

The first measure of location, called the *sample mean*, is just the average of the values and is generally labeled $\bar{X}$. The notation $\bar{X}$ is read as X bar. In summation notation,

$$\bar{X} = \frac{1}{n}\sum X_i.$$

Example 1

A commercial trout farm wants to advertise and as part of their promotion plan they want to tell customers how much their typical trout weighs. To keep things simple for the moment, suppose they catch five trout having weights 1.1, 2.3, 1.7, 0.9 and 3.1 pounds. The trout farm does not want to report all five weights to the public but rather one number that conveys the typical weight among the five trout caught. For these five trout, a measure of the typical weight is the sample mean,

$$\bar{X} = \frac{1}{5}(1.1 + 2.3 + 1.7 + 0.9 + 3.1) = 1.82.$$

Example 2

You sample ten married couples and determine the number of children they have. The results are 0, 4, 3, 2, 2, 3, 2, 1, 0, 8. The sample mean is $\bar{X} = (0+4+3+2+2+3+2+1+0+8)/10 = 2.5$. Of course, nobody has 2.5 children. The intention is to provide a number that is centrally located among the 10 observations with the goal of conveying what is typical. The sample mean is frequently used for this purpose, in part because it greatly simplifies technical issues related to methods covered in subsequent chapters. In some cases, the sample mean suffices as a summary of data, but it is important to keep in mind that for various reasons, it can highly unsatisfactory. One of these reasons is illustrated next (and other practical concerns are described in subsequent chapters).

Example 3

Imagine an investment firm is trying to recruit you. As a lure, they tell you that among the 11 individuals currently working at the company, the average salary, in thousands of dollars, is 88.7. However, on closer inspection, you find that the salaries are

$$30, 25, 32, 28, 35, 31, 30, 36, 29, 200, 500,$$

where the two largest salaries correspond to the vice president and president, respectively. The average is 88.7, as claimed, but an argument can be made that this is hardly typical because the salaries of the president and vice president result in a sample mean that gives a distorted sense of what is typical. Note that the sample mean is considerably larger than 9 of the 11 salaries.

Example 4

Pedersen et al. (1998) conducted a study, a portion of which dealt with the sexual attitudes of undergraduate students. Among other things, the students were asked how many sexual partners they desired over the next 30 years. The responses of 105 males are shown in table 2.3. The sample mean is $\bar{X} = 64.9$. But this is hardly typical because 102 of the 105 males gave a response less than the sample mean.

Outliers are values that are unusually large or small. In the last example, one participant responded that he wanted 6,000 sexual partners over the

Table 2.3 Responses by males in the sexual attitude study

6 1 1 3 1 1 1 1 1 1 6 1 1 1 4
5 3 9 1 1 1 5 12 10 4 2 1 1 4 45
8 5 0 1 150 13 19 2 1 18 3 1 3 1 11
1 2 1 1 1 12 1 1 2 6 1 1 1 1 4
1 150 6 40 4 30 10 1 1 0 3 4 1 4 7
1 10 0 19 1 9 1 1 1 5 0 1 1 15 4
1 4 1 1 11 1 1 30 12 6000 1 0 1 1 15

next 30 years, which is clearly unusual compared to the other 104 students. Also, two gave the response 150, which again is relatively unusual. An important point made by these last two examples is that the sample mean can be highly influenced by one or more outliers. That is, care must be exercised when using the sample mean because its value can be highly atypical and therefore potentially misleading. Also, outliers are not necessarily mistakes or inaccurate reflections of what was intended. For example, it might seem that nobody would seriously want 6,000 sexual partners, but a documentary on the outbreak of AIDS made it clear that such individuals do exist. Moreover, similar studies conducted within a wide range of countries confirm that generally a small proportion of individuals will give a relatively extreme response.

The median

Another important measure of location is called the *sample median*. The basic idea is easily described using the example based on the weight of trout. The observed weights were

$$1.1, 2.3, 1.7, 0.9, 3.1.$$

Putting the values in ascending order yields

$$0.9, 1.1, \mathbf{1.7}, 2.3, 3.1.$$

Notice that the value 1.7 divides the observations in the middle in the sense that half of the remaining observations are less than 1.7 and half are larger. If instead the observations are

$$0.8, 4.5, 1.2, 1.3, 3.1, 2.7, 2.6, 2.7, 1.8,$$

we can again find a middle value by putting the observations in order yielding

$$0.8, 1.2, 1.3, 1.8, \mathbf{2.6}, 2.7, 2.7, 3.1, 4.5.$$

Then 2.6 is a middle value in the sense that half of the observations are less than 2.6 and half are larger. This middle value is an example of what is called a *sample median*.

Notice that there are an odd number of observations in the last two illustrations; the last illustration has $n = 9$. If instead we have an even number of observations, there is no middle value, in which case the most common strategy is to average the two middle values to get the so-called sample median. For the last illustration, suppose we eliminate the value 1.2, so now $n = 8$ and the observations, written in ascending order, are

$$0.8, 1.3, 1.8, \mathbf{2.6}, \mathbf{2.7}, 2.7, 3.1, 4.5.$$

The sample median in this case is taken to be the average of 2.6 and 2.7, namely $(2.6+2.7)/2 = 2.65$. In general, with n odd, the median is a value in your sample, but with n even this is not necessarily the case.

A more formal description of the sample median helps illustrate some commonly used notation. Recall that the notation $X_1, \ldots, X_n$ is typically used to represent the observations associated with n individuals or objects. Consider again the trout example where $n = 5$ and the observations are $X_1 = 1.1$, $X_2 = 2.3$, $X_3 = 1.7$, $X_4 = 0.9$ and $X_5 = 3.1$ pounds. That is, the first trout that is caught has weight 1.1 pounds, the second has weight 2.3 pounds, and so on. The notation $X_{(1)}$ is used to indicate the smallest observation. In the illustration, the smallest of the five observations is 0.9, so $X_{(1)} = 0.9$. The smallest of the remaining four observations is 1.1, and this is written as $X_{(2)} = 1.1$. The smallest of the remaining three observations is 1.7, so $X_{(3)} = 1.7$, the largest of the five values is 3.1, and this is written as $X_{(5)}$. More generally,

$$X_{(1)} \leq X_{(2)} \leq X_{(3)} \leq \cdots \leq X_{(n)}$$

is the notation used to indicate that n values are to be put in ascending order.

The *sample median* is computed in one of two ways:

1. If the number of observations, n, is odd, compute $m = (n+1)/2$. Then the sample median is

$$M = X_{(m)},$$

the mth value after the observations are put in order.

2. If the number of observations, n, is even, compute $m = n/2$. Then the sample median is

$$M = (X_{(m)} + X_{(m+1)})/2,$$

the average of the mth and (m+1)th observations after putting the observed values in ascending order.

Example 5

Seven individuals are given a test that measures depression. The observed scores are

$$34, 29, 55, 45, 21, 32, 39.$$

Because the number of observations is $n = 7$, which is odd, $m = (7+1)/2 = 4$. Putting the observations in order yields

$$21, 29, 32, \mathbf{34}, 39, 45, 55.$$

The fourth observation is $X_{(4)} = 34$, so the sample median is $M = 34$.

Example 6

We repeat the last example, only with six test scores

$$29, 55, 45, 21, 32, 39.$$

Because the number of observations is $n = 6$, which is even, $m = 6/2 = 3$. Putting the observations in order yields

$$21, 29, \mathbf{32}, \mathbf{39}, 45, 55.$$

The third and fourth observations are $X_{(3)} = 32$ and $X_{(4)} = 39$, so the sample median is $M = (32 + 39)/2 = 35.5$.

Example 7

Consider again the data in example 3 dealing with salaries. We saw that the sample mean is 88.7. In contrast, the sample median is $M = 31$, providing a substantially different impression of the typical salary earned. This illustrates that the sample median is relatively insensitive to outliers, for the simple reason that the smallest and largest values are trimmed away when it is computed. For this reason, the median is called a *resistant measure of location*. The sample mean is an example of a measure of location that is not resistant to outliers.

Example 8

As previously noted, the sample mean for the sexual attitude data in table 2.3 is $\bar{X} = 64.9$. But the median is $M = 1$, which provides a substantially different perspective on what is typical.

With the sample mean and median in hand, we can now be a bit more formal and precise about what is meant by a measure of location.

Definition A summary of data, based on the observations $X_1, \ldots, X_n$, is called a *measure of location* if it satisfies two properties. First, its value must lie somewhere between the smallest and largest values observed. In symbols, the measure of location must have a value between $X_{(1)}$ and $X_{(n)}$, inclusive. Second, if all observations are multiplied by some constant c, then the measure of location is multiplied by c as well.[1]

Example 9

You measure the height, in feet, of ten women yielding the values 5.2, 5.9, 6.0, 5.11, 5.0, 5.5, 5.6, 5.7, 5.2, 5.8. The sample mean is $\bar{X} = 5.501$. Notice that the mean cannot be less than the smallest value and it cannot be greater than the largest value. That is, it satisfies the first criterion for being a measure of location. We could get the mean in inches by multiplying each value by 12 and recomputing the average, but it is easier to simply multiply the mean by 12 yielding 66.012. Similarly, the median is 5.55 in feet, and in inches it is easily verified that the median is $12 \times 5.55 = 66.6$. More generally, if M is the median, and if each value is multiplied by some number c, the median becomes cM. This illustrates that both the mean and median satisfy the second condition in the definition of a measure location.

The practical point being made here is that when a statistician refers to a measure of location, this does not necessarily imply that this measure reflects what is typical. We have already seen that the sample mean can be very atypical, yet it is generally referred to as a measure of location.

1. Readers interested in more mathematical details about the definition of a measure of location are referred to Staudte and Sheather (1990).

The sample mean versus the sample median

How do we choose between the mean and median? It might seem that because the median is resistant to outliers and the mean is not, use the median. But the issue is not this simple. Indeed, for various reasons outlined later in this book, both the mean and median can be highly unsatisfactory. What is needed is a good understanding of their relative merits, which includes issues covered in subsequent chapters. To complicate matters, even when the mean and median have identical values, it will be seen that for purposes beyond merely describing the data, the choice between these two measures of location can be crucial. It is also noted that although the median can better reflect what is typical, in some situations its resistance to outliers can be undesirable.

Example 10

Imagine someone invests \$200,000 and reports that the median amount earned per year, over a 10-year-period, is \$100,000. This sounds great, but now imagine that the earnings for each year are: \$100,000, \$200,000, \$200,000, \$200,000, \$200,000, \$200,000, \$200,000, \$300,000, \$300,000, \$-1,900,000. So at the end of 10 years this individual has earned nothing and in fact lost the \$200,000 initial investment. (The sample mean is 0.) Certainly the long-term total amount earned is relevant in which case the sample mean provides a useful summary of the investment strategy that was followed.

Quartiles

As already explained and illustrated, the sample median divides the data into two parts: the lower half and the upper half after putting the observations in ascending order. *Quartiles* are measures of location aimed at dividing data into four parts. This is done with two additional measures of location called the lower and upper quartiles. (The median is sometimes called the middle quartile.) Roughly, the *lower quartile* is the median of the smaller half of the data. And the *upper quartile* is the median of the upper half. So it will be approximately the case that a fourth of the data lies below the lower quartile, a fourth will lie between the lower quartile and the median, a fourth will lie between the median and the upper quartile, and a fourth will lie above the upper quartile.

There are, in fact, many suggestions about how the lower and upper quartiles should be computed. Again let $X_{(1)} \leq \cdots \leq X_{(n)}$ denote the observations written in ascending order. A simple approach is to take the lower quartile to be $X_{(j)}$, where $j = n/4$. If $n = 16$, for example, then $j = 4$ and a fourth of the values will be less than or equal to $X_{(4)}$, and using $X_{(4)}$ is consistent with how the lower quartile is defined. But when $n = 10$, this simple approach is unsatisfactory. Should we use $j = 10/4$ rounded down to the the value 2, or should we use j rounded up to the value 3? Here we deal with this issue using a method that is relatively simple and which has been found to be well suited for another problem considered later in this chapter. The method is based on what are called the *ideal fourths*. To explain, let j be the integer portion of $(n/4) + (5/12)$, meaning that j is $(n/4) + (5/12)$ rounded down to the nearest integer, and let

$$h = \frac{n}{4} + \frac{5}{12} - j.$$

The lower quartile is taken to be

$$q_1 = (1-h)X_{(j)} + hX_{(j+1)}. \tag{2.1}$$

Letting $k = n - j + 1$, the upper quartile is

$$q_2 = (1-h)X_{(k)} + hX_{(k-1)}. \tag{2.2}$$

Example 10

Consider the values

$$-29.6, -20.9, -19.7, -15.4, -12.3, -8.0, -4.3, 0.8, 2.0, 6.2, 11.2, 25.0.$$

There are twelve values, so $n = 12$, and

$$\frac{n}{4} + \frac{5}{12} = 3.41667.$$

Rounding this last quantity down to the nearest integer gives $j = 3$. That is, j is just the number to the left of the decimal. Also, $h = 3.416667 - 3 = .41667$. That is, h is the decimal portion of 3.41667. Because $X_{(3)} = -19.7$ and $X_{(4)} = -15.4$, the lower quartile is

$$q_1 = (1 - .41667)(-19.7) + .41667(-15.4) = -17.9.$$

In a similar manner, the upper quartile is

$$q_2 = (1 - .41667)(6.2) + .41667(2) = 4.45.$$

An important feature of the lower quartile is that it is insensitive to the smallest values among the data under study. In modern terminology, it is resistant to outliers. In the last example, the smallest value is -29.6. If the value -29.6 is lowered to -100, or even $-1{,}000{,}000$, the lower quartile does not change and the upper quartile is unchanged as well. In a similar manner, the upper quartile is resistant to outliers as well. (This property will be exploited in section 2.4.)

Five number summary of data

The term *five number summary* refers to five numbers used to characterize data: (1) the lowest observed value, (2) the lower quartile, (3) the median, (4) the upper quartile, and (5) the largest observed value. (Software packages typically have a function that computes all five values.)

Problems

4. Find the mean and median of the following sets of numbers. (a) −1, 03, 0, 2, −5. (b) 2, 2, 3, 10, 100, 1,000.
5. The final exam scores for 15 students are 73, 74, 92, 98, 100, 72, 74, 85, 76, 94, 89, 73, 76, 99. Compute the mean and median.
6. The average of 23 numbers is 14.7. What is the sum of these numbers?

7. Consider the ten values 3, 6, 8, 12, 23, 26, 37, 42, 49, 63. The mean is $\bar{X} = 26.9$. (a) What is the value of the mean if the largest value, 63, is increased to 100? (b) What is the mean if 63 is increased to 1,000? (c) What is the mean if 63 is increased to 10,000?

8. Repeat the previous problem, only compute the median instead.

9. In general, how many values must be altered to make the sample mean arbitrarily large?

10. In general, approximately how many values must be altered to make the sample median arbitrarily large?

11. For the values 0, 23, −1, 12, −10, −7, 1, −19, −6, 12, 1, −3, compute the lower and upper quartiles (the ideal fourths).

12. For the values −1, −10, 2, 2, −7, −2, 3, 3, −6, 12, −1, −12, −6, 8, 6, compute the lower and upper quartiles (the ideal fourths).

13. Approximately how many values must be altered to make q_2 arbitrarily large?

14. Argue that the smallest observed value, $X_{(1)}$, as well as the the lower and upper quartiles, satisfy the definition of a measure of location.

2.3 Measures of variation

Often, measures of location are of particular interest. But measures of variation play a central role as well. Indeed, it is variation among responses that motivates many of the statistical methods covered in this book.

For example, imagine that a new diet for losing weight is under investigation. Of course, some individuals will lose more weight than others, and conceivably, some might actually gain weight instead. How might we take this variation into account when trying to assess the efficacy of this new diet? When a new drug is being researched, the drug might have no detrimental effect for some patients, but it might cause liver damage in others. What must be done to establish that the severity of liver damage is small? When asked whether they approve of how a political leader is performing, some will say they approve and others will give the opposite response. How can we take this variability into account when trying to assess the proportion of individuals who approve? The first step toward answering these questions is to introduce measures of variation, which play a central role when summarizing data. (The manner in which these measures are used to address the problems just described will be covered in subsequent chapters.)

The range

The *range* is just the difference between the largest and smallest observations. In symbols, it is $X_{(n)} - X_{(1)}$. In table 2.1, the largest value is 31, the smallest is −34, so the range is $31 - (-34) = 65$. Although the range provides some useful information about the data, relative to other measures that might be used, it plays a minor role at best. One reason has to do with technical issues that are difficult to describe at this point.

The variance and standard deviation

Another approach to measuring variation, one that plays a central role in applied work, is the sample variance. The basic idea is to measure the typical distance observations have from the mean. Imagine we have n numbers labeled $X_1, \ldots, X_n$. *Deviation scores* are just the difference between an observation and the sample mean. For example, the deviation score for the first observation, X_1, is $X_1 - \bar{X}$. In a similar manner, the deviation score for the second observation is $X_2 - \bar{X}$.

Example 1

For various reasons, a high fiber diet is thought to promote good health. Among cereals regarded to have high fiber, is there much variation in the actual amount of fiber contained in one cup? For 11 such cereals, the amount of fiber (in grams), written in ascending order, is

$$7.5, 8.0, 8.0, 8.5, 9.0, 11.0, 19.5, 19.5, 28.5, 31.0, 36.0.$$

The sample mean is $\bar{X} = 17$, so the deviation scores are

$$-9.5, -9.0, -9.0, -8.5, -8.0, -6.0, 2.5, 2.5, 11.5, 14.0, 19.0.$$

Deviation scores reflect how far each observation is from the mean, but often it is convenient and desirable to find a single numerical quantity that summarizes the amount of variation in our data. An initial suggestion might be to simply average the deviation scores. That is, we might use

$$\frac{1}{n}\sum(X_i - \bar{X}).$$

But using the rules for summation already described, it can be seen that this average difference is always zero, so this approach is unsatisfactory. Another possibility is to use the average of the absolute deviation scores:

$$\frac{1}{n}\sum|X_i - \bar{X}|.$$

This is reasonable, but it makes certain theoretical developments difficult. It turns out that using the squared differences instead greatly reduces certain mathematical problems related to methods covered in subsequent chapters. That is, use what is called the *sample variance*, which is

$$s^2 = \frac{1}{n-1}\sum(X_i - \bar{X})^2.$$

In other words, use the average squared difference from the mean. The sample *standard deviation* is the (positive) square root of the variance, s.

Notice that when computing the sample mean, we divide by n, the number of observations, but when computing the sample variance, s^2, we divide by $n - 1$. When first encountered, this usually seems strange, but it is too soon to explain why this is done. We will return to this issue in chapter 5.

Example 2

Imagine you sample 10 adults ($n = 10$), ask each to rate the performance of the president on a 10-point scale, and that their responses are:

$$3, 9, 10, 4, 7, 8, 9, 5, 7, 8.$$

The sample mean is $\bar{X} = 7$, $\sum(X_i - \bar{X})^2 = 48$, so the sample variance is $s^2 = 48/9 = 5.33$. Consequently, the standard deviation is $s = \sqrt{5.33} = 2.31$.

Another way to summarize the calculations is as follows.

	i	X_i	$X_i - \bar{X}$	$(X_i - \bar{X})^2$
	1	3	−4	16
	2	9	2	4
	3	10	3	9
	4	4	−3	9
	5	7	0	0
	6	8	1	1
	7	9	2	4
	8	5	−2	4
	9	7	0	0
	10	8	1	1
$\sum$			0	48

The sum of the observations in the last column is $\sum(X_i - \bar{X})^2 = 48$. So again, $s^2 = 48/9 = 5.33$.

The interpretation and practical utility of the sample variance, s^2, is unclear at this point. For now, the main message is that for some purposes it is very useful, as will be seen. But simultaneously, there are variety of situations where it can highly unsatisfactory. What is needed is a basic understanding of when it performs well, and when and why it can yield highly misleading results. One of the main reasons it can be unsatisfactory is its sensitivity to outliers.

Example 3

Consider the 10 values 50, 50, 50, 50, 50, 50, 50, 50, 50, 50. As is evident, the sample mean is $\bar{X} = 50$, and because all values are equal to the sample mean, $s^2 = 0$. Suppose we decrease the first value to 45 and increase the last to 55. Now $s^2 = 5.56$. If we decrease the first value to 20 and increase the last to 80, $s^2 = 200$. The point is that the sample variance can be highly influenced by unusually large or small values, even when the bulk of the values are tightly clustered together. Put another way, the sample variance can be small only when *all* of the values are tightly clustered together. If even a single value is unusually large or small, the sample variance will tend to be large, regardless of how bunched together the other values might be. This property can wreak havoc on methods routinely used to analyze data, as will be seen. Fortunately, many new methods have been derived that deal effectively with this problem.

The interquartile range

For some purposes, it is important to measure the variability of the centrally located values. If, for example, we put the observations in ascending order, how much variability is there among the central half of the data? The last example illustrated that the sample variance can be unsatisfactory in this regard. An alternative approach, which has practical importance, is the *interquartile range*, which is just $q_2 - q_1$, the difference between the upper and lower quartiles.

Notice that the interquartile range is insensitive to the more extreme values under study. As previously noted, the upper and lower quartiles are resistant to outliers, which means that the most extreme values do not affect the values of q_1 and q_2. Consequently, the interquartile range is resistant to outliers as well.

Example 4

Consider again the 10 values 50, 50, 50, 50, 50, 50, 50, 50, 50, 50. The interquartile range is zero. If we decrease the first value to 20 and increase the last to 80, the interquartile range is still zero because it measure the variability of the central half of the data, while ignoring the upper and lower fourth of the observations. Indeed, no matter how small we make the first value, and no matter how much we increase the last value, the interquartile range remains zero.

Problems

15. The height of 10 plants is measured in inches and found to be 12, 6, 15, 3, 12, 6, 21, 15, 18 and 12. Verify that $\sum(X_i - \bar{X}) = 0$.

16. For the data in the previous problem, compute the range, variance and standard deviation.

17. Use the rules of summation notation to show that it is always the case that $\sum(X_i - \bar{X}) = 0$.

18. Seven different thermometers were used to measure the temperature of a substance. The readings in degrees Celsius are −4.10, −4.13, −5.09, −4.08, −4.10, −4.09 and −4.12. Find the variance and standard deviation.

19. A weightlifter's maximum bench press (in pounds) in each of six successive weeks was 280, 295, 275, 305, 300, 290. Find the standard deviation.

2.4 Detecting outliers

The detection of outliers is important for a variety of reasons. One rather mundane reason is that they can help identify erroneously recorded results. We have already seen that even a single outlier can grossly affect the sample mean and variance, and of course we do not want a typing error to substantially alter or color our perceptions of the data. Such errors seem to be rampant in applied work, and the subsequent cost of such errors can be enormous (De Veaux and Hand, 2005). So it can be prudent to check for outliers, and if any are found, make sure they are valid.

But even if data are recorded accurately, it cannot be stressed too strongly that modern outlier detection techniques suggest that outliers are more the rule rather than the exception. That is, unusually small or large values occur naturally in a wide range of situations. Interestingly, in 1960, the renowned statistician John Tukey (1915–2000) predicted that in general we should expect outliers. What is fascinating about his prediction is that it was made before good outlier detection techniques were available.

A simple approach to detecting outliers is to merely look at the data. And another possibility is to inspect graphs of the data described in chapter 3. But for various purposes (to be described), these two approaches are unsatisfactory. What is needed are outlier detection techniques that have certain properties, the nature of which, and why they are important, is impossible to appreciate at this point. But one basic goal is easy to understand. A fundamental requirement of any outlier detection technique is that it does not suffer from what is called masking. An outlier detection technique is said to suffer from *masking* if the very presence of outliers causes them to be missed.

A classic outlier detection method

A classic outlier detection technique illustrates the problem of masking. This classic technique declares the value X an outlier if

$$\frac{|X - \bar{X}|}{s} \geq 2. \tag{2.3}$$

(The value 2 in this last equation is motivated by results covered in chapter 4.)

Example 1

Consider the values

$$2, 2, 2, 2, 2, 3, 3, 3, 3, 3, 4, 4, 4, 4, 4, 1{,}000.$$

The sample mean is $\bar{X} = 65.94$, the sample standard deviation is $s = 249.1$,

$$\frac{|1{,}000 - 65.94|}{249.1} = 3.75,$$

3.75 is greater than 2, so the value 1,000 is declared an outlier. In this particular case, the classic outlier detection method is performing in a reasonable manner; it identifies what is surely an usual value.

Example 2

Now consider the values

$$2, 2, 3, 3, 3, 4, 4, 4, 100{,}000, 100{,}000.$$

The sample mean is $\bar{X} = 20{,}002.5$, the sample standard deviation is $s = 42{,}162.38$,

$$\frac{|100{,}000 - 20{,}002.5|}{42{,}162.38} = 1.897,$$

and so the classic method would not declare the value 100,000 an outlier even though certainly it is highly unusual relative to the other eight values. The problem is that both the sample mean and the sample standard deviation are sensitive to outliers. That is, the classic method for detecting outliers suffers from masking. It is left as an exercise to show that even if the two values 100,000 in this example are increased to 10,000,000, the value 10,000,000 is not declared an outlier.

In some cases the classic outlier detection rule will detect the largest outlier but miss other values that are clearly unusual. Consider the sexual attitude data in table 2.3. It is evident that the response 6,000 is unusually large. But even the response 150 seems very large relative to the majority of values listed, yet the classic rule does not flag it as an outlier.

The boxplot rule

One of the earliest improvements on the classic outlier detection rule is called the *boxplot rule*. It is based on the fundamental strategy of avoiding masking by replacing the mean and standard deviation with measures of location and dispersion that are relatively insensitive to outliers. In particular, the *boxplot rule* declares the value X an outlier if

$$X < q_1 - 1.5(q_2 - q_1) \tag{2.4}$$

or

$$X > q_2 + 1.5(q_2 - q_1). \tag{2.5}$$

So the rule is based on the lower and upper quartiles, as well as the interquartile range, which provide resistance to outliers.

Example 3

Consider the values

$$1, 2, 3, 4, 5, 6, 7, 8, 9, 10, 11, 12, 13, 14, 100, 500.$$

A little arithmetic shows that the lower quartile is $q_1 = 4.417$, the upper quartile is $q_2 = 12.583$, so $q_2 + 1.5(q_2 - q_1) = 12.583 + 1.5(12.583 - 4.417) = 24.83$. That is, any value greater than 24.83 is declared an outlier. In particular, the values 100 and 500 are labeled outliers.

Example 4

For the sexual attitude data in table 2.3, the classic outlier detection rule declares only one value to be an outlier: the largest response, 6,000. In contrast, the boxplot rule labels all values 15 and larger as outliers. So of the 105 responses, the classic outlier detection rule finds only one outlier, and the boxplot rule finds 12.

Problems

20. For the values

$$20, 121, 132, 123, 145, 151, 119, 133, 134, 130,$$

use the classic outlier detection rule to determine whether any outliers exist.

21. Apply the boxplot rule for outliers to the values in the preceding problem.
22. Consider the values

$$0, 121, 132, 123, 145, 151, 119, 133, 134, 130, 250.$$

Are the values 0 and 250 declared outliers using the classic outlier detection rule?

23. Verify that for the data in the previous problem, the boxplot rule declares the values 0 and 250 outliers.
24. Consider the values

$$20, 121, 132, 123, 145, 151, 119, 133, 134, 240, 250.$$

Verify that no outliers are found using the classic outlier detection rule.

25. Verify that for the data in the previous problem, the boxplot rule declares the values 20, 240, and 250 outliers.
26. What do the last three problems suggest about the boxplot rule versus the classic rule for detecting outliers?

2.5 Some modern advances and insights

During the last half-century, and particularly during the last twenty years, there have been major advances and insights relevant to the most basic methods covered in an introductory statistics course. Most of these advances cannot be covered here, but it is very important to at least alert students to some of the more important advances and insights and to provide a glimpse of why more modern techniques have practical value. The material covered here will help achieve this goal.

Means, medians and trimming

The *mean* and *median* are the two best-known measures of location, with the mean being used in a large proportion of applied investigations. There are circumstances where using a mean gives satisfactory results. Indeed, there are conditions where it is optimal (versus any other measure of location that might be used.) But recent advances and insights have made it clear that both the mean and median can be highly unsatisfactory for a wide range of practical situations. Many new methods have been developed for dealing with known problems, some of which are based in part on using measures of location other than the mean and median. One of the simpler alternatives is introduced here.

The sample median is an example of what is called a *trimmed mean*; it trims all but one or two values. Although there are circumstances where this extreme amount of trimming can be beneficial, for various reasons covered in subsequent chapters, this extreme amount of trimming can be detrimental. The sample mean represents the other extreme: zero trimming. We have already seen that this can result in a measure of location that is a rather poor reflection of what is a typical observation. But even when it provides a good indication of the typical value, many basic methods based on the mean suffer from other fundamental concerns yet to be described. One way of reducing these problems is to use a compromise amount of trimming. That is, trim some values, but not as many as done by the median. No specific amount of trimming is always best, but for various reasons, 20% trimming is often a good choice. This means that the smallest 20%, as well as the largest 20%, are trimmed and the average of the remaining data is computed. In symbols, first compute $.2n$, round down to the nearest integer, call this result g, in which case the 20% trimmed mean is given by

$$\bar{X}_t = \frac{1}{n-2g}(X_{(g+1)} + \cdots + X_{(n-g)}). \tag{2.6}$$

Example 1

Consider the values

$$46, 12, 33, 15, 29, 19, 4, 24, 11, 31, 38, 69, 10.$$

Putting these values in ascending order yields,

$$4, 10, 11, 12, 15, 19, 24, 29, 31, 33, 38, 46, 69.$$

The number of observations is $n = 13$, $0.2(n) = 0.2(13) = 2.6$, and rounding this down to the nearest integer yields $g = 2$. That is, trim the two smallest values, 4 and 10, trim the two largest values, 46 and 69, and average the numbers that remain yielding

$$\bar{X}_t = \frac{1}{9}(11 + 12 + 15 + 19 + 24 + 29 + 31 + 33 + 38) = 23.56.$$

Example 2

Imagine a figure skating contest that uses nine judges who rate a skater on a six-point scale. Suppose the nine ratings are

$$5.1, 5.3, 5.3, 5.5, 5.0, 5.1, 5.4, 4.2, 5.2.$$

A natural concern is that some raters might not be fair under certain circumstances, or they might provide a poor reflection of how most raters would judge the skater, which in turn might make a difference in a competition. From a statistical point of view, we do not want an unusual rating to overly influence our measure of the typical rating a skater would receive. For the data at hand, the sample mean is 5.1, but notice that the rating 4.2 is unusually small compared to the remaining eight. To guard against unusually high or low ratings, it is common in skating competitions to throw out the highest and lowest scores and average those that remain. Here, $n = 9$, $0.2n = 1.8$, so $g = 1$. That is, a 20% trimmed mean corresponds to throwing out the lowest and highest scores and averaging the ratings that remain, yielding $\bar{X}_t = 5.2$.

Other measures of location

Yet another approach when measuring location is to check for outliers, remove any that are found, and then average the remaining values. There are, in fact, several variations of this strategy. There are circumstances where this approach has practical value, but the process of removing outliers creates certain technical problems that require advanced techniques that go beyond the scope of this book.[2] Consequently, this approach to measuring location is not discussed further.

Winsorized data and the winsorized variance

When using a trimmed mean, certain types of analyses, to be covered later, are not done in an intuitively obvious manner based on standard training. To illustrate how technically correct methods are applied, we will need to know how to Winsorize data and how to compute the Winsorized variance.

The process of Winsorizing data by 20% is related to 20% trimming. When we compute a 20% trimmed mean, we compute g as previously described, remove the g smallest and largest observations, and average the remaining values. Winsorizing the data by 20% means that the g smallest values are not trimmed, but rather, they are set equal to the smallest value not trimmed. Similarly, the g largest values are set equal to the largest value not trimmed.

Example 3

Suppose the reaction times of individuals are measured yielding

$$2, 3, 4, 5, 6, 7, 8, 9, 10, 50.$$

There are $n = 10$ values, $0.2(10) = 2$, so $g = 2$. Here, 20% Winsorizing of the data means that the two smallest values are set equal to 4. Simultaneously the two largest observations, 10 and 50, are set equal to 9, the largest value not trimmed. That is, 20% Winsorizing of the data yields

$$4, 4, 4, 5, 6, 7, 8, 9, 9, 9.$$

In symbols, the observations $X_1, \ldots, X_n$ are Winsorized by first putting the observations in order yielding $X_{(1)} \leq X_{(2)} \leq \cdots \leq X_{(n)}$. Then the g smallest observations are replaced by $X_{(g+1)}$, and the g largest observations are replaced by $X_{(n-g)}$.

Example 4

To Winsorize the values

$$10, 8, 22, 35, 42, 2, 9, 18, 27, 1, 16, 29$$

using 20% Winsorization, first note that there are $n = 12$ observations, $.2 \times 12 = 2.4$, and rounding down gives $g = 2$. Putting the values in order yields

$$1, 2, 8, 9, 10, 16, 18, 22, 27, 29, 35, 42.$$

Then the two smallest values are replaced by $X_{(g+1)} = X_{(3)} = 8$, the two largest values are replaced by $X_{(n-g)} = X_{(10)} = 29$, and the resulting Winsorized values are

$$8, 8, 8, 9, 10, 16, 18, 22, 27, 29, 29, 29.$$

2. The technical problems are related to methods for testing hypotheses, a topic introduced in chapter 7.

The *Winsorized sample variance* is just the sample variance based on the Winsorized values and will be labeled s_w^2. In symbols, if $W_1, \ldots, W_n$ are the Winsorized values,

$$s_w^2 = \frac{1}{n-1} \sum (W_i - \overline{W})^2, \qquad (2.7)$$

where

$$\overline{W} = \frac{1}{n} \sum W_i,$$

the average of the Winsorized values. The sample mean of the Winsorized values, $\overline{W}$, is called the *sample Winsorized mean*. The *Winsorized sample standard deviation* is the square root of the Winsorized sample variance, s_w.

Example 5

To compute the 20% Winsorized mean and variance for the observations

$$1, 2, 8, 9, 10, 16, 18, 22, 27, 29, 35, 42,$$

first Winsorize these values yielding

$$8, 8, 8, 9, 10, 16, 18, 22, 27, 29, 29, 29.$$

The mean of these Winsorized values is the Winsorized mean given by

$$\overline{X}_w = \frac{8+8+8+9+10+16+18+22+27+29+29+29}{12} = 17.75.$$

The Winsorized sample variance is

$$s_w^2 = \frac{(8-17.75)^2 + (8-17.75)^2 + \cdots + (29-17.75)^2}{12-1} = 82.57.$$

The Winsorized sample standard deviation is $s_w = \sqrt{82.57} = 9.1$.

For the observations in the last example, the sample mean is $\bar{X} = 18.25$ and the sample variance is $s^2 = 170.57$, which is about twice as large as the sample Winsorized variance, $s_w^2 = 82.57$. Notice that the Winsorized variance is less sensitive to extreme observations and roughly reflects the variation for the middle portion of your data. In contrast, the sample variance, s^2, is highly sensitive to extreme values. This difference between the sample variance and the Winsorized sample variance will be seen to be important.

Example 6

For the data in the last example, suppose we increase the largest value, 42, to 60. Then the sample mean increases from 18.25 to 19.75 and the sample variance, s^2, increases from 170.57 to 275.3. In contrast, the Winsorized sample mean and variance do not increase at all, they are still equal to 17.75 and 82.57, respectively. The sample Winsorized variance provides resistance to outliers because its value does not increase as we increase the largest observation, a property that will turn out to have great practical value.

A Summary of Some Key Points

- Several measures of location were introduced. How and when should one measure of location be preferred over another? It is much too soon to discuss this issue in a satisfactory manner. An adequate answer depends in part on concepts yet to be described. For now, the main point is that different measures of location vary in how sensitive they are to outliers.
- The sample mean can be highly sensitive to outliers. For some purposes, this is desirable, but in many situations this creates practical problems, as will be demonstrated in subsequent chapters.
- The median is highly insensitive to outliers. This plays an important role in some situations, but the median has some negative characteristics yet to be described.
- In terms of sensitivity to outliers, the 20% trimmed mean lies between two extremes: no trimming (the mean) and the maximum amount of trimming (the median).
- The sample variance also is highly sensitive to outliers. We saw that this property creates difficulties when checking for outliers (it results in masking), and some additional concerns will become evident later in this book.
- The interquartile range measures variability without being sensitive to the more extreme values. This property makes it well suited to detecting outliers.
- The 20% Winsorized variance also measures variation without being sensitive to the more extreme values. But it is too soon to explain why it has practical importance.

Problems

27. What is the typical pulse rate (beats per minute) among adults? Imagine that you sample 21 adults, measure their pulse rate and get

$$80, 85, 81, 75, 77, 79, 74, 86, 79, 55$$

$$82, 89, 73, 79, 83, 82, 88, 79, 77, 81, 82.$$

Compute the 20% trimmed mean.

28. For the observations

$$21, 36, 42, 24, 25, 36, 35, 49, 32$$

verify that the sample mean, trimmed mean and median are $\bar{X} = 33.33$, $\bar{X}_t = 32.9$ and $M = 35$.

29. The largest observation in the last problem is 49. If 49 is replaced by the value 200, verify that the sample mean is now $\bar{X} = 50.1$ but the trimmed mean and median are not changed.

30. For the last problem, what is the minimum number of observations that must be altered so that the trimmed mean is greater than 1,000?

31. Repeat the previous problem but use the median instead. What does this illustrate about the resistance of the mean, median and trimmed mean?

32. For the observations

$$6, 3, 2, 7, 6, 5, 8, 9, 8, 11$$

verify that the sample mean, trimmed mean and median are $\bar{X} = 6.5$, $\bar{X}_t = 6.7$ and $M = 6.5$.

33. In general, when you have n observations, what proportion of the values must be altered to make the 20% trimmed mean as large as you want.

34. A class of fourth graders was asked to bring a pumpkin to school. Each of the 29 students counted the number of seeds in their pumpkin and the results were

$$250, 220, 281, 247, 230, 209, 240, 160, 370, 274, 210, 204, 243, 251, 190,$$

$$200, 130, 150, 177, 475, 221, 350, 224, 163, 272, 236, 200, 171, 98.$$

Compute the sample mean, trimmed mean and median.

35. Compute the 20% Winsorized values for the observations

$$21, 36, 42, 24, 25, 36, 35, 49, 32.$$

36. For the observations in the previous problem, compute the sample 20% Winsorized variance.

37. In the previous problem, would you expect the sample variance to be larger or smaller than 51.4? Verify your answer.

38. In general, will the Winsorized sample variance, s_w^2, be less than the sample variance, s^2?

39. For the observations

$$6, 3, 2, 7, 6, 5, 8, 9, 8, 11$$

verify that the sample variance and Winsorized variance are 7.4 and 1.8, respectively.

40. Consider again the number of seeds in 29 pumpkins given in problem 34. Compute the 20% Winsorized variance.

41. Snedecor and Cochran (1967) report results from an experiment dealing with weight gain in rats as a function of source and amount of protein. One of the groups was fed beef with a low amount of protein. The weight gains were

$$90, 76, 90, 64, 86, 51, 72, 90, 95, 78.$$

Compute the 20% trimmed mean and the 20% Winsorized variance.

3

GRAPHICAL SUMMARIES OF DATA AND SOME RELATED ISSUES

This chapter covers some of the more basic methods for graphically summarizing data and it comments on some related issues. Included are some modern insights regarding when these graphical methods perform well, and when and why they might be unsatisfactory. A few comments about more modern methods are provided.

3.1 Relative frequencies

The notation f_x is used to denote the *frequency* or number of times the value x occurs. To be concrete, imagine that 100 individuals are asked to rate a recently released movie using a 10-point scale, where a 1 means the movie received a poor rating and a 10 indicates an excellent film. The results, which for convenience are written in ascending order, are shown in table 3.1. In table 3.1, the value 2 occurs five times, which is written as $f_2 = 5$, the number 3 occurs 18 times, so $f_3 = 18$. In a similar manner $f_4 = 24, f_5 = 25, f_6 = 15$, $f_7 = 9$ and $f_9 = 4$. The *relative frequency* is just the frequency divided by the sample size, which here is $n = 100$. That is, the relative frequency associated with the value x is f_x/n, the proportion of times the value x occurs among the n observations. The value having the largest frequency is called the *mode*. The plot used to indicate the relative frequencies consists of the observed x values along the x-axis with the height of spikes used to indicate the relative frequencies. The relative frequencies, taken as whole, are called an *empirical distribution*. Figure 3.1 shows a plot of the relative frequencies (the empirical distribution) for the data in table 3.1. The mode is 5.

Plots of relative frequencies help add perspective on the sample variance, mean and median introduced in chapter 2. Moreover, relative frequencies help convey some basic principles covered in chapter 4. But first it is noted how the sample mean is computed using relative frequencies. Observe that the sum of the frequencies yields the sample size, n. That is,

$$n = \sum f_x,$$

Table 3.1 One hundred ratings of a film

2	2	2	2	2	3	3	3	3	3	3	3	3	3	3	3	3	3	3	3	3	3	3	4	4	4	4	4	4	4	4	4	4	4	4	4	4
4	4	4	4	4	4	4	4	4	4	5	5	5	5	5	5	5	5	5	5	5	5	5	5	5	5	5	5	5	5	5	5	5	5	5	6	6
6	6	6	6	6	6	6	6	6	6	6	6	6	7	7	7	7	7	7	7	7	7	8	8	8	8											

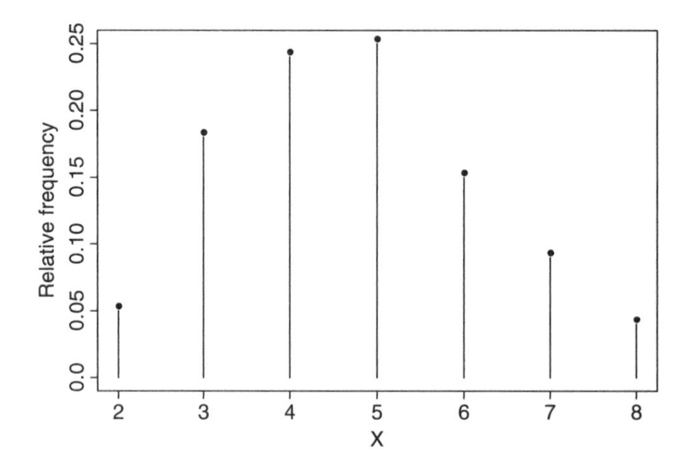

Figure 3.1 Relative frequencies for the data in table 3.1.

where now the summation is over all possible values of x. For table 3.1,

$$\sum f_x = f_1 + f_2 + f_3 + f_4 + f_5 + f_6 + f_7 + f_8 + f_9 + f_{10}$$

$$= 0 + 5 + 18 + 24 + 25 + 15 + 9 + 40 + 0 + 0 = 100.$$

The sample mean is

$$\bar{X} = \frac{1}{n} \sum x f_x = \sum x \frac{f_x}{n}. \tag{3.1}$$

So if we know the relative frequencies (f_x/n), it is a simple matter to compute the mean even when the number of observations, n, is large. (And writing the sample mean in terms of relative frequencies helps elucidate a basic principle covered in chapter 4.) The sample variance is

$$s^2 = \frac{n}{n-1} \sum \frac{f_x}{n} (x - \bar{X})^2. \tag{3.2}$$

Example 1

One million couples are asked how many children they have. For illustrative purposes, suppose that the maximum number of possible children is 5 and that the relative frequencies are $f_0/n = 0.10$, $f_1/n = 0.20$, $f_2/n = 0.25$, $f_3/n = 0.29$, $f_4/n = 0.12$ and $f_5/n = 0.04$. Then the sample mean is

$$\bar{X} = 0(.10) + 1(.20) + 2(.25) + 3(.29) + 4(.12) + 5(.04) = 2.25.$$

To compute the sample variance, first compute

$$0.10(0 - 2.25)^2 + 0.20(1 - 2.25)^2 + 0.25(2 - 2.25)^2 + 0.29(3 - 2.25)^2$$

$$+0.12(4 - 2.25)^2 + 0.04(5 - 2.25)^2 = 1.6675.$$

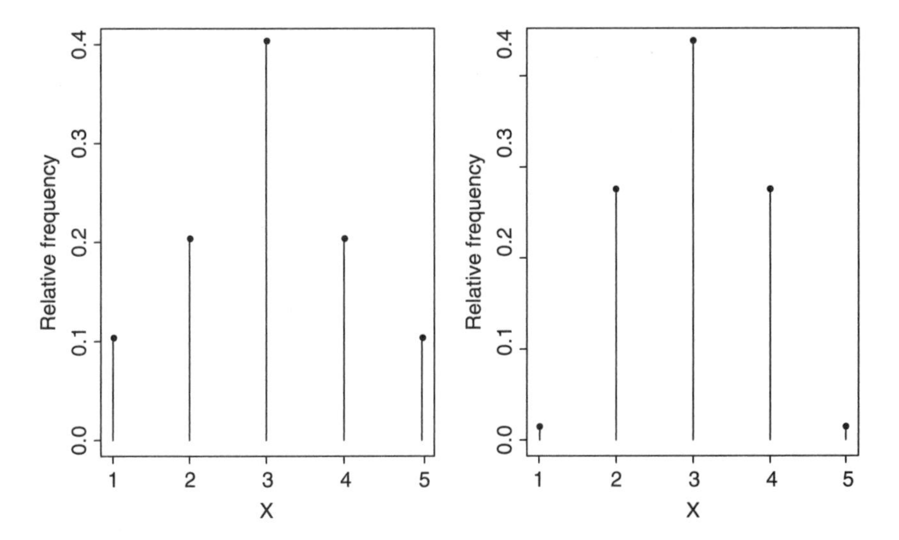

Figure 3.2 Relative frequencies that are symmetric about a central value. For this special case, the mean and median have identical values, the middle value, which here is 3. The relative frequencies in the left panel are higher for the more extreme values, versus the right panel, indicating that the variance associated with the left panel is higher.

Because $n/(n-1) = 1{,}000{,}000/999{,}999 = 1.000001$, the sample variance is $1.000001(1.6675) = 1.667502$.

Look at figure 3.2. The left panel shows five relative frequencies where the middle spike is the highest and the other spikes are symmetric about this middle value. When this happens, the mode, median, mean and 20% trimmed mean all have the same value, which is equal to the middle value. Here this middle value is 3. The right panel shows another plot of relative frequencies symmetric about the value 3 (again the mean), but now the relatively frequencies associated with the values 1 and 3 are much smaller, and the relatively frequencies associated with the values 2 and 4 are higher, indicating that the sample variance is smaller in the right panel versus the left. In the left panel, the data used to create the plot has a sample variance of 1.21 and in the right panel it is 0.63.

The *cumulative relative frequency distribution* refers to the proportion of observations less than or equal to a given value. Note that for each possible x value, a certain proportion of the observed values will be less than or equal to x, which will be denoted by $F(x)$. These proportions, taken as a whole, are sometimes called the *empirical cumulative distribution function*.

Example 2

In the previous example, the smallest observed value is 0, the corresponding relative frequency is $f_0/n = 0.10$, so the proportion of values less than or equal to 0 is $F(0) = 0.10$. The next largest observation is 1, its relative frequency is $f_1/n = 0.20$, so the corresponding cumulative relative frequency is $F(1) = 0.10 + 0.20 = 0.30$. In a similar manner, $F(2) = 0.45$, $F(3) = 0.84$, $F(4) = 0.96$ and $F(5) = 1.00$.

Problems

1. Based on a sample of 100 individuals, the values 1, 2, 3, 4, 5 are observed with relative frequencies 0.2, 0.3, 0.1, 0.25, 0.15. Compute the mean, variance and standard deviation.

2. Fifty individuals are rated on how open minded they are. The ratings have the values 1, 2, 3, 4 and the corresponding relative frequencies are 0.2, 0.24, 0.4, 0.16, respectively. Compute the mean, variance and standard deviation.

3. For the values 0, 1, 2, 3, 4, 5, 6 the corresponding relative frequencies based on a sample of 10,000 observations are 0.015625, 0.093750, 0.234375, 0.312500, 0.234375, 0.093750, 0.015625, respectively. Determine the mean, median, variance, standard deviation and mode.

4. For a local charity, the donations in dollars received during the last month were 5, 10, 15, 20, 25, 50 having the frequencies 20, 30, 10, 40, 50, 5. Compute the mean, variance and standard deviation.

5. The values 1, 5, 10, 20 have the frequencies 10, 20, 40, 30. Compute the mean, variance and standard deviation.

3.2 Histograms

An important feature of the data previously used to illustrate plots of relative frequencies is that a few values occur many times. In figure 3.2, for example, the responses are limited to five values. But when there are many values, with most values occurring a small number of times, plots of relatively frequencies can be rather uninteresting. If each value occurs only once, a plot of the relatively frequencies would consist of n spikes, each having height $1/n$. A histogram is one way of trying to deal with this problem. It is similar to plots of relatively frequencies, the main difference being that values are binned together to get a more useful plot. That is, a histogram simply groups the data into categories and plots the corresponding frequencies.

Example 1

A histogram is illustrated with data from a heart transplant study conducted at Stanford University between October 1, 1967 and April 1, 1974. Of primary concern is whether a transplanted heart will be rejected by the recipient. With the goal of trying to address this issue, a so-called T5 mismatch score was developed by Dr. C. Bieber. It measures the degree of dissimilarity between the donor and the recipient tissue with respect to HL-A antigens. Scores less than 1 represent a good match and scores greater than 1 a poor match. The

Table 3.2 T5 mismatch scores from a heart transplant study

0.00	0.12	0.16	0.19	0.33	0.36	0.38	0.46	0.47	0.60	0.61	0.61	0.66	0.67	0.68
0.69	0.75	0.77	0.81	0.81	0.82	0.87	0.87	0.87	0.91	0.96	0.97	0.98	0.98	1.02
1.06	1.08	1.08	1.11	1.12	1.12	1.13	1.20	1.20	1.32	1.33	1.35	1.38	1.38	1.41
1.44	1.46	1.51	1.58	1.62	1.66	1.68	1.68	1.70	1.78	1.82	1.89	1.93	1.94	2.05
2.09	2.16	2.25	2.76	3.05										

Table 3.3 Frequencies and relative frequencies for grouped T5 scores, $n = 65$

Test score (x)	Frequency	Relative frequency
–0.5–0.0	1	$1/65 = .015$
0.0–0.5	8	$8/65 = .123$
0.5–1.0	20	$20/65 = .308$
1.0–1.5	18	$18/65 = .277$
1.5–2.0	12	$12/65 = .138$
2.0–2.5	4	$4/65 = .062$
2.5–3.0	1	$1/65 = .015$
3.0–3.5	1	$1/65 = .015$

T5 scores, written in ascending order, are shown in table 3.2 and are taken from Miller (1976). Suppose we group the T5 values into eight categories: (1) values between -0.5 and 0.0, (2) values greater than 0.0 but less than or equal to 0.5, (3) values greater than 0.5 but less than or equal to 1.0, and so on. The beginning and end of each interval are called *boundaries* or *class interval* and the point midway between any to boundaries is called the *class mark* or *midpoint*. So here, the first interval has boundaries -0.5 and 0.0 and the corresponding class mark or midpoint is $(-0.5+0)/2 = -0.25$. Similarly, the second interval has boundaries 0.0 and 0.5, so the class mark is $(0.0+0.5)/2 = 0.25$. Note that all of the categories have the same length, which is a feature routinely used. The frequency and relative frequency associated with each of these intervals is shown in table 3.3. For example, there are eight T5 mismatch scores in the interval extending from 0.0 to 0.5 and the proportion of all scores belonging to this interval is 0.123. Figure 3.3 shows the resulting histogram.

How many bins should be used when constructing a histogram and how should the length of the bins be chosen? The general goal is to choose the number of bins so as to get an informative plot of the data. If we have one bin only, this tells us little about the data, and too many bins suffer from the same problem. There are simple rules for choosing the number of bins. One is called

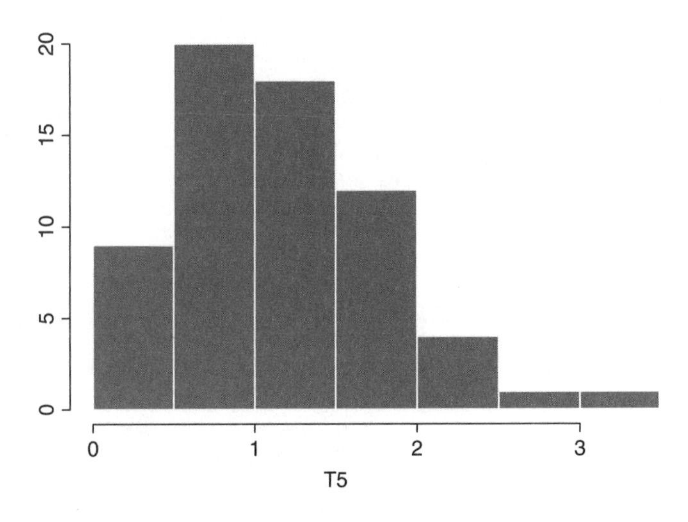

Figure 3.3 A histogram of the heart transplant data in table 3.5.

Sturges's rule, which is commonly used by statistical software, but no details are given here. The main point is that standard methods for choosing the number of bins can result in a rather unsatisfactory summary of the data, as will be illustrated. The good news is that substantially better methods are now available, some of which are outlined in the final section of this chapter.

What do histograms tell us?

Like so many graphical summaries of data, histograms attempt, among other things, to tell us something about the shape of the data. One issue of some concern is whether data are reasonably symmetric about some central value. In figure 3.2, we see exact symmetry, but often data are highly skewed, and this can be a serious practical problem when dealing with inferential techniques yet to be described. The left panel of figure 3.4 shows data that are not symmetric, but rather *skewed to the right*. The right panel shows data that are *skewed to the left*. In recent years, skewness, roughly referring to a lack of symmetry, has been found to be a much more serious problem than once thought for reasons that are best postponed for now. But one important point that should be stressed here is that when distributions are skewed, generally the mean, median and 20% trimmed mean will differ. In some cases they differ by very little, but in other situations these measures of location can differ substantially, as was illustrated in chapter 2. Moreover, even when these measures of location are virtually identical, subsequent chapters will demonstrate that often the choice for a measure of location can make a practical difference when addressing common problems yet to be described.

Example 2

In an unpublished study (by M. Earleywine) was performed that generally dealt with the effects of consuming alcohol. A portion of the was concerned with

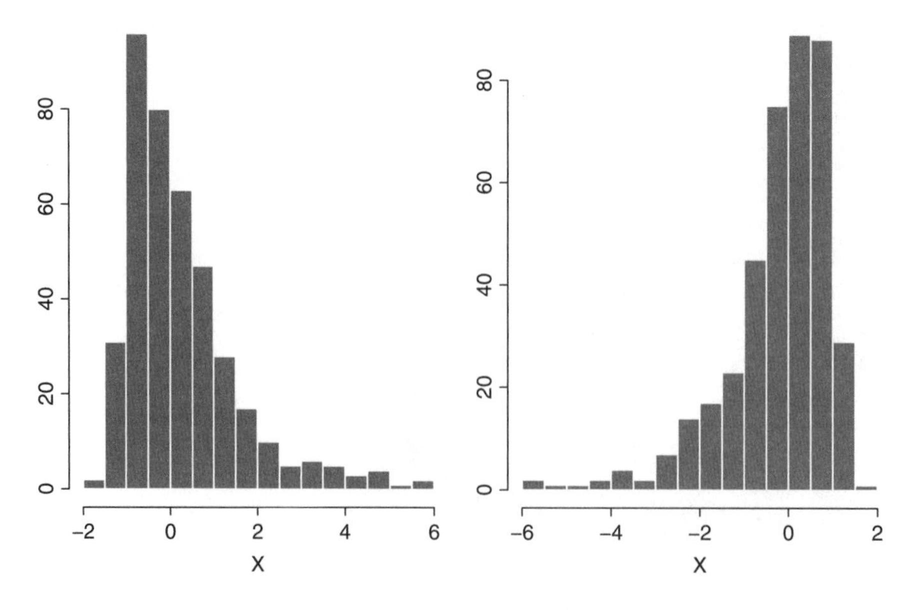

Figure 3.4 The left panel is an example of a histogram that is said to be skewed to the right. In the right panel, it is skewed to the left.

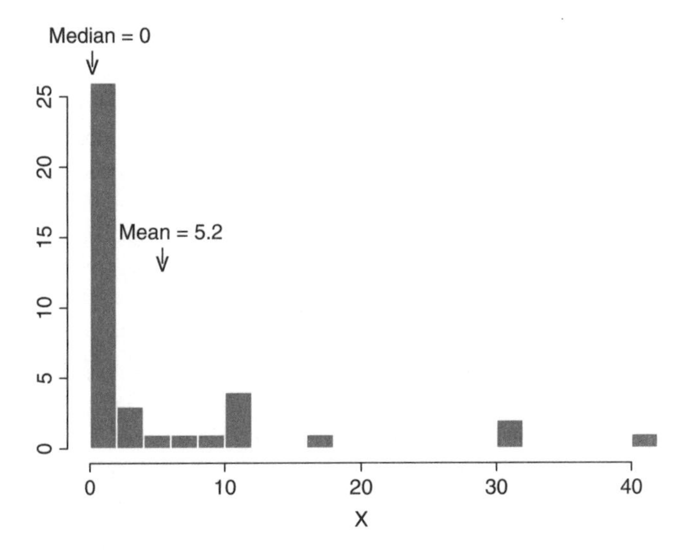

Figure 3.5 A histogram of the measurement of hangover symptoms.

measuring hangover symptoms after consuming a specific amount of alcohol in a laboratory setting. The resulting measures, written in ascending order, were

0 0 0 0 0 0 0 0 0 0 0 0 0 0 0 0 0 0 0 0 0 0 0 0 0 1 2 2 2 3 3 3 6 8 9 11 11 11 12 18 32 32 41.

Figure 3.5 shows a histogram based on these data. As is evident, the histogram is skewed to the right with a fairly large difference between the median and mean. Note that the rightmost portion indicates three bins that are separated from the rest of the plot. This might suggest that any value greater than 16 should be flagged an outlier, but a boxplot rule indicates that only values greater than or equal to 32 are outliers. Although histograms can provide a useful summary of data, as an outlier detection method, it is not very satisfactory. One fundamental problem is that, when using a histogram, no precise rule is available for telling us when a value should be declared an outlier. Without a precise rule, there can be no agreement on the extent to which masking (described in chapter 2) is avoided.[1] (Also, examples will be seen where extremely unlikely values occur, yet a histogram does not suggest that they are unusual.) As a general rule, histograms might suggest that certain values are outliers, but when making a decision about which values are outliers, it is better to use methods specifically designed for this task, such as the boxplot rule.

Populations, samples, and potential concerns about histograms

Histograms are routinely taught in an introductory statistics course and in some cases they provide useful information about data. But like so many methods covered in this book, it is important to understand not only when histograms perform well,

1. Outlier detection methods are typically designed to have other properties not covered here. Again, without a precise rule for deciding what constitutes an outlier, it is impossible to determine whether a specific method achieves the properties desired.

but also when, and in what sense, they might be highly misleading. Without a basic understanding of the relative merits of a method, there is the potential of drawing erroneous conclusions, as well as missing interesting results. The immediate goal is to illustrate a fundamental concern about histograms, and in the final section of this chapter, methods aimed at correcting known problems are briefly indicated.

As mentioned in chapter 1, there is an important distinction between samples of individuals or things versus a population of individuals or things. Samples represent a subset of the population under study. Consider, for example, the last example dealing with hangover symptoms. There were 40 participants who represent only a small proportion of the individuals who might have taken part in this study. Ideally, the available participants will provide a reasonably accurate reflection of the histogram we would obtain if all participants could be measured. In some cases, histograms satisfy this goal. But an important practical issue is whether they can be highly unsatisfactory. And if they can be unsatisfactory, is there some strategy that might give substantially better results? In turns out that they can indeed by unsatisfactory, and fundamental improvements are now available (e.g., Silverman, 1986). The only goal here is to illustrate what might go wrong and provide information about where to look for better techniques.

First we consider a situation where the histogram tends to perform tolerably well. Imagine that the population consists of one million individuals and that if we could measure everyone, the resulting histogram would appear as in figure 3.6. (Here, the y-axis indicates the relative frequencies.) Now imagine that 100 individuals are selected from the one million individuals in the population, with every individual having the same probability of being chosen. An issue of fundamental importance is the extent to which a histogram based on a sample of only 100 individuals will reflect the histogram we would get if all individuals could be measured. Mimicking this process on a computer resulted in the histogram shown in figure 3.7. So in this particular case, the histogram provides a reasonable reflection of the population histogram, roughly capturing its bell shape. A criticism of this illustration is that maybe we just got lucky. That is, in general, perhaps with only 100 individuals, the histogram will not accurately reflect the population.

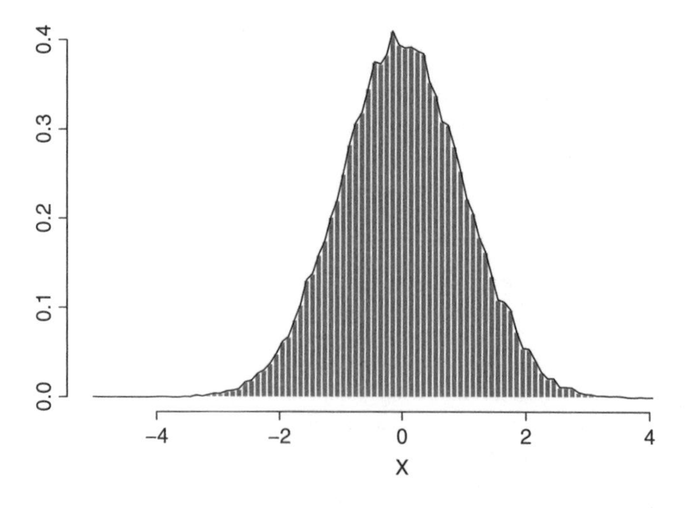

Figure 3.6 A histogram of an entire population that is approximately symmetric about 0 with relatively light tails, meaning outliers tend to be rare.

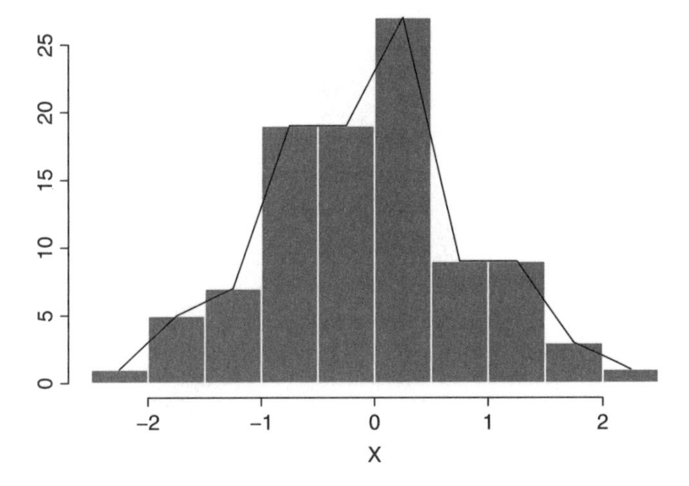

Figure 3.7 A histogram based on a sample of 100 observations generated from the histogram in figure 3.6.

This might indeed happen, but generally it gives a reasonable sense of the shape of the population histogram.

Now consider the population histogram in figure 3.8. This histogram has the same bell shape as in figure 3.6, but the tails extend out a bit farther. This reflects the fact that for this particular population, there are more outliers or extreme values. Now look at figure 3.9, which is based on 100 individuals sampled from the population histogram in figure 3.8. As is evident, it provides a poor indication of what the population histogram looks like. Figure 3.9 also provides another illustration that the histogram can perform rather poorly as an outlier detection rule. It suggests that values greater than 10 are highly unusual, which turns out to be true based on how the data were generated. But values less than −5 are also highly unusual, which is less evident here. The fact that the histogram can miss outliers limits its ability to deal with problems yet to be described.

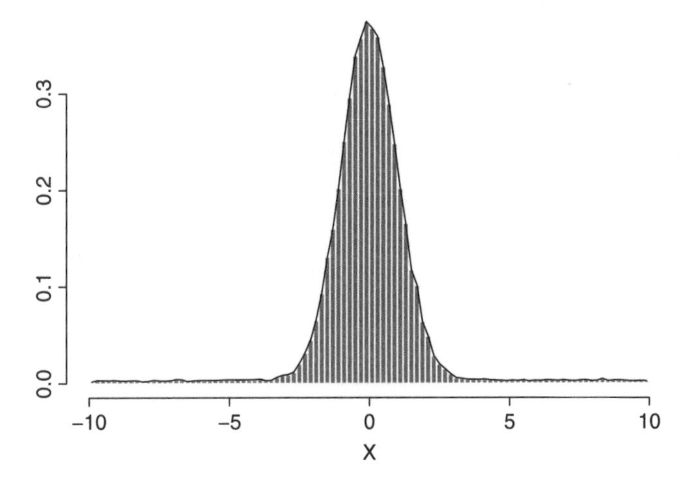

Figure 3.8 A histogram of an entire population that is approximately symmetric about 0 with relatively heavy tails, meaning outliers tend to be common.

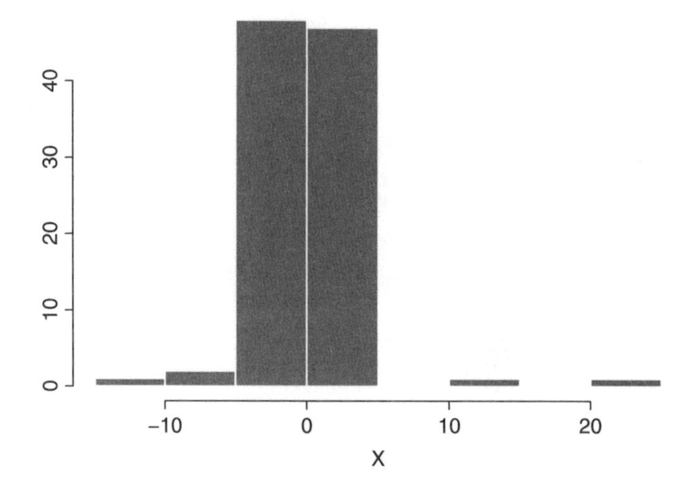

Figure 3.9 A histogram based on a sample of 100 observations generated from the histogram in figure 3.8.

A rough characterization of the examples just given is that when the population histogram is symmetric and bell-shaped, and outliers tend to be rare, it performs tolerably well with 100 observations, in terms of indicating the shape of the population histogram. But when outliers are relatively common, the reverse is true.

Problems

6. For the data in table 2.1, dealing with changes in cholesterol levels, suppose a histogram is to be created with bins defined as follows: −40 −30 −20 −10 0 10 20 30 40. That is, the first bin has boundaries −40 and −30, the next bin contains all values greater than −30 but less than or equal to −20, and so on. Determine the frequencies for each bin and construct a histogram.

7. For the data in table 2.2, suppose a histogram is to be created with bins defined as follows: −50 −40 −30 −20 −10 0 10 20 30 40 50 60 70 80. Determine the frequencies for each bin and construct a histogram.

8. The heights of 30 male Egyptian skulls from 4000 BC were reported by Thomson and Randall-Maciver (1905) to be

   121 124 129 129 130 130 131 131 132 132 132 133 133 134 134 134 134 135 135 136 136 136 136 137 137 138 138 138 140 143.

   Create a histogram with bins extending from 120–125, 125–130, and so on. Based on this histogram, does the largest value, 143, appear to be an outlier?

9. For the data in the previous problem, does the boxplot rule (described in chapter 2) indicate that 143 is an outlier?

10. What do the last two problems suggest about using a histogram to detect outliers?

Table 3.4 Word identification scores

58 58 58 58 58 64 64 68 72 72 72 75 75 77 77 79 80 82 82
82 82 82 84 84 85 85 90 91 91 92 93 93 93 95 95 95 95 95
95 95 95 98 98 99 101 101 101 102 102 102 102 102 103 104 104 104 104
104 105 105 105 105 105 107 108 108 110 111 112 114 119 122 122 125 125 125
127 129 129 132 134

3.3 Boxplots and stem-and-leaf displays

A *stem-and-leaf display* is another method of gaining some overall sense of what data are like. The method is illustrated with measures taken from a study aimed at understanding how children acquire reading skills. A portion of the study was based on a measure that reflects the ability of children to identify words. (These data were supplied by L. Doi.) Table 3.4 lists the observed scores in ascending order.

The construction of a stem-and-leaf display begins by separating each value into two components. The first is the *leaf* which, in this example, is the number in the ones position (the single digit just to the left of the decimal place). For example, the leaf corresponding to the value 58 is 8. The leaf for the value 64 is 4 and the leaf for 125 is 5. The digits to the left of the leaf are called the *stem*. Here the stem of 58 is 5, the number to the left of 8. Similarly, 64 has a stem of 6 and 125 has a stem of 12. We can display the results for all 81 children as follows:

Stems	Leaves
5	88888
6	448
7	22255779
8	0222224455
9	011233355555555889
10	1112222234444455555788
11	01249
12	22555799
13	24

There are five children who have the score 58, so there are five scores with a leaf of 8, and this is reflected by the five 8s displayed to the right of the stem 5 and under the column headed by Leaves. Two children got the score 64, and one child got the score 68. That is, for the stem 6, there are two leaves equal to 4 and one equal to 8, as indicated by the list of leaves in the display. Now look at the third row of numbers where the stem is 7. The leaves listed are 2, 2, 2, 5, 5, 7, 7 and 9. This indicates that the value 72 occurred three times, the value 75 occurred two times, as did the value 77, and the value 79 occurred once. Notice that the display of the leaves gives us some indication of the values that occur most frequently and which are relatively rare. Like the histogram, the stem-and-leaf display gives us an overall sense of what the values are like.

The leaf always consists of the numbers corresponding to a specified digit. For example, the leaf might correspond to tenths digit, meaning that the leaf is the first number to the right of the decimal, in which case the stem consists of all the numbers to the left of the leaf. So for the number 158.234, the leaf would be 2 and the stem would be 158. If we specify the leaf to be the hundredth digit, the leaf would now be 3 and the stem would be 158.2. The choice of which digit is to be used as the leaf depends in part on which digit provides a useful graphical summary of the data. But details about how to address this problem are not covered here. Suffice it to say that algorithms have been proposed for deciding which digit should be used as the leaf and determining how many lines a stem-and-leaf display should have (for example, Emerson and Hoaglin, 1983).

Example 1

Chapter 1 mentioned the software S-PLUS. When its version of a stem-and-leaf display is applied to the T5 mismatch scores, the result is

```
Decimal point is at the colon
   0 : z122344
   0 : 556667777788889999
   1 : 00000111111223334444
   1 : 5566777788999
   2 : 0122
   2 : 8
   3 : 0
```

The z in the first row stands for zero. So this plot suggests that the data are reasonably symmetric, with maybe a hint of being skewed to the right. Also, there are no values visibly separated from the overall plot suggesting that there are no outliers. (The boxplot rule, described in chapter 2, also finds no outliers.)

Boxplot

Proposed by Tukey (1977), a boxplot is a commonly used graphical summary of data, an example of which is shown in figure 3.10. As indicated, the ends of the rectangular box mark the lower and upper quartiles. That is, the box indicates where the middle half of the data lie. The horizontal line inside the box indicates the position of the median. The lines extending out from the box are called *whiskers*.

Boxplots determine whether values are outliers using the boxplot rule described in chapter 2. (See equations 2.4 and 2.5.) Figure 3.11 shows a boxplot with two outliers. The ends of the whiskers are called *adjacent values*. They are the smallest and largest values not declared outliers.

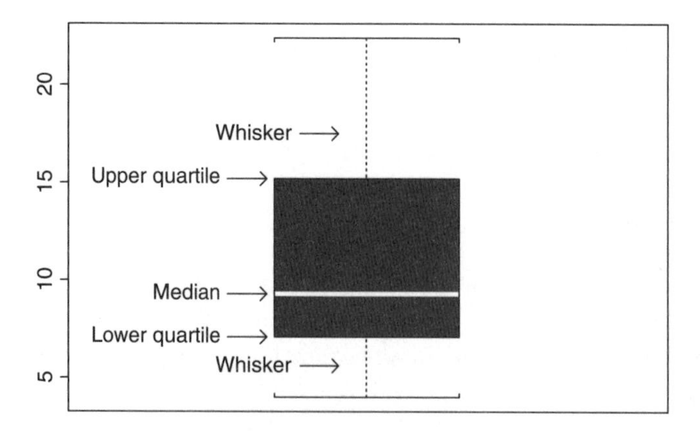

Figure 3.10 An example of a boxlpot with no outliers.

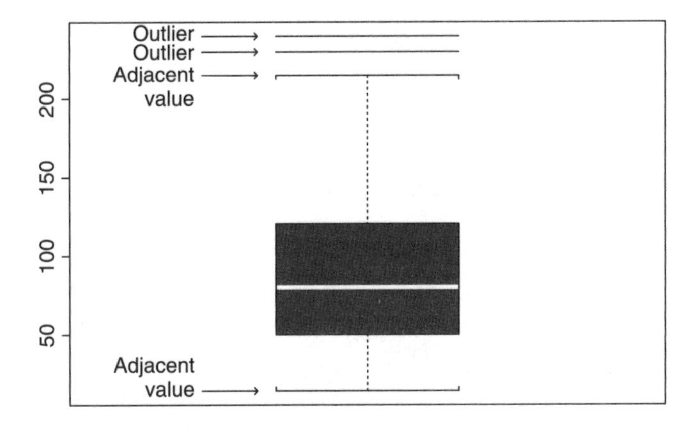

Figure 3.11 An example of a boxplot with outliers.

All of the statistical software mentioned in chapter 1 contain routines for creating boxplots. In case it helps, here is an outline of how they are constructed using the data in figure 3.11. First, compute the lower and upper quartiles using the method in chapter 2 (based on the ideal fourths). This yields $q_1 = 49.8$ and $q_2 = 120.75$, which correspond to the lower and upper ends of the box, respectively. The median is $M = 80$ and determines where the line within the box is placed. Next, using the boxplot rule in chapter 2, determine how small a value must be to be declared an outlier. Here, this value is -56.5. The smallest value not declared an outlier determines the end of the lower whisker, which is 14. Again using the boxplot rule, any value greater than 227 is declared an outlier. The largest value not declared an outlier is 215, so this value marks the end of the upper whisker. There are two values greater than 227, which correspond to the horizontal lines at the top of figure 3.11.

Problems

11. Table 3.5 shows the exam scores for 27 students. Create a stem-and-leaf display using the digit in the ones position as the stem.
12. If the leaf is the hundredths digit, what is the stem for the number 34.679?
13. Consider the values 5.134, 5.532, 5.869, 5.809, 5.268, 5.495, 5.142, 5.483, 5.329, 5.149, 5.240, 5.823. If the leaf is taken to be the tenths digit, why would this make an uninteresting stem-and-leaf display?
14. For the boxplot in figure 3.11, determine, approximately, the quartiles, the interquartile range, and the median. Approximately how large is the largest value not declared an outlier?
15. In figure 3.11, about how large must a value be to be declared an outlier? How small must it be?

Table 3.5 Examination Scores

83	69	82	72	63	88	92	81	54
57	79	84	99	74	86	71	94	71
80	51	68	81	84	92	63	99	91

16. Create a boxplot for the data in table 3.1.
17. Create a boxplot for the data in table 3.2.

3.4 Some modern trends and developments

We have seen that in terms of providing information about the shape of the population histogram, a histogram based on 100 observations can be relatively ineffective in certain situations. There is a vast literature on how this problem might be addressed using what are called *kernel density estimators*. There are in fact many variations of this approach, some of which appear to perform very well over a fairly broad range of situations. Some of these methods come with the software R and S-PLUS mentioned in chapter 1. The computational details go well beyond the scope of this book, but an illustration might help motivate their use.

Example 1

Consider again the data used to create the histogram shown in figure 3.9. Recall that the 100 observations were sampled from a population having the symmetric histogram shown in figure 3.8, yet the histogram in figure 3.9 suggests a certain amount of asymmetry. In particular, the right tail differs from the left; values in the right tail appear to be outliers and the values in the left tail seem to have a low relatively frequency, but otherwise there is no sense that they are unusually far from the central values. One of the seemingly better methods for improving on the histogram is called an adaptive kernel density estimator. Figure 3.12 shows a plot of the data in figure 3.9 using this method.[2] The plot in figure 3.12 does not capture the exact symmetry of the population histogram, but typically it does a better job of indicating its shape versus the histogram.

A Summary of Some Key Points

- No single graphical summary of data is always best. Different methods provide different and potentially interesting perspectives. What is required is some familiarity with the various methods to help you choose which one to use.
- However, the choice of method is not always academic. The histogram is a classic, routinely taught method, but it is suggested that kernel density estimators be given serious consideration. Perhaps the most important point to keep in mind is that the histogram performs rather poorly as an outlier detection technique.
- The boxplot is one of the more useful graphical tools for summarizing data. It conveys certain important features that were described and illustrated, but kernel density estimators can help add perspective.
- The stem-and-leaf display can be useful when trying to understand the overall pattern of the data. But with large sample sizes, it can be highly unsatisfactory.

2. The S-PLUS function akerd was used, which belongs to a library of S-PLUS functions mentioned in chapter 1.

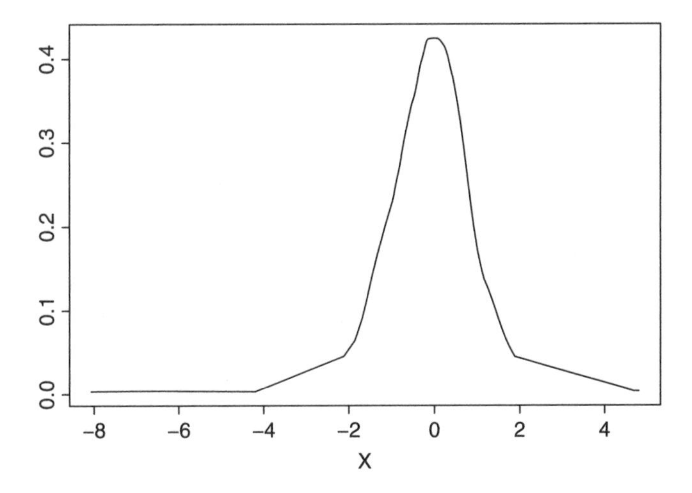

Figure 3.12 An example of a kernel density plot based on the same 100 observations generated for figure 3.8 and used in figure 3.9. Note how the kernel density plot does a better job of capturing the shape of the population histogram in figure 3.8.

Problems

18. Describe a situation where the sample histogram is likely to give a good indication of the population histogram based on 100 observations.
19. Comment generally on how large a sample size is needed to ensure that the sample histogram will likely provide a good indication of the population histogram?
20. When trying to detect outliers, discuss the relative merits of using a histogram versus a boxplot.
21. A sample histogram indicates that the data are highly skewed to the right. Is this a reliable indication that if all individuals of interest could be measured, the resulting histogram would also be highly skewed?

4

PROBABILITY AND RELATED CONCEPTS

As indicated in chapters 1 and 3, one of the main goals is to cover basic methods that are aimed at making inferences about a population of individuals or things based on a sample or subset of the population that is available. Basic probability plays a key role when addressing this issue. This chapter covers the fundamentals of probability and some related concepts that will be needed.

4.1 The meaning of probability

Coming up with an acceptable definition of the term *probability* turns out to be a nontrivial task. For example, imagine that we have a bowl with 100 marbles, 50 are blue and the other 50 are green. When someone picks a marble from the bowl, without looking, what is the probability that the chosen marble will be blue? A seemingly natural response is 0.5 because half the marbles are blue. But suppose all of the green marbles are on the bottom of the bowl and the person picking a marble has a penchant for picking marbles from those on top. Then the probability of a blue marble is virtually 1; it will happen with near certainty unless this individual decides to reach deeply into the bowl. Of course we can mix the marbles in the bowl so that not all of the blue marbles are at the top. But when is the bowl of marbles sufficiently mixed so that the probability of a blue marble is 0.5? A response might be that all marbles must have the same probability of being chosen. But we cannot use the term probability as if we know what it means in our attempt to define probability.

Another seemingly natural way of defining probability is in terms of a long series of events. For example, if you flip a coin, what is the probability of a head? You might flip the coin a hundred times, a thousand times, and we can conceive of flipping the coin billions of times with the idea that the proportion of times we observe a head is equal to the probability of a head on any one flip. There is a famous result (called the law of large numbers) telling us the conditions under which the proportion of heads will indeed converge to the true probability of a head. This result says that if the probability of a head is the same on every flip, as we increase the number of flips, the proportion of times we observe a head will converge to the true probability. But again, we have used the term probability as if we already know what it means.

Probability functions

The typical way of dealing with probability, at least for the problems covered in this book, is to view probability in terms of what are called *probability functions*, which must obey certain rules that are about to be described. This approach does not define probability, but rather makes precise the rules of probability that are needed to make progress. Before explaining what this means, some new terms and notation are needed.

A *random variable* refers to a measurement or observation that cannot be known in advance. If someone with heart disease is put on a new medication, will they have a heart attack during the next 12 months? How much weight will someone lose if they follow Aunt Sally's weight reduction method for 4 weeks? Consistent with previous chapters, usually some upper case Roman letter is used to represent a random variable, the most common letter being X. That is, X is a generic term representing whatever we want to measure. A lower case x is used to represent an observed value corresponding to the random variable X. So the notation $X = x$ means that the observed value of X is x. For instance, we might let a 1 indicate that someone has a heart attack and no heart attack is indicated by a 0. So here, $X = 1$ indicates that a heart attack occurred and $X = 0$ means that it did not occur. This is an example of what is called a *discrete random variable*, meaning that there are gaps between any value and the next possible value. Here there are only two possible values, the next possible value after 0 is 1, no value between 0 and 1 is possible, so X is discrete. In terms of Aunt Sally's weight reduction method, X represents how much weight was lost and $X = 12$ refers to the event that someone lost 12 pounds. That is $X = 12$. In principle, this is a *continuous random variable*, meaning that for any two outcomes, any value between these two values is possible. For example, someone might lose 12 pounds, 13 pounds, or any amount between 12 and 13.

The set of all possible outcomes or values of X we might observe is called the *sample space*. For the heart disease example, the sample space consists of two outcomes: heart attack or no heart attack, or in numerical terms, 1 and 0. For the weight loss example, the sample space consists all possible results we might observe in terms of how much weight was lost or gained. Unless stated otherwise, the elements of the sample space are assumed to be *mutually exclusive*. That is, one, and only one, element of the sample space can occur. In the heart disease example, $X = 0$ and $X = 1$ are two mutually exclusive events simply because if an individual did not have a heart attack ($X = 0$), this eliminates the possibility that a heart attack occurred ($X = 1$). In a similar manner, in the weight loss example, all possible outcomes are mutually exclusive.

Example 1

An investment strategy for buying stocks has been recommended to you. You plan to try out the new strategy by buying 10 stocks and observing how many stocks gain value after 6 months. Here, X represents the number of stocks that gain value and the sample space is 0, 1, 2, 3, 4, 5, 6, 7, 8, 9 and 10. This is an example of a discrete random variable.

Next, we introduce the notion of a probability function, which for convenience is described in terms of a discrete random variable, still assuming that the elements of the sample space are mutually exclusive.

Definition A *probability function* is a rule, denoted by $p(x)$, that assigns numbers to elements of the sample space with the following properties:

1. $p(x) \geq 0$
2. For any two distinct elements in the sample space, say x and y, $p(x \text{ or } y) = p(x) + p(y)$. In words, the probability of observing the value x or y is equal to the sum of the individual probabilities (assuming that these two events are mutually exclusive).
3. The sum of the probabilities associated with all of the elements in the sample space is one. (That is, $\sum p(x) = 1$).

The definition of a probability function is somewhat abstract, but it is easy to understand if we think of probabilities in terms of proportions. To be concrete, imagine that 100 students are asked to rate their college experience on a 5-point scale: 1, 2, 3, 4, 5. Further imagine that among the 100 students, 10 respond 1, 20 respond 2, 35 respond 3, 30 respond 4 and 5 respond 5. So the proportions corresponding to 1, 2, 3, 4, 5 are .1, .2, .35, .3 and .05, respectively. Now think of these proportions as probabilities. It is evident that all proportions are greater than or equal to 0. Consider any two responses, say 2 and 4. The number of students responding 2 or 4 is $20 + 30 = 50$, so the proportion responding 2 or 4 is .5. In symbols, letting $P(2 \text{ or } 4)$ be the probability of a 2 or 4, $P(2 \text{ or } 4) = p(2) + p(4) = .2 + .3 = .5$. That is, condition 2 in the definition of a probability function is satisfied for these two values and for any other two values we might choose. And condition 3 is satisfied because the proportions sum to 1.

Example 2

Imagine that the sample space consists of the values

$$x : 0, 1, 2, 3, 4$$

and consider

$$p(x) : .1, .2, .25, .3, .25, .2.$$

That is, $p(0) = .1$, $p(1) = .2$ and so on. Then $p(x)$ does not qualify as a probability function because the sum of all five $p(x)$ values is greater than 1.

Problems

1. If the possible values for x are 0, 1, 2, 3, 4, 5, and the corresponding values for $p(x)$ are .2, .2, .15, .3, .35, .2, .1, respectively, does $p(x)$ qualify as a probability function?
2. If the possible values for x are 2, 3, 4 and the corresponding values for $p(x)$ are .2, -0.1, .9, respectively, does $p(x)$ qualify as a probability function?
3. If the possible values for x are -1, 2, 3, 4, and the corresponding values for $p(x)$ are .1, .15, .5, .25, respectively, does $p(x)$ qualify as a probability function?

4. If the possible values for x are 2, 3, 4, 5, and the corresponding values for $p(x)$ are .2, .3, .4, .1, respectively, what is the probability of observing a value less than or equal to 3.4?
5. In problem 4, what is the probability of observing a value less than or equal to 1?
6. In problem 4, what is the probability of observing a value greater than 3?
7. In problem 4, what is the probability of observing a value greater than or equal to 3?
8. If the probability of observing a value less than or equal to 6 is .3, what is the probability of observing a value greater than 6?

4.2 Expected values

The notion of expected values is important for two general reasons. First, it provides a precise definition of quantities associated with populations that play a fundamental role in methods to be described. Second, it provides a link between samples and populations in a sense explained in chapter 5.

Consider any random variable X with probability function $p(x)$. The *expected value* of X is

$$E(X) = \sum xp(x). \tag{4.1}$$

In words, multiply every value in the sample space by its probability and sum the results. The expected value of a random variable plays such a fundamental role, it has been given a special name: the population mean, which is typically written as μ (a lower case Greek mu). That is, $\mu = E(X)$.

The notion of expected value might be made clearer by noting that it is similar to how the sample mean was defined in chapter 3. Recall that if we collect data and observe the values x having relative frequencies f_x/n, the sample mean is

$$\bar{X} = \sum x\frac{f_x}{n}.$$

Now imagine that an entire population consists of N individuals and that the relative frequencies are f_x/N. Further assume that we view these relative frequencies as probabilities. That is, $p(x)$ is taken to be f_x/N. Then the average value for the population of individuals is

$$\mu = E(X) = \sum x\frac{f_x}{N}.$$

That is, the population mean is just the average of all the individuals in the population, if only they could be measured.

Example 1

For

$$x: 1, 5, 10$$

and

$$p(x): .2, .5, .3$$

the population mean is

$$\mu = 1(.2) + 5(.5) + 10(.3) = 5.7.$$

Example 2

A carnival game costs 2 dollars to play and for each play, a contestant can win 1, 2 or 3 dollars. Imagine that the probabilities associated with these three outcomes are .25, .40 and .35, respectively. That is, $p(1) = .25$, $p(2) = .4$ and $p(3) = .35$. Then the expected winnings are

$$\sum xp(x) = 1(.25) + 2(.40) + 3(.35) = 2.10.$$

This says that on average, a contestant would win \$2.10 and, because it costs \$2.00 to play, on average the carnival would lose 10 cents.

Example 3

Imagine that when buying a car, you are offered an extended warranty for \$200. To keep the example simple, further imagine that four outcomes are possible: no repairs are needed, or one of three repairs can occur costing \$50, \$150 and \$250. If the probabilities corresponding to these four outcomes are .7, .15, .10 and .05, respectively, the expected cost of a repair is

$$0(.7) + 50(.15) + 150(.10) + 200(.05) = 30.$$

This says that on average, the cost to customers who buy this warranty is $\$200 - \$30 = \$170$.

Example 4

Consider a population of one million individuals who have played a particular video game. Further assume that if asked to rate the game on a 5 point scale, the proportion who would give a rank of 1, 2, 3, 4 and 5 would be .1, .15, .35, .30 and .1, respectively. If we view these proportions as probabilities, the mean or expected rating is

$$\mu = 1(.1) + 2(.15) + 3(.35) + 4(.30) + 5(.1) = 3.15.$$

A population mean is an example of what is called a *population parameter*, which is a quantity that is generally unknown because it is impossible to measure everyone in the population. The sample mean is an example of an *estimator*, a value computed based on data available to us and based on a sample taken from the population. As is probably evident, the sample mean $\bar{X}$ is intended as an estimate of the population mean, μ. A fundamental issue is how well the sample mean $\bar{X}$ estimates the population mean, μ, an issue we will begin to address in chapters 5 and 6.

Population variance

We have seen that there are two types of means: a population mean, which is the average, μ, that we would get if all individuals or things could be measured, and a sample mean, which is just the average based on a sample from the population. In a similar manner there is both a sample variance, s^2, introduced in chapter 2, and a population variance, which is the average squared distance between the population mean and all the measures associated with all individuals or things in the population under study. More formally, the *population variance* is

$$\sigma^2 = \sum (x - \mu)^2 p(x), \tag{4.2}$$

where σ is a lower case Greek sigma. The (positive) square root of the population variance variance, σ, is called the *population standard deviation.*

Example 5

Consider the following probability function.

x:	0	1	2	3
$p(x)$:	.1	.3	.4	.2

The population mean is $\mu = 1.7$. So for the value 0, its squared distance from the population mean is $(0 - 1.7)^2 = 2.89$ and reflects how far away the value 0 is from the population mean. Moreover, the probability associated with this squared difference is .1, the probability of observing the value 0. In a similar manner, the squared difference between 1 and the population mean is .49, and the probability associated with this squared difference is .3, the same probability associated with the value 1. Continuing in this manner, we can compute the population variance. In particular,

$$\sigma^2 = (0 - 1.7)^2(.1) + (1 - 1.7)^2(.3) + (2 - 1.7)^2(.4) + (3 - 1.7)^2(.2) = .81.$$

Example 6

For a five-point scale measuring anxiety, the probability function for all adults living in New York City is

x:	1	2	3	4	5
$p(x)$:	.05	.1	.7	.1	.05

The population mean is

$$\mu = 1(.05) + 2(.1) + 3(.7) + 4(.1) + 5(.05) = 3,$$

so the population variance is

$$\sigma^2 = (1 - 3)^2(.05) + (2 - 3)^2(.1) + (3 - 3)^2(.7) + (4 - 3)^2(.1) + (5 - 3)^2(.05) = .6,$$

and the population standard deviation is $\sigma = \sqrt{.6} = .775$.

Problems

9. For the probability function

$$x : 0, 1$$

$$p(x) : .7, .3$$

verify that the mean and variance are .3 and .21, respectively. What is the probability of getting a value less than the mean?

10. Imagine that an auto manufacturer wants to evaluate how potential customers will rate handling for a new car being considered for production. Also suppose that if all potential customers were to rate handling on a 4-point scale, 1 being poor and 4 being excellent, the corresponding probabilities associated with these ratings would be $p(1) = .2$, $p(2) = .4$, $p(3) = .3$, and $p(4) = .1$. Determine the population mean, variance and standard deviation.

11. If the possible values for x are 1, 2, 3, 4, 5 with probabilities .2, .1, .1, .5, .1, respectively, what are the population mean, variance and standard deviation?

12. In problem 11, determine the probability of getting a value within one standard deviation of the mean. That is, determine the probability of getting a value between $\mu - \sigma$ and $\mu + \sigma$.

13. If the possible values for x are 1, 2, 3 with probabilities .2, .6 and .2, respectively, what is the mean and standard deviation?

14. In problem 13, suppose the possible values for x are now 0, 2, 4 with the same probabilities as before. Will the standard deviation increase, decrease, or stay the same? Verify your answer.

15. For the probability function

$$x: 1, 2, 3, 4, 5$$

$$p(x): .15, .2, .3, .2, .15$$

determine the mean, the variance, and the probability that the a value is less than the mean.

16. For the probability function

$$x: 1, 2, 3, 4, 5$$

$$p(x): .1, .25, .3, .25, .1$$

would you expect the variance to be larger or smaller than the variance associated with the probability function used in the previous exercise? Verify your answer by computing the variance for the probability function given here.

17. For the probability function

$$x: 1, 2, 3, 4, 5$$

$$p(x): .2, .2, .2, .2, .2$$

would you expect the variance to be larger or smaller than the variance associated with the probability function used in the previous exercise? Verify your answer by computing the variance.

4.3 Conditional probability and independence

Conditional probability refers to the probability of some event given that some other event has occurred. For example, what is the probability of developing heart disease

Table 4.1 Hypothetical probabilities for getting a flu shot and getting the flu

Get a shot	Get the flu		
	Yes	No	
Yes	.25	.20	.45
No	.28	.27	.55
	.53	.47	1.00

given that your cholesterol level is 250? What is the probability of winning one million dollars or more given that you buy 100 lottery tickets?

A convenient way of illustrating how conditional probabilities are computed is in terms of what are called *contingency tables*, an example of which is shown in table 4.1. In the contingency table are the probabilities associated with four mutually exclusive groups: individuals who (1) receive a flu shot and get the flu, (2) do not receive a flu shot and get the flu, (3) receive a flu shot and do not get the flu, and (4) do not receive a flu shot and do not get the flu. The last column shows what are called the *marginal probabilities*. For example, the probability of getting a flu shot is $0.25 + 0.20 = 0.45$. Put another way, it is the sum of the probabilities associated with two mutually exclusive events. The first event is getting the flu and simultaneously getting a flu shot, which has probability .25. The second event is not getting the flu and simultaneously getting a flu shot, which has probability .2. The last line of table 4.1 shows the marginal probabilities associated with getting or not getting the flu. For example, the probability of getting the flu is $0.25 + 0.28 = 0.53$.

Now consider the probability of someone getting the flu given that they receive a flu shot, and for convenience, view probabilities as proportions. So among all individuals we might observe, according to table 4.1, the proportion of people who get a flu shot is 0.45. Among the people who got a flu shot, the proportion who got the flu is $0.25/0.45 = 0.56$. That is, the probability of getting the flu, given that an individual received a flu shot, is 0.56.

Notice that a conditional probability is determined by altering the sample space. In the illustration, the proportion of all people who got a flu shot is 0.45. But restricting attention to individuals who got a flu shot means that the sample space has been altered. More precisely, the contigency table reflects four possible outcomes, but by focusing exclusively on individuals who got a flu shot, the sample space is reduced to two outcomes.

In a more general notation, if A and B are any two events, and if we let $P(A)$ represent the probability of event A and $P(A \text{ and } B)$ represent the probability that events A and B occur simultaneously, then the conditional probability of A, given that B has occurred, is

$$P(A|B) = \frac{P(A \text{ and } B)}{P(B)}. \tag{4.3}$$

In the illustration, A is the event of getting the flu and B is the event of getting a flu shot. So according to table 4.1, $P(A \text{ and } B) = 0.25$, $P(B) = 0.45$, so $P(A|B) = 0.25/0.45$, as previously indicated.

Example 1

From table 4.1, the probability that someone does not get the flu, given that they get a flu shot, is

$$.20/.45 = .44.$$

Independence and dependence

Roughly, two events are *independent* if the probability associated with the first event is not altered when the second event is known. If the probability is altered, the events are *dependent*.

Example 2

According to table 4.1, the probability that someone gets the flu is 0.53. The event that someone gets the flu is independent of the event that someone gets a flu shot if among the individuals getting a shot, the probability of getting the flu remains 0.53. We have seen, however, that the probability of getting the flu, given that the person gets a shot, is 0.56, so these two events are dependent.

Consider any two variables, say X and Y, and let x and y be any two possible values corresponding to these variables. We say that the variables X and Y are independent if for any x and y we might pick,

$$P(Y = y|X = x) = P(Y = y). \tag{4.4}$$

Otherwise they are said to be dependent.

Example 3

Imagine that married couples are asked to rate the the extent marital strife is reduced by following the advice in a book on having a happy marriage. Assume that both husbands and wives rate the effectiveness of the book with the values 1, 2, and 3, where the values stand for fair, good, and excellent, respectively. Further assume that the probabilities associated with the possible outcomes are as shown in table 4.2. We see that the probability a wife (Y) gives a rating of 1 is 0.2. In symbols, $P(Y = 1) = .2$. Furthermore, $P(Y = 1|X = 1) = .02/.1 = .2$, where $X = 1$ indicates that the wife's husband gave a rating of 1. So the event $Y = 1$ is independent of the event $X = 1$. If the probability had changed, we could stop and say that X and Y are dependent. But to say that they

Table 4.2 Hypothetical probabilities for rating a book

	Husband (X)			
Wife (Y)	1	2	3	
1	.02	.10	.08	0.2
2	.07	.35	.28	0.7
3	.01	.05	.04	0.1
	0.1	0.5	0.4	

are independent requires that we check all possible outcomes. For example, another possible outcome is $Y = 1$ and $X = 2$. We see that $P(Y = 1|X = 2) = .1/.5 = .2$, which again is equal to $P(Y = 1)$. Continuing in this manner, it can be seen that for any possible values for Y and X, the corresponding events are independent, so we say that X and Y are independent. That is, they are independent regardless of what their respective values might be.

Now the notion of *dependence* is described in a slightly more general context that contains contingency tables as a special case. A common and fundamental question in applied research is whether information about one variable influences the probabilities associated with another variable. For example, in a study dealing with diabetes in children, one issue of interest was the association between a child's age and the level of serum C-peptide at diagnosis. For convenience, let X represent age and Y represent C-peptide concentration. For any child we might observe, there is some probability that her C-peptide concentration is less than 3, or less than 4, or less than c, where c is any constant we might pick. The issue at hand is whether information about X (a child's age) alters the probabilities associated with Y (a child's C-peptide level). That is, does the conditional probability of Y, given X, differ from the probabilities associated with Y when X is not known or ignored. If knowing X does not alter the probabilities associated with Y, we say that X and Y are independent. Equation (4.4) is one way of providing a formal definition of independence. An alternative way is to say that X and Y are independent if

$$P(Y \le y|X = x) = P(Y \le y) \tag{4.5}$$

for any x and y values we might pick. Equation (4.4) implies equation (4.5). Yet another way of describing independence is that for any x and y values we might pick,

$$\frac{P(Y = y \text{ and } X = x)}{P(X = x)} = P(Y = y), \tag{4.6}$$

which follows from equation (4.4). From this last equation it can be seen that if X and Y are independent, then

$$P(X = x \text{ and } Y = y) = P(X = x)P(Y = y). \tag{4.7}$$

Equation (4.7) is called the *product rule* and says that if two events are independent, the probability that they occur simultaneously is equal to the product of their individual probabilities.

Example 4

In table 4.1, if getting a flu shot is independent of getting the flu, then the probability of both getting a flu shot and getting the flu is $.45 \times .53 = 0.2385$.

Example 5

Consider again the diabetes in children study where one of the variables of interest was C-peptide concentrations at diagnosis. Suppose that for all children we might measure, the probability of having a C-peptide concentration less than or equal to 3 is $P(Y \le 3) = .4$. Now consider only children who are 7 years old and imagine that for this subpopulation of children, the probability of having a C-peptide concentration less than 3 is 0.2. In symbols, $P(Y \leq 3|X = 7) = 0.2$. Then C-peptide concentrations and age are said to be dependent because knowing that the child's age is 7 alters the probability that the child's C-peptide concentration is less than 3. If instead $P(Y \leq 3|X = 7) = 0.4$, the events $Y \leq 3$ and $X = 7$ are independent. More generally, if, for any x and y we pick, $P(Y \leq y|X = x) = P(Y = y)$, then C-peptide concentrations and age are independent.

Problems

18. For the following probabilities

	Income		
Age	High	Medium	Low
< 30	.030	.180	.090
30–50	.052	.312	.156
> 50	.018	.108	.054

determine (a) the probability someone is under 30, (b) the probability that someone has a high income given that they are under 30, (c) the probability of someone having a low income given that they are under 30, and (d) the probability of a medium income given that they are over 50.

19. For the previous exercise, are income and age independent?

20. Coleman (1964) interviewed 3,398 schoolboys and asked them about their self-perceived membership in the "leading crowd." Their response was either yes, they were a member, or no they were not. The same boys were also asked about their attitude concerning the leading crowd. In particular, they were asked whether membership meant that it does not require going against one's principles sometimes or whether they think it does. Here, the first response will be indicated by a 1, while the second will be indicated by a 0. The results were as follows:

	Attitude	
Member	1	0
Yes	757	496
No	1071	1074

The sample size is 3,398. So, for example, the relative frequency of the event (yes, 1) is 757/3398. Treat the relative frequencies as probabilities and determine (a) the probability that an arbitrarily chosen boy responds yes, (b) $P(\text{yes}|1)$, (c) $P(1|\text{yes})$, (d) whether the response yes is independent of the attitude 0, (e) the probability of a (yes and 1) or a (no and 0) response, (f) the probability of not responding (yes and 1), (g) the probability of responding yes or 1.

21. Let Y be the cost of a home and let X be a measure of the crime rate. If the variance of the cost of a home changes with X, does this mean that cost of a home and the crime rate are dependent?

22. If the probability of $Y < 6$ is .4 given that $X = 2$, and if the probability of $Y < 6$ is .3 given that $X = 4$, does this mean that X and Y are dependent?
23. If the range of possible Y values varies with X, does this mean that X and Y are dependent?

4.4 The Binomial probability function

The most important discrete distribution is the binomial. It arises in situations where only two possible outcomes are possible when making a single observation. The outcomes might be, for example, yes and no, success and failure, agree and disagree. Such random variables are called *binary*. Typically the number 1 is used to represent a success and a failure is represented by 0. A common convention is to let p represent the probability of success and to let $q = 1 - p$ be the probability of a failure. When dealing with the binomial probability function, the random variable X represents the total number of successes among n observations.

The immediate goal is to describe how to compute the probability of x successes among n trials or observations. In symbols, we want to evaluate $P(X = x)$. For example, imagine that five people under go a new type of surgery and you observe whether the disorder is successfully treated. If the probability of a success is $p = .6$, what is the probability that exactly three of the five surgeries will be a success?

Assuming that the outcomes are independent and that the probability of success is the same every time the surgery is performed, there is a convenient formula for solving this problem based on the *binomial probability function*. It says that among n observations, the probability of exactly x successes, $P(X = x)$, is given by

$$p(x) = \binom{n}{x} p^x q^{n-x}. \tag{4.8}$$

Here, $n = 5$, $x = 3$, and the goal is to determine $p(3)$. The first term on the right side of this equation, called the binomial coefficient, is defined to be

$$\binom{n}{x} = \frac{n!}{x!(n-x)!},$$

where $n!$ represents n factorial. That is,

$$n! = 1 \times 2 \times 3 \times \cdots \times (n-1) \times n.$$

For example, $1! = 1, 2! = 2$, and $3! = 6$. By convention, $0! = 1$.

Example 1

For the situation just described regarding whether a surgical procedure is a success, the probability of exactly three successes, can be determined as follows.

$$\begin{aligned} n! &= 1 \times 2 \times 3 \times 4 \times 5 = 120, \\ x! &= 1 \times 2 \times 3 = 6, \\ (n-x)! &= 2! = 2, \end{aligned}$$

in which case

$$p(3) = \frac{120}{6 \times 2}(.6^3)(.4^2) = .3456.$$

Example 2

Imagine 10 couples who recently got married. What is the probability that four of the ten couples will report that they are happily married at the end of 1 year? Assuming responses by couples are independent, and that the probability of success is $p = .3$, the probablity that exactly $x = 4$ couples will report that they are happily married is

$$p(4) = \frac{10!}{4! \times 6!}(.3^4)(.7^6) = .2001.$$

Often attention is focused on the probability of *at least x* successes in n trials or *at most x* successes, rather than the probability of getting *exactly x* successes. In the last illustration, you might want to know the probability that four couples or fewer are happily married as opposed to exactly four. The former probability consists of five mutually exclusive events, namely, $x = 0$, $x = 1$, $x = 2$, $x = 3$, and $x = 4$. Thus, the probability that four couples or fewer are happily married is

$$P(X \leq 4) = p(0) + p(1) + p(2) + p(3) + p(4).$$

In summation notation,

$$P(X \leq 4) = \sum_{x=0}^{4} p(x).$$

More generally, the probability of k successes or less in n trials is

$$P(X \leq k) = \sum_{x=0}^{k} p(x)$$

$$= \sum_{x=0}^{k} \binom{n}{x} p^x q^{n-x}.$$

For any k between 0 and n, table 2 in appendix B gives the value of $P(X \leq k)$ for various values of n and p. Returning to the illustration where $p = .3$ and $n = 10$, table 2 reports that the probability of four successes or less is .85. Notice that the probability of five successes or more is just the complement of getting four successes or less, so

$$P(X \geq 5) = 1 - P(X \leq 4) = 1 - .85$$

$$= .15.$$

In general,

$$P(X \geq k) = 1 - P(X \leq k - 1),$$

so $P(X \geq k)$ is easily evaluated with table 2.

Expressions like

$$P(2 \leq x \leq 8),$$

meaning you want to know the probability that the number of successes is between 2 and 8, inclusive, can also be evaluated with table 2 by noting that

$$P(2 \leq x \leq 8) = P(x \leq 8) - P(x \leq 1).$$

In words, the event of eight successes or less can be broken down into the sum of two mutually exclusive events: the event that the number of successes is less than or equal to 1 and the event that the number of successes is between 2 and 8, inclusive. Rearranging terms yields the last equation. The point is that $P(2 \le x \le 8)$ can be written in terms of two expressions that are easily evaluated with table 2 in appendix B.

Example 3

Assume $n = 10$ and $p = .5$. From table 2 in appendix B, $P(X \le 1) = .011$ and $P(X \le 8) = .989$, so

$$P(2 \le X \le 8) = .989 - .011 = .978.$$

A related problem is determining the probability of one success or less or nine successes or more. The first part is simply read from table 2 and can be seen to be .011. The probability of nine successes or more is the complement of eight successes or less, so $P(X \ge 9) = 1 - P(X \le 8) = 1 - .989 = .011$, again assuming that $n = 10$ and $p = .5$. Thus, the probability of one success or less or nine successes or more is $.011 + .011 = .022$. In symbols,

$$P(X \le 1 \text{ or } X \ge 9) = .022.$$

There are times when you will need to compute the mean and variance of a binomial probability function once you are given n and p. It can be shown that the (population) mean and variance are given by

$$\mu = E(X)$$
$$= np,$$

and

$$\sigma^2 = npq.$$

Example 4

If $n = 16$ and $p = .5$, the mean of the binomial probability function is $\mu = np = 16(.5) = 8$. That is, on average, eight of the 16 observations will be a success, while the other eight will not. The variance is $\sigma^2 = npq = 16(.5)(.5) = 4$, so the standard deviation is $\sigma = \sqrt{4} = 2$. If instead, $p = .3$, *then* $\mu = 16(.3) = 4.8$. That is, the average number of successes is 4.8.

In most situations, p, the probability of a success, is not known and must be estimated based on x, the observed number of successes. The estimate typically used is x/n, the proportion of observed successes. Often this estimator is written as

$$\hat{p} = \frac{x}{n},$$

where $\hat{p}$ is read p hat. It can be shown that

$$E(\hat{p}) = p.$$

That is, if you were to repeat an experiment millions of times (and in theory infinitely many times), each time sampling n observations, the average of the resulting $\hat{p}$ values would be p. It can also be shown that the variance of $\hat{p}$ is

$$\sigma^2_{\hat{p}} = \frac{pq}{n}.$$

Example 5

If you sample 25 people and the probability of success is .4, the variance of $\hat{p}$ is

$$\sigma^2_{\hat{p}} = \frac{.4 \times .6}{25} = .098.$$

The characteristics and properties of the binomial probability function can be summarized as follows:

- The experiment consists of exactly n independent trials.
- Only two possible outcomes are possible on each trial, usually called 'success' and 'failure'.
- Each trial has the same probability of success, p.
- $q = 1 - p$ is the probability of a failure.
- There are x successes among the n trials.
- $p(x) = \binom{n}{x} p^x q^{n-x}$ is the probability of x successes in n trials, $x = 0, 1, \ldots, n$.
- $\binom{n}{x} = \frac{n!}{x!(n-x)!}$.
- You estimate p with $\hat{p} = \frac{x}{n}$, where x is the total number of successes.
- $E(\hat{p}) = p$.
- The variance of $\hat{p}$ is $\sigma^2 = \frac{pq}{n}$.
- The average or expected number of successes in n trials is $\mu = E(X) = np$.
- The variance of X is $\sigma^2 = npq$.

Example 6

Two basketball teams are contending for the title of national champion. The first team to win four games is declared champion. If we assume that the probability of a win is the same each time they play, and that the outcome for any two games are independent, can we use the binomial probability function to determine the probability of four wins? The answer is no. The problem here is that the number of games to be played, n, is not fixed. If the two teams played exactly seven games, then the binomial probability function would apply. But the total number of games to be played is 4, 5, 6 or 7.

Problems

24. For a binomial with $n = 10$ and $p = .4$, use table 2 in appendix B to determine (a) $p(0)$, the probability of exactly 0 successes, (b) $P(X \leq 3)$, (c) $P(X < 3)$, (d) $P(X > 4)$, (e) $P(2 \leq X \leq 5)$.

25. For a binomial with $n = 15$ and $p = .3$, use table 2 in appendix B to determine (a) $p(0)$, the probability of exactly 0 successes, (b) $P(X \leq 3)$, $P(X < 3)$, (c) $P(X > 4)$, (d) $P(2 \leq X \leq 5)$.

26. For a binomial with $n = 15$ and $p = .6$, use table 2 to determine the probability of exactly 10 successes.

27. For a binomial with $n = 7$ and $p = .35$, what is the probability of exactly 2 successes.

28. For a binomial with $n = 18$ and $p = .6$, determine the mean and variance of X, the total number of successes.

29. For a binomial with $n = 22$ and $p = .2$, determine the mean and variance of X, the total number of successes.

30. For a binomial with $n = 20$ and $p = .7$, determine the mean and variance of $\hat{p}$, the proportion of observed successes.

31. For a binomial with $n = 30$ and $p = .3$, determine the mean and variance of $\hat{p}$.

32. For a binomial with $n = 10$ and $p = .8$, determine (a) the probability that $\hat{p}$ is less than or equal to .7, (b) the probability that $\hat{p}$ is greater than or equal to .8, (c) the probability that $\hat{p}$ is exactly equal to .8.

33. A coin is rigged so that when it is flipped, the probability of a head is .7. If the coin is flipped three times, which is the more likely outcome, exactly three heads, or two heads and a tail?

34. Imagine that the probability of head when flipping a coin is given by the binomial probability function with $p = .5$. (So the outcomes are independent.) If you flip the coin nine times and get nine heads, what is the probability of head on the tenth flip?

35. The Department of Agriculture of the United States reports that 75% of all people who invest in the futures market lose money. Based on the binomial probability function, with $n = 5$, determine

    (a) the probability that all 5 lose money.
    (b) the probability that all 5 make money.
    (c) the probability that at least 2 lose money.

36. If for a binomial, $p = .4$ and $n = 25$, determine (a) $P(X < 11)$, (b) $P(X \leq 11)$, (c) $P(X > 9)$ and (d) $P(X \geq 9)$

37. In the previous problem, determine the mean of X, the variance of X, the mean of $\hat{p}$), and the variance of $\hat{p}$.

4.5 The normal curve

This section explains how probabilities are computed when dealing with continuous variables that follow what is called the normal curve. In contrast to discrete variables, probabilities associated with continuous variables are given by the area under a curve. The equation for this curve is called a *probability density function*, which is typically labeled $f(x)$. The normal curve, or normal distribution, plays a central role in a wide range of statistical techniques and is routinely used in a plethora of disciplines including physics, astronomy, manufacturing, economics, meteorology, medicine, biology, agriculture, sociology, geodesy, anthropology, communications, accounting, education, and psychology. Many published papers make it clear that for a wide range of situations, the normal distribution provides a very convenient and highly accurate method for analyzing data. But there is also a wide range of situations where it performs poorly. So in terms of building a good foundation for understanding data, a crucial

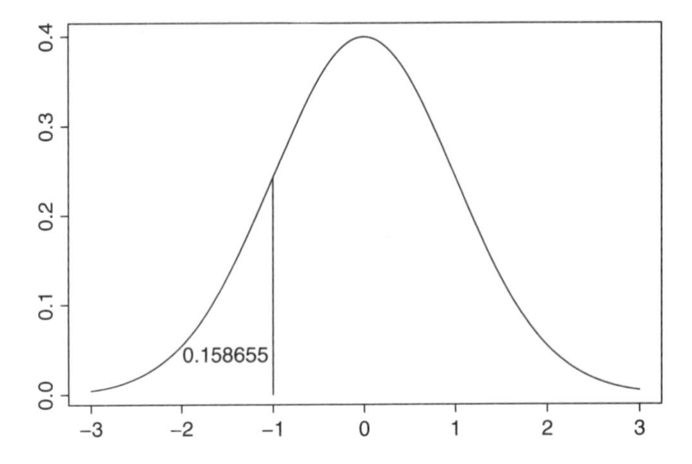

Figure 4.1 A normal distribution mean $\mu = 0$. The area under the curve and to the left of -1 is 0.158655, which is the probability that an observation is less than or equal to -1.

component is understanding when the normal distribution serves us well, and when it can be misleading and highly unsatisfactory.

An example of a normal distribution is shown in figure 4.1. Note the vertical line at -1. It can be shown that the area under the normal curve and to the left of -1 is 0.158655. This means that if a random variable X has the normal distribution shown in figure 4.1, the probability that X is less than or equal to -1 is 0.158655. In symbols, $P(X \leq -1) = 0.158655$.

Normal distributions have the following important properties:

1. The total area under the curve is 1. (This is a requirement of any probability density function.)
2. All normal distributions are bell-shaped and symmetric about their mean, μ.
3. Although not indicated in figure 4.1, all normal curves extend from $-\infty$ to ∞ along the x-axis.
4. If the variable X has a normal distribution, the probability that X has a value within one standard deviation of the mean is .68 as indicated in figure 4.2. In symbols, if X has a normal distribution, $P(\mu - \sigma < X < \mu + \sigma) = .68$ regardless of what the population mean and variance happen to be. The probability of being within two standard deviations is approximately .954. In symbols, $P(\mu - 2\sigma < X < \mu + 2\sigma) = .954$. The probability of being within three standard deviations is $P(\mu - 3\sigma < X < \mu + 3\sigma) = .9975$.
5. The probability density function of a normal distribution is

$$f(x) = \frac{1}{\sigma\sqrt{2\pi}} \exp\left[-\frac{(x-\mu)^2}{2\sigma^2}\right], \tag{4.9}$$

where as usual, μ and σ^2 are the mean and variance. This equation does not play a direct role in this book and is reported simply for informational purposes. The main point is that for any distribution to qualify as a normal distribution, its probability density function must be given by this last equation. It turns out that many distributions are symmetric and bell-shaped, yet they do not qualify as a normal distribution. That is, the

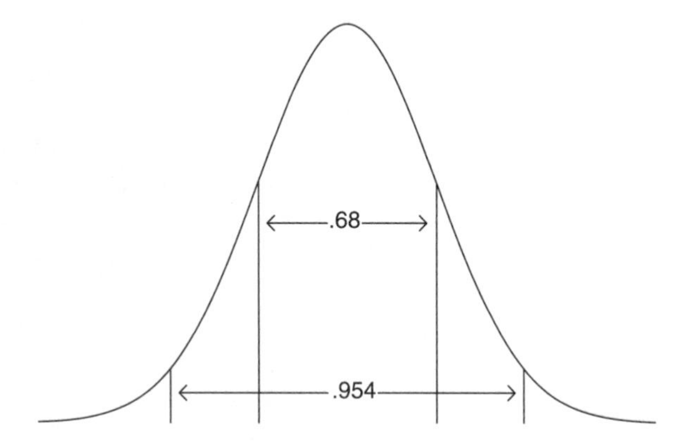

Figure 4.2 For all normal distributions, the probability that an observation is within one standard deviation of the mean is .68. The probability of being within two standard deviations is .954.

equation for these distributions do not conform to the equation for a normal curve. Another important point is that the probability density function is determined by the mean and variance. If, for example, we want to determine the probability that a variable is less than 25, this probability is completely determined by the mean and variance *if* we assume normality.

Figure 4.3 shows three normal distributions, two of which have equal means of zero but standard deviations $\sigma = 1$ and $\sigma = 1.5$. The other distribution again has standard deviation $\sigma = 1$, but now the mean is $\mu = 2$. There are two things to notice. First, if two normal distributions have equal variances but unequal means, the two probability curves are centered around different values but otherwise they are identical. Second, for *normal* distributions, there is a distinct and rather noticeable difference between the two curves when the standard deviation increases from 1 to 1.5.

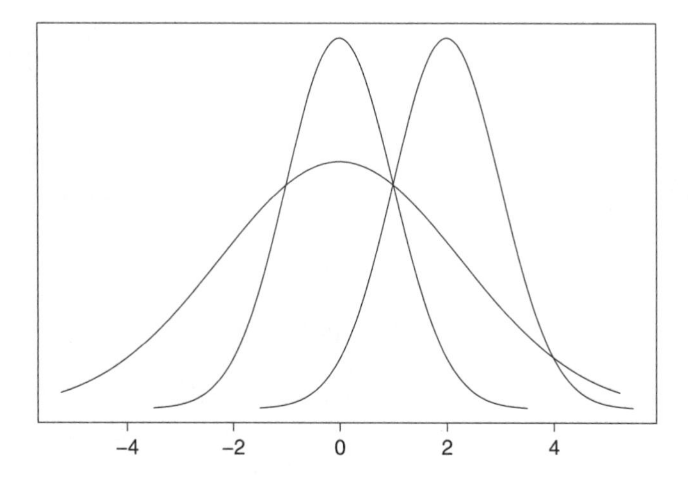

Figure 4.3 The left two distributions have the same mean and standard deviations 1 and 1.5. The right distribution has a mean of 2 and standard deviation 1.

Historical remarks

Some brief comments about the history of the normal distribution will help put its relative merits in perspective. Consider a binomial probability function with $n = 1{,}000$ and $p = .5$. To be concrete, imagine we flip a coin 1,000 times and that unknown to use, the probability of a head is .5. Note that if we get 450 heads, we would estimate the probability of a head to be .45. In a similar manner, if we get 550 heads, we would estimate the probability of a head to be .55. Abraham de Moivre (1667–1754) asked the following question: what is the probability of getting between 450 and 550 heads? That is, what is the probability that the estimate of p, the probability of success in a binomial distribution, will be between .45 and .55 when in fact $p = .5$? Put another way, what is the probability that the estimated probability of a success will be close to its true value?

The problem, of course, is that without a computer, the calculations needed to answer this question are prohibitive. Even calculating the probability of exactly 450 heads is a tremendously difficult task, and the problem requires that one also compute the probability of exactly 451 heads, 452 heads, and so on, until we get to 550 heads, and then these 101 values would need to be summed. So de Moivre set out to find a reasonably accurate approximation for solving the problem. Over 12 years later, he had a solution, which he announced in 1733, and which was based on the normal distribution.

Initially, de Moivre's derivation of the normal distribution generated little interest. Many years later, Laplace was searching for a family of distributions that might be used to solve some of the basic problems covered in this book. But his best attempt led to an approach that proved to be highly impractical in terms of both theoretical and computational details. Karl Gauss realized that a slight modification of Laplace's approach greatly simplified the theoretical and computational details needed to solve practical problems. His slight modification resulted in the normal distribution. Gauss's work was so influential, the normal curve is sometimes called the Gaussian distribution. But although the normal curve proved to be extremely convenient from a mathematical point of view, both Laplace and Gauss were concerned about how to justify the use of the normal distribution when addressing practical problems. In 1809, Gauss described his first attempt at solving this problem, which proved to be highly unsatisfactory. In 1810, Laplace announced an alternative approach, the so-called central limit theorem, the details of which are covered in chapter 5. Laplace's result forms the basis for assuming normality for an extremely wide range of problems. We will see that in some instances, this justification is extremely successful, but in others it is not.

Another issue should be discussed: If we could measure all the individuals (or things) who constitute a population of interest, will a plot of the observations follow a normal distribution? The answer is rarely, and some would argue never. The first study to determine whether data follow normal distribution was conducted by Wilhelm Bessel (1784–1846) in the year 1818. His astronomical data appeared to be reasonably bell-shaped, consistent with a normal curve, but Bessel made an important observation: The distribution of his data had thicker or heavier tails than what would be expected for a normal curve. Bessel appears to have made an extremely important observation. Roughly, heavy-tailed distributions tend to generate outliers, which turn out to create practical problems to be illustrated. But it would be about another 150 years before the practical importance of Bessel's observation would be fully appreciated.

During the nineteenth century, various researchers came to the conclusion that data follow a normal curve, a view that stems from results covered in chapter 5. Indeed, the term normal distribution stems from the first paper ever published by Karl Pearson. He was so convinced that the bell-shaped distribution used by Laplace and Gauss applied to data, he named it the normal distribution, meaning the distribution we should expect. To his credit, Pearson examined data from actual studies to see whether his belief could be confirmed, and eventually he concluded that his initial speculation was incorrect. He attempted to deal with this problem by introducing a larger family of distributions for describing data, but his approach does not play a role today when analyzing data. More modern investigations confirm Pearson's conclusion that data generally do not follow a normal curve (e.g., Micceri, 1989), but this does not mean that the normal curve has no practical value. What is important is whether it provides an adequate approximation, and as already mentioned, in some cases it does, and in others it does not.

4.6 Computing probabilities associated with normal curves

For the moment, we ignore any practical limitations associated with the normal curve and focus on some basic issues when the normal distribution is used. To be concrete, assume that human infants have birth weights that are normally distributed with a mean of 3,700 grams and a standard deviation of 200 grams. What is the probability that a baby's birth weight will be less than or equal to 3,000 grams? As previously explained, this probability is given by the area under the normal curve, but simple methods for computing this area are required. Today the answer is easily obtained on a computer. But for pedagogical reasons a more traditional method is covered here. We begin by considering the special case where the mean is zero and the standard deviation is one ($\mu = 0, \sigma = 1$) after which we illustrate how to compute probabilities for any mean and standard deviation.

The standard normal distribution

The *standard normal distribution* is a normal distribution with mean $\mu = 0$ and standard deviation $\sigma = 1$; it plays a central role in many areas of statistics. As is typically done, Z is used to represent a variable that has a standard normal distribution. Our immediate goal is to describe how to determine the probability that an observation randomly sampled from a standard normal distribution is less than any constant c we might choose.

These probabilities are easily determined using table 1 in appendix B which reports the probability that a standard normal random variable has probability less than or equal to c for $c = -3.00, -2.99, -2.98, \ldots, -0.01, 0, .01, \ldots 3.00$. The first entry in the first column shows -3. The column next to it gives the corresponding probability, .0013. That is, the probability that a standard normal random variables is less than or equal to -3 is $P(Z \leq -3) = .0013$. Going down the first column we see the entry -2.08, and the column next to it indicates that the probability of a standard normal variable being less than or equal to -2.08 is .0188. Looking at the last entry in the third column, we see -1.55, the entry just to the right, in the fourth column, is .0606, so $P(Z \leq -1.55) = .0606$. This probability corresponds to the area in the left portion of figure 4.4. Because the standard normal curve is symmetric about zero, the probability

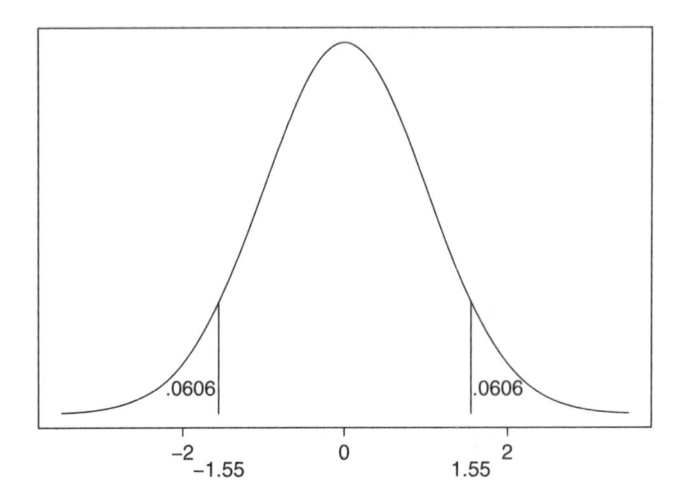

Figure 4.4 The left tail indicates that for a standard normal distribution, the probability of a value less than or equal to −1.55 is .0606, and the probability of a value greater or equal to 1.55 is .0606 as well.

that X is greater than 1.55 is also .0606, which is shown in the right portion of figure 4.4. Again, looking at the first column of table 1 in appendix B, we see the value $c = 1.53$, and next to it is the value .9370 meaning that $P(Z \leq 1.53) = .9370$.

In applied work, there are three types of probabilities that need to be determined:

1. $P(Z \leq c)$, the probability that a standard normal random variable is less than or equal to c,
2. $P(Z \geq c)$, the probability that a standard normal random variable is greater than or equal to c, and
3. $P(a \leq Z \leq b)$, the probability that a standard normal random variable is between the values a and b.

The first of these is determined from table 1 in appendix B, as already indicated. Because the area under the curve is one, the second is given by

$$P(Z \geq c) = 1 - P(Z \leq c).$$

The third is given by

$$P(a \leq Z \leq b) = P(Z \leq b) - P(Z \leq a).$$

Example 1

Determine $P(Z \geq 1.5)$, the probability that a standard normal random variable is greater than 1.5. From table 1 in appendix B, $P(Z \leq 1.5) = .9332$. Therefore, $P(Z \geq 1.5) = 1 - .9332 = .0668$.

Example 2

Next we determine $P(-1.96 \leq Z \leq 1.96)$, the probability that a standard normal random variable is between −1.96 and 1.96. From table 1 in appendix B, $P(Z \leq 1.96) = .975$. Also, $P(Z \leq -1.96) = .025$, so $P(-1.96 \leq Z \leq 1.96) = .975 - .025 = .95$.

In some situations it is necessary to use table 1 (in appendix B) backwards. That is, we are given a probability and the goal is to determine c. For example, if we are told that $P(Z \le c) = .99$, what is c? We simply find where .99 happens to be in table 1 under the columns headed by $P(Z \le c)$, and then read the number to the left, under the column headed by c. The answer is 2.33.

Before continuing, this is a convenient moment to illustrate the notion of a *quantile*. If for any variable X, $P(X \le c) = .5$, then c is said to be the .5 quantile. The .5 quantile corresponds to the *population median*, which is estimated with the sample median described in chapter 2. If $P(X \le c) = .2$, c is the .2 quantile, and if $P(X \le c) = .75$, c is the .75 quantile. Percentiles are just quantiles multiplied by 100.

Example 3

We have seen that for a standard normal distribution, $P(Z \le 2.33) = .99$. This says that 2.33 is the .99 quantile. From table 1 in appendix B, $P(Z \le -1.96) = .025$. Said another way, -1.96 is the .025 quantile of a standard normal distribution. To determine the .4013 quantile of a standard normal, we simply go to table 1 in appendix B and find entry the value z such that $P(Z \le z) = .4013$. The answer is $-.25$.

When using table 1 in appendix B, two related problems also arise. The first is determining c given the value of

$$P(Z \ge c).$$

A solution is obtained by noting that the area under the curve is one, so $P(Z \ge c) = 1 - P(Z \le c)$, which involves a quantity we can determine from table 1. That is, you compute $d = 1 - P(Z \ge c)$ and then determine c such that

$$P(Z \le c) = d.$$

Example 4

To determine c if $P(Z \ge c) = .9$, first compute $d = 1 - P(Z \le c) = 1 - .9 = .1$. Then c is given by $P(Z \le c) = .1$. Referring to table 1 in appendix B, $c = -1.28$.

The other type of problem is determining c given

$$P(-c \le Z \le c).$$

Letting $d = P(-c \le Z \le c)$, the answer is given by

$$P(Z \le c) = \frac{1+d}{2}.$$

Example 5

To determine c if $P(-c \le Z \le c) = .9$, let $d = P(-c \le Z \le c) = .9$ and then compute $(1+d)/2 = (1+.9)/2 = .95$. Then c is given by $P(Z \le c) = .95$. Referring to table 1 in appendix B, $c = 1.645$.

Solution for any normal distribution

Now consider any normal random variable having mean μ and standard deviation σ. The next goal is to describe how to determine the probability of an observation being less than c, where as usual, c is any constant that might be of interest. The solution is based on *standardizing* a normal random variable, which means that we subtract the population mean μ and divide by the standard deviation, σ. In symbols, we standardize a normal random variable X by transforming it to

$$Z = \frac{X - \mu}{\sigma}. \tag{4.10}$$

The quantity Z is often called a Z score; it reflects how far the value X is from the mean in terms of the standard deviation.

Example 6

If $Z = .5$, then X is a half standard deviation away from the mean. If $Z = 2$, then X is two standard deviations away from the mean.

It can be shown that if X has a normal distribution, then the distribution of Z is standard normal. In particular, the probability that a normal random variable X is less than or equal to c is

$$P(X \leq c) = P\left(Z \leq \frac{c - \mu}{\sigma}\right). \tag{4.11}$$

Example 7

Someone claims that the cholesterol levels in adults have a normal distribution with mean $\mu = 230$ and standard deviation $\sigma = 20$. If this is true, what is the probability that an adult will have a cholesterol level less than or equal to 200? Referring to equation (4.11), the answer is

$$P(X \leq 200) = P\left(Z \leq \frac{200 - 230}{20}\right) = P(Z < -1.5) = .0668,$$

where .0668 is read from table 1 in appendix B. This means that the probability of an adult having a cholesterol level less than 200 is .0668.

In a similar manner, we can determine the probability that an observation is greater than or equal to 240 or between 210 and 250. More generally, for any constant c that is of interest, we can determine the probability that an observation is greater than c with the equation

$$P(X \geq c) = 1 - P(X \leq c),$$

the point being that the right side of this equation can be determined with equation (4.11). In a similar manner, for any two constants a and b,

$$P(a \leq X \leq b) = P(X \leq b) - P(X \leq a).$$

Example 8

Continuing the last example, determine the probability of observing an adult with a cholesterol level greater than or equal to 240. We have that

$$P(X \geq 240) = 1 - P(X \leq 240).$$

Referring to equation (4.11),

$$P(X \leq 240) = P\left(Z \leq \frac{240 - 230}{20}\right) = P(Z < .5) = .6915,$$

so

$$P(X \geq 240) = 1 - .6915 = .3085.$$

In words, the probability of an adult having a cholesterol level greater than or equal to 240 is .3085.

Example 9

Continuing the cholesterol example, we determine

$$P(210 \leq X \leq 250).$$

We have that

$$P(210 \leq X \leq 250) = P(X \leq 250) - P(X \leq 210).$$

Now

$$P(X \leq 250) = P\left(Z < \frac{250 - 230}{20}\right) = P(Z \leq 1) = .8413$$

and

$$P(X \leq 210) = P\left(Z < \frac{210 - 230}{20}\right) = P(Z \leq -1) = .1587,$$

so

$$P(210 \leq X \leq 250) = .8413 - .1587 = .6826,$$

meaning that the probability of observing a cholesterol level between 210 and 250 is .6826.

Problems

38. Given that Z has a standard normal distribution, use table 1 in appendix B to determine (a) $P(Z \geq 1.5)$, (b) $P(Z \leq -2.5)$, (c) $P(Z < -2.5)$, (d) $P(-1 \leq Z \leq 1)$.

39. If Z has a standard normal distribution, determine (a) $P(Z \leq .5)$, (b) $P(Z > -1.25)$, (c) $P(-1.2 < Z < 1.2)$, (d) $P(-1.8 \leq Z < 1.8)$.

40. If Z has a standard normal distribution, determine (a) $P(Z < -.5)$, (b) $P(Z < 1.2)$, (c) $P(Z > 2.1)$, (d) $P(-.28 < Z < .28)$.

41. If Z has a standard normal distribution, find c such that (a) $P(Z \leq c) = .0099$, (b) $P(Z < c) = .9732$, (c) $P(Z > c) = .5691$, (d) $P(-c \leq Z \leq c) = .2358$.

42. If Z has a standard normal distribution, find c such that (a) $P(Z > c) = .0764$, (b) $P(Z > c) = .5040$, (c) $P(-c \leq Z < c) = .9108$, (d) $P(-c \leq Z \leq c) = .8$.

43. If X has a normal distribution with mean $\mu = 50$ and standard deviation $\sigma = 9$, determine (a) $P(X \leq 40)$, (b) $P(X < 55)$, (c) $P(X > 60)$, (d) $P(40 \leq X \leq 60)$.

44. If X has a normal distribution with mean $\mu = 20$ and standard deviation $\sigma = 9$, determine (a) $P(X < 22)$, (b) $P(X > 17)$, (c) $P(X > 15)$, (d) $P(2 < X < 38)$.

45. If X has a normal distribution with mean $\mu = .75$ and standard deviation $\sigma = .5$, determine c is (a) $P(X < .25)$, (b) $P(X > .9)$, $(a)P(X < c) = .1587$, (b) P(X > c)=.382, (c) $P(.5 < X < 1)$, (d) $P(.25 < X < 1.25)$.

46. If X has a normal distribution, determine c such that

$$P(\mu - c\sigma < X < \mu + c\sigma) = .95.$$

Hint: Convert the above expression so that the middle term has a standard normal distribution.

47. If X has a normal distribution, determine c such that

$$P(\mu - c\sigma < X < \mu + c\sigma) = .8.$$

48. Assuming that the scores on a math achievement test are normally distributed with mean $\mu = 68$ and standard deviation $\sigma = 10$, what is the probability of getting a score greater than 78?

49. In the previous problem, how high must someone score to be in the top 5%? That is, determine c such that $P(X > c) = .05$.

50. A manufacturer of car batteries claims that the life of their batteries is normally distributed with mean $\mu = 58$ months and standard deviation $\sigma = 3$. Determine the probability that a randomly selected battery will last at least 62 months.

51. Assume that the income of pediatricians is normally distributed with mean μ =\$100,000 and standard deviation $\sigma = 10,000$. Determine the probability of observing an income between \$85,000 and \$115,000.

52. Suppose the winnings of gamblers at Las Vegas are normally distributed with mean $\mu = -300$ (the typical person loses \$300), and standard deviation $\sigma = 100$. Determine the probability that a gambler does not lose any money.

53. A large computer company claims that their salaries are normally distributed with mean \$50,000 and standard deviation 10,000. What is the probability of observing an income between \$40,000 and \$60,000?

54. Suppose the daily amount of solar radiation in Los Angeles is normally distributed with mean 450 calories and standard deviation 50. Determine the probability that for a randomly chosen day, the amount of solar radiation is between 350 and 550.

55. If the cholesterol levels of adults are normally distributed with mean 230 and standard deviation 25, what is the probability that a randomly sampled adult has a cholesterol level greater than 260?

56. If after one year, the annual mileage of privately owned cars is normally distributed with mean 14,000 miles and standard deviation 3,500, what is the probability that a car has mileage greater than 20,000 miles?

4.7 Some modern advances and insights

Four types of distributions are important when trying to assess and understand the relative merits of methods covered in this book: symmetric with outliers rarely occurring, symmetric with outliers commonly occurring, asymmetric with outliers rarely occurring, and asymmetric with with outliers commonly occurring. (Symmetric distributions with outliers rarely occurring includes the normal distribution as a special case.) Most of the details must be postponed for the moment, but we will see that the latter three types of distributions are a serious concern when analyzing data with commonly used methods based on means. However, some comments can be made now regarding situations where distributions are symmetric with outliers commonly occurring. A fundamental issue is this: If a plot of the data indicates that the underlying distribution is bell-shaped like a normal distribution, is it safe to assume normality? For some purposes the answer is yes, but for others the answer is an emphatic no. Normal distributions typically result in relatively few outliers. But when outliers tend to occur more frequently, some of the basic properties of a normal distribution are no longer true.

The so-called contaminated or mixed normal distribution is a classic way of illustrating some of the consequences associated with symmetric, bell-shaped distributions where outliers tend to be common. Consider a situation where we have two populations of individuals or things. Assume each population has a normal distribution, but they differ in terms of their means, or variances, or both. When we mix the two populations together we get what is called a *mixed* or *contaminated normal.* Generally, mixed normals fall outside the class of normal distributions. That is, for a distribution to qualify as normal, the equation for its curve must have the form given by equation (4.9), and the mixed normal does not satisfy this requirement. When the two normals mixed together have a common mean, but unequal variances, the resulting probability curve is again symmetric about the mean, but even now the mixed normal is not a normal curve.

To provide a more concrete description of the mixed normal, consider the entire population of adults living around the world and let X represent the amount of weight they have gained or lost during the last year. Imagine that we divide the population of adults into two groups: those who have tried some form of dieting to lose weight and those that have not. For illustrative purposes, assume that for adults who have not tried to lose weight, the distribution of their weight loss is standard normal (so $\mu = 0$ and $\sigma = 1$). As for adults who have dieted to lose weight, assume that their weight loss is normally distributed again with mean $\mu = 0$ but with standard deviation $\sigma = 10$. Finally, suppose that 10% of all adults went on a diet last year to lose weight. That is, there is a 10% chance of selecting an observation from a normal distribution having standard deviation ten, so there is a 90% chance of selecting an observation from a normal curve having a standard deviation of one.

Now, if we mix these two populations of adults together, the exact distribution can be derived and is shown in figure 4.5. Also shown is the standard normal distribution, and as is evident there is little separating the two curves. Note that in figure 4.5, the tails of the mixed normal lie above the tails of the normal. For this reason, the mixed normal is often described as being *heavy-tailed.* Because the area under the extreme portions of a heavy-tailed distribution is larger than the area under a normal curve, extreme values or outliers are more likely when sampling from the mixed normal.

Here is the point: Very small departures from normality can greatly influence the value of the population variance. For the standard normal the variance is 1, but for the mixed normal it is 10.9. A related implication is that slight changes in any distribution, not just the normal distribution, can have a big impact on the population variance. And as for the sample variance, its value can be drastically altered by only a few outliers no matter how large the sample size might be. The full implications of this result are

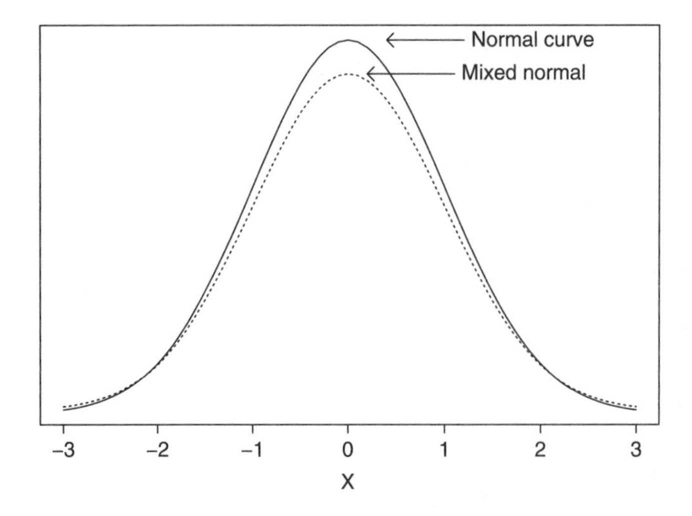

Figure 4.5 Shown is the normal and the mixed normal described in the text.

impossible to appreciate at this point, but they will become clear in subsequent chapters. But two implications can be illustrated here.

We have seen that for any normal distribution, the probability of being within one standard deviation of the mean is .68, as illustrated in figure 4.2. In symbols, if X has a normal distribution,

$$P(\mu - \sigma \leq X \leq \mu + \sigma) = .68.$$

More generally, if a distribution is symmetric and bell-shaped, but not a normal distribution, is it approximately true that the probability of being within one standard deviation of the mean is .68? The answer is no, not necessarily. For the contaminated normal considered here, the probability exceeds .925.

A criticism of this last example is that perhaps we never encounter a contaminated normal in practice, but this misses the point. The contaminated normal illustrates a basic principle: When sampling from distributions where outliers tend to be common, certain interpretations of the standard deviation that are reasonable under normality can be highly inaccurate. And we will see that other uses of the standard deviation can result in serious practical problems yet to be described.

Here is another implication worth mentioning. As previously pointed out, normal curves are completely determined by their mean and variance, and figure 4.3 illustrated that under normality, increasing the variance from 1 to 1.5 results in a very noticeable difference in the graphs of the probability curves. If we assume that curves are normal, or at least approximately normal, this might suggest that in general, if two distributions have equal variances, they will will be very similar in shape. But this is not necessarily true even when the two curves are symmetric about the population mean and are bell-shaped. Figure 4.6 provides another illustration that two curves can have equal means and variances yet differ substantially.

The illustration just given is not intended to suggest that the variance be abandoned when trying to understand data. Rather, the main message is that when learning basic statistical techniques, it is important to be aware of when the variance provides accurate and useful information, and when and why it might be unsatisfactory. That is, basic

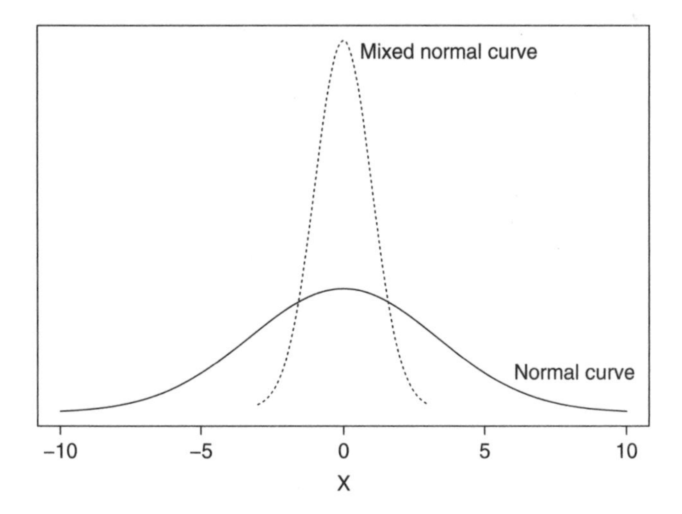

Figure 4.6 Two distributions with equal means and variances.

training should include concepts and results that help you avoid reading more into data than is warranted.

Skewness

Heavy-tailed distributions are one source of concern when using means. Another is skewness, which generally refers to distributions that are not exactly symmetric. It is too soon to discuss all the practical problems associated with skewed distributions, but one of the more fundamental issues can be described here.

Consider how we might choose a single number to represent the typical individual or thing under study. A seemingly natural approach is to use the population mean. If a distribution is symmetric about its mean, as is the case when a distribution is normal, there is general agreement that the population mean is indeed a reasonable reflection of what is typical. But when distributions are skewed, at some point doubt begins to arise as to whether the mean is a good choice. Consider, for example, the distribution

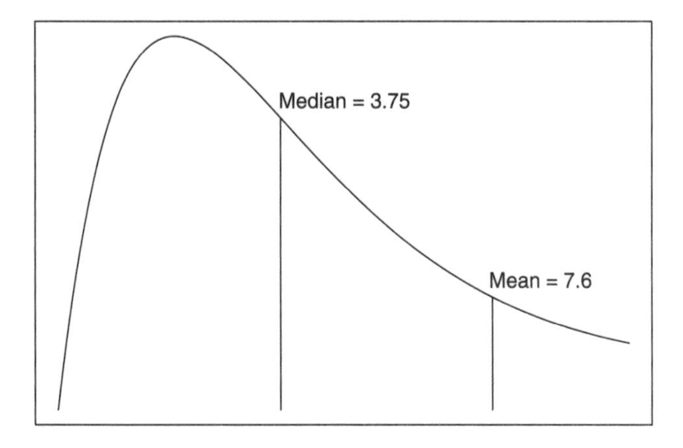

Figure 4.7 For skewed distributions, the population mean and median can differ tremendously.

shown in figure 4.7, which is skewed to the right. In this particular case the population mean is located in the extreme right portion of the curve. In fact, the probability that an observation is less than the population mean is 0.74. So from a probabilistic point of view, the population mean is rather atypical. In contrast, the median is located near the more likely outcomes and would seem to better reflect what is typical.

Comments on skewness and transforming data

A classic suggestion regarding how to deal with skewed distributions is to transform the data. One of the easiest transformations that is often suggested is to take logarithms. It is true that in some cases, this creates a more symmetric looking plot of the data, but even when using more complex transformations, a plot of the data can remained skewed. Another important point is that this is a relatively ineffective way of dealing with outliers. Although simple transformations might reduce outliers, often the number of outliers remains the same and in some instances the number of outliers actually increases. There are more effective methods for dealing with skewness, most of which are not covered in this book. (But a few comments on dealing with skewness will be made in subsequent chapters.)

Example 1

The left panel of figure 4.8 shows a plot of 100 observations generated on a computer. Using the boxplot rule, five of the values are declared outliers. The right panel shows a plot of the same data after taking the logarithm of each value. The plot appears to be more symmetric, but an important point is that the same five outliers are again declared outliers.

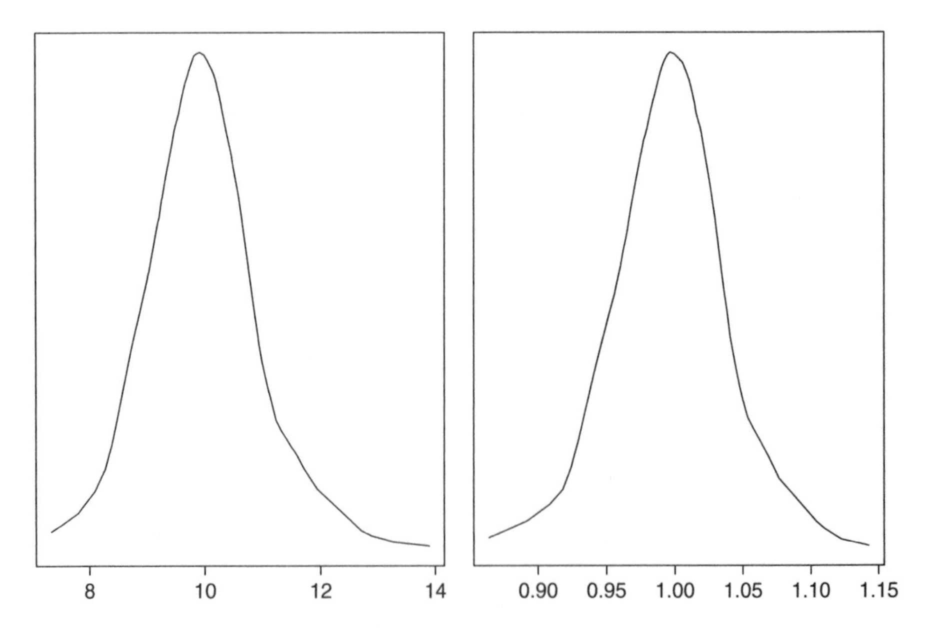

Figure 4.8 Taking logarithms sometimes results in a plot of the data being more symmetric, as illustrated here, but outliers can remain.

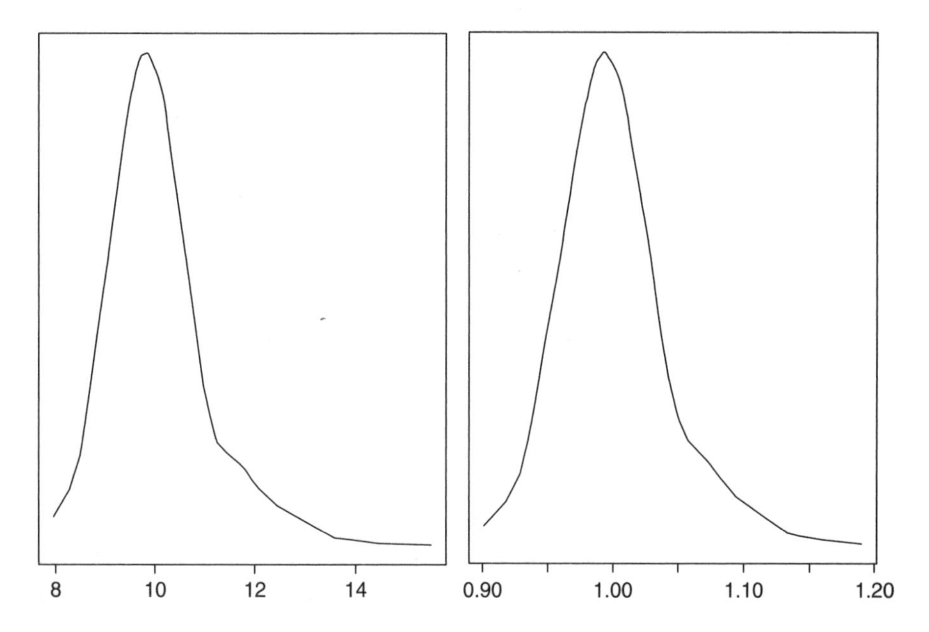

Figure 4.9 Taking logarithms sometimes reduces skewness but does not eliminate it, as illustrated here.

Example 2

The left panel of figure 4.9 shows a plot of another 100 observations generated on a computer, only this time the data were generated from a distribution a bit more skewed than the distribution used in figure 4.8. Using the boxplot rule, six of the values are declared outliers. The right panel shows a plot of the same data after taking the logarithm of each value. This time the plot remains noticeably skewed. The number of outliers is reduced to four, this is better than six, but the more salient point is that the presence of outliers is not eliminated.

This next example illustrates yet one more point worth mentioning. Consider the following values:

$$1, 2, 3, 4, 5, 6, 7, 8, 9, 10, 200, 500$$

The sample mean is 62.9 and the median is 6.5. If we take logarithms (to the base 10), the values are now

$$0.0000000, 0.3010300, 0.4771213, 0.6020600, 0.6989700, 0.7781513,$$
$$0.8450980, 0.9030900, 0.9542425, 1.0000000, 2.3010300, 2.6989700.$$

The mean of these value is 0.8116246. Does this somehow provide an estimate of the mean before we transformed the data? In general, transforming data makes it difficult to estimate the mean of the original values. If we simply transform back by raising 10 to the power 0.8116246, we get $10^{0.8116246} = 6.48$, a value close to the median of the original values, but quite different from the mean. This illustrates a common situation: Using the mean of values after taking logarithms often results in using a measure of location that is close to the median of the original values, rather than the mean. This does not necessarily mean that transformations are bad. But it is important to be aware of this property.

A Summary of Some Key Points

- This chapter introduced the normal distribution and described some of its properties. One important property has to do with the fact that the standard deviation determines the probability that an observation is close to the mean. For example, the probability that an observation is within one standard deviation of the mean is .68, as was illustrated.
- In various situations, the normal distribution has considerable practical value, but this is not always the case. An important goal in this chapter was to provide a foundation for understanding one reason why it can be unsatisfactory. This was done with the mixed normal distribution, which illustrates that even when a distribution is bell-shaped, certain properties of the normal distribution are no longer true. For example, the probability that an observation is within one standard deviation of the mean exceeds .925 when dealing with the mixed normal distribution. The practical implications associated with this result will become clear in the next three chapters.
- Even when an entire population of individuals is measured, the mean, median and 20% trimmed mean can differ substantially.
- Transforming data might reduce skewness, but a fair amount of skewness can remain, and under fairly general conditions, this represents an unsatisfactory approach when dealing with outliers.

Problems

57. Can small changes in the tails of a distribution result in large changes in the population mean, μ, relative to changes in the median?

58. Explain in what sense the population variance is sensitive to small changes in a distribution.

59. For normal random variables, the probability of being within one standard deviation of the mean is .68. That is, $P(\mu - \sigma \leq X \leq \mu + \sigma) = .68$ if X has a normal distribution. For nonnormal distributions, is it safe to assume that this probability is again .68? Explain your answer.

60. If a distribution appears to be bell-shaped and symmetric about its mean, can we assume that the probability of being within one standard deviation of the mean is .68?

61. Can two distributions differ by a large amount yet have equal means and variances?

62. If a distribution is skewed, is it possible that the mean exceeds the .85 quantile?

5

SAMPLING DISTRIBUTIONS

Recall that the population mean μ represents the average of all individuals or things under study. But typically, not all individuals can be measured. Rather, we have only a small subset of all individuals available to us, and the average response based on this sample, $\bar{X}$, is used to estimate the population mean, μ. An issue of fundamental importance is how well the sample mean, $\bar{X}$, estimates the population mean, μ. If the sample mean is $\bar{X} = 23$, we estimate that the population mean is 23, but generally this estimate will be wrong. So what is needed is some method that can be used to assess the precision of this estimate. That is, based on the available data, if $\bar{X} = 23$, can we be reasonably certain that the population mean is less than 42? Can we be reasonably certain that the population mean is greater than 16? In a similar manner, when we compute the sample median, M, how well does it estimate the population median? Given some data, is there some method that allows us to conclude, for example, that the population median is between 8 and 10? When working with the binomial, if we observe 67 successes among 100 observations, we estimate the probability of success to be .67. But how might we assess the accuracy of this estimate? A key component when trying to address these problems is the notion of a sampling distribution.

5.1 Sampling distribution of a binomial random variable

The notion of a sampling distribution is perhaps easiest to explain and illustrate when working with the binomial distribution. But first the binomial distribution is described in a slightly different manner, and the notion of random sampling needs to be made more precise.

As done in chapter 2, imagine we have n observations, which we label $X_1, \ldots, X_n$. To be concrete, suppose we want to determine the proportion of adults over the age of 40 who suffer from arthritis. For the first person in our study, set $X_1 = 1$ if this individual has arthritis, otherwise set $X_1 = 0$. Similarly, for the second person, set $X_2 = 1$ if this individual has arthritis, otherwise set $X_2 = 0$. We repeat this process n times, so each of the variables $X_1, \ldots, X_n$ has the value 0 or 1. These n values are said to be a *random sample* if two conditions are met:

- Any two observations are independent,
- Each observation has the same probability function.

For the binomial, the second condition merely means that for every observation made, the probability of getting a 1 is the same, which we label p. In symbols, for the n observations available,

$$P(X_1 = 1) = P(X_2 = 1) = \cdots = P(X_n = 1) = p.$$

When working with the binomial, the sample mean of these n values is typically denoted by $\hat{p}$. That is,

$$\hat{p} = \frac{1}{n} \sum X_i,$$

which is just the proportion of ones among the n observations.

Example 1

Imagine you want to know the percentage of marriages that end in divorce among couples living in Iceland. You do not have the resources to check all records, so you want to estimate this percentage based on available data. To keep the illustration simple, suppose we have data on ten couples:

$$X_1 = 1, X_2 = 0, X_3 = 0, X_4 = 0, X_5 = 1,$$

$$X_6 = 0, X_7 = 0, X_8 = 0, X_9 = 0, X_{10} = 1.$$

That is, the first couple got a divorce, the next three couples did not get a divorce, the fifth couple got a divorce, and so on. The number of divorces among these ten couples is

$$\sum X_i = 1+0+0+0+1+0+0+0+0+1 = 3,$$

so the estimated probability of a divorce is

$$\hat{p} = \frac{3}{10} = .3.$$

Notice that for the binomial, if we knew the true probability of a divorce, p, we could compute the probability of getting $\hat{p} = .3$ based on a sample of size ten. When $n = 10$, it is just the probability of observing 3 divorces. Referring to equation (4.8), this probability is

$$p(3) = \binom{10}{3} p^3 q^7,$$

where $q = 1 - p$. If, for example, $p = .4$, then $p(3) = .215$. That is, the probability of getting $\hat{p} = .3$ is .215. More generally, if we observe x successes, the estimate of p is $\hat{p} = x/n$, which occurs with probability $p(x)$, where $p(x)$ is the binomial probability function given by equation (4.8).

Now imagine a collection of research teams and suppose each team estimates the divorce rate based on the records of ten married couples. By chance, different research teams will get different results. For example, the first team might get $\hat{p} = .5$, the second team might get $\hat{p} = .1$, the third team might get $\hat{p} = .3$, and so on. The *sampling distribution* of $\hat{p}$ refers to the distribution of the $\hat{p}$ values we would get if millions of research teams were to conduct the same study. Put another way, the sampling distribution of $\hat{p}$ is just the probabilities associated with all possible values for $\hat{p}$ that we might observe. But for the case of the binomial, and assuming random sampling, the distribution of $\hat{p}$ is simple; it corresponds to the binomial probability function. That is, we can determine the probability of getting a particular value for $\hat{p}$ among the many studies that might be performed once we are given the sample size, n, and the probability of success p. Moreover, from chapter 4, the average of the $\hat{p}$ values among the many studies that might be conducted is p and the variance of the $\hat{p}$ values is $p(1-p)/n$.

Example 2

Various research teams plan to conduct a study aimed at estimating the occurrence of tooth decay among adults living in a particular geographic region. Assume random sampling, that each team plans to base their estimate on five individuals, and that unknown to them, the proportion of people with tooth decay is .3. So $n = 5$ and $p = .3$. Then for each research team that might investigate this issue, the possible values for $\hat{p}$ are 0/5, 1/5, 2/5, 3/5, 4/5, 5/5, and the corresponding probabilities are 0.16807, 0.36015, 0.30870, 0.13230, 0.02835 and 0.00243, respectively, which correspond to the probability of a 0, 1, 2, 3, 4, and 5, based on the binomial probability function. So from chapter 4, the average value of $\hat{p}$, among the many research teams, is $p = .3$, and the variance of the $\hat{p}$ values is $p(1-p)/n = .3(.7)/5 = .042$.

Example 3

A college president claims that the proportion of students at her institution with an IQ greater than 120 is .4. If various individuals plan to sample 20 students, with the goal of estimating the proportion who have an IQ greater than 120, what is the probability that an investigator will get an estimate less than or equal to 4/20 if the claim is correct? An estimate less than or equal to 4/20 corresponds to getting 0 or 1 or 2 or 3 or 4 students with an IQ greater than 120. If the claim $p = .4$ is correct, then from table 2 in appendix B (with $n = 20$), the probability of getting four or fewer students with an IQ greater than 120 is .051. That is, if the president's claim is correct, it would be rather unusual to get $\hat{p} \leq 4/20$, suggesting that perhaps the claim is wrong.

Problems

1. For a binomial with $n = 25$ and $p = .5$, determine (a) $P(\hat{p} \leq 15/25)$, (b) $P(\hat{p} > 15/25)$, (c) $P(10/25 \leq \hat{p} \leq 15/25)$.
2. Many research teams intend to conduct a study regarding the proportion of people who have colon cancer. If a random sample of ten individuals could be obtained, and if the probability probability of having colon cancer is .05, what is the probability that a research team will get $\hat{p} = .1$?
3. In the previous problem, what is the probability of $\hat{p} = .05$?

4. Someone claims that the probability of losing money, when using an investment strategy for buying and selling commodities, is .1. If this claim is correct, what is the probability of getting $\hat{p} \leq .05$ based on a random sample of 25 investors?

5. You interview a married couple and ask the wife whether she supports the current leader of their country. Her husband is asked the same question. Describe why it might be unreasonable to view these two responses as a random sample.

6. Imagine that a thousand research teams draw a random sample from a binomial distribution with $p = .4$, with each study based on a sample size of 30. So this would result in 1,000 $\hat{p}$ values. If these 1,000 values were averaged, what, approximately, would be the result?

7. In the previous problem, if you computed the sample variance of the $\hat{p}$ values, what, approximately, would be the result?

5.2 Sampling distribution of the mean under normality

When working with the binomial, we have seen that the sampling distribution of $\hat{p}$ is fairly easy, because we merely use the binomial probability function as described in chapter 4. The goal in this section is to extend the notion of a sampling distribution to situations where data have a normal distribution. So again we have n observations $X_1, \ldots, X_n$, but rather than having a value of 0 or 1, these variables have values that are continuous with mean μ and standard deviation σ, and if all individuals could be measured, a plot of the data would have the normal distribution described in chapter 4.

As with the binomial, we imagine that many research teams plan to conduct the same study, or that the same research team plans to repeat their study many times. To be concrete, imagine the goal is to estimate how many additional hours of sleep an individual gets after taking a particular drug. Further suppose that the drug is tried on 20 individuals yielding a sample mean of $\bar{X} = .8$ hours. But if the study were repeated with another 20 participants, chances are we would get a different result. This time we might get $\bar{X} = 1.3$. And repeating the study yet again might yield $\bar{X} = -0.2$. In statistical terms, there will be variation among the sample means. The goal is to be able to determine the probability that the sample mean is less than .5, less than 1, less than 1.3, and more generally, less than c, where c is any constant we might choose. In symbols, we want to be able to determine $P(\bar{X} \leq c)$ for any constant c.

Of course, we could solve this problem simply by repeating the study many times, but this is impractical. It turns out that three key results provide a solution based on a single study when n observations are randomly sampled from a population having a population mean μ, and a population standard deviation σ. These results are:

- Under random sampling, the average value of the sample mean, over millions of studies (and in theory over infinitely many studies) can be shown to be equal to μ, the population mean. In symbols, $E(\bar{X}) = \mu$. Said another way, in any given study, chances are that the sample mean will not be equal to the population mean. But on average (over many studies), the sample mean provides a correct estimate of the population mean.
- Under random sampling, the variance of the sample mean, over millions of studies (and in theory over infinitely many studies) can be shown to be σ^2/n.

That is, the average squared difference between the sample mean and the population mean is σ^2/n. In symbols, the variance of the sample mean is $E[(\bar{X} - \mu)^2] = \sigma^2/n$.

- When observations are randomly sampled from a normal distribution, the sample mean also has a normal distribution. Put more succinctly, when n observations are randomly sampled from a normal distribution with mean μ and variance σ^2, the sample mean has a normal distribution with mean μ and variance σ^2/n. The practical implication is that the probability of getting a value for $\bar{X}$ less than or equal to 1, 3, or c, for any c we might choose, can be determined under normality when the mean, variance and sample size are known, as will be illustrated.

Note that the first two results require the assumption of random sampling only—normality is not required. The third result assumes normality. As previously remarked, normality is rarely if ever true, so there is the issue of how to deal with the more realistic situation where data do not follow a normal curve. For the moment this issue is ignored, but it will be discussed in detail at various points.

The variance of the sample mean is called the *squared standard error* of the sample mean. Often this variance is written as $\text{VAR}(\bar{X})$ or $\sigma^2_{\bar{X}}$. To be a bit more concrete, imagine we randomly sample 25 observations where, unknown to us, the population mean is 1.5 and the variance is 2 ($\sigma^2 = 2$). We might get a sample mean of $\bar{X} = 1.45$. Further imagine that we repeat the study many times yielding the sample means

$$1.45, 1.53, 1.90, 1.43, 2.72, 1.70, 1.13, 1.94, 1.23, \ldots.$$

It can be shown that if the study is repeated a very large number of times, the average of these sample means will be equal to the population mean, 1.5, and that if we were to compute the sample variance based on these values, we would get $\sigma^2/n = 2/25$. That is, the variance of the sample means is equal to the variance of the distribution from which the observations were sampled, divided by the sample size, assuming random sampling only. The (positive) square root of the squared standard error, $\sigma_{\bar{X}} = \sigma/\sqrt{n}$, is called the *standard error* of the mean. In practice, the variance (σ^2) is rarely known, but it can be estimated with the sample variance, s^2, as previously noted. This, in turn, provides an estimate of the squared standard error, namely, s^2/n, and an estimate of the standard error is $s/\sqrt{n}$.

Example 1

Ten randomly sampled batteries from a particular manufacturing company are found to have the following lifetimes (in months):

$$55, 69, 77, 53, 63, 71, 58, 62, 80, 61.$$

The sample variance is $s^2 = 82.54$, so an estimate of σ^2/n, the squared standard error of the sample mean, is $82.54/10 = 8.254$ and an estimate of the standard error is $\sqrt{8.254} = 2.87$.

Example 2

Sixteen observations are randomly sampled from a normal distribution having mean 10 and standard deviation 1. That is, $n = 16$, $\mu = 10$ and $\sigma = 1$. By chance

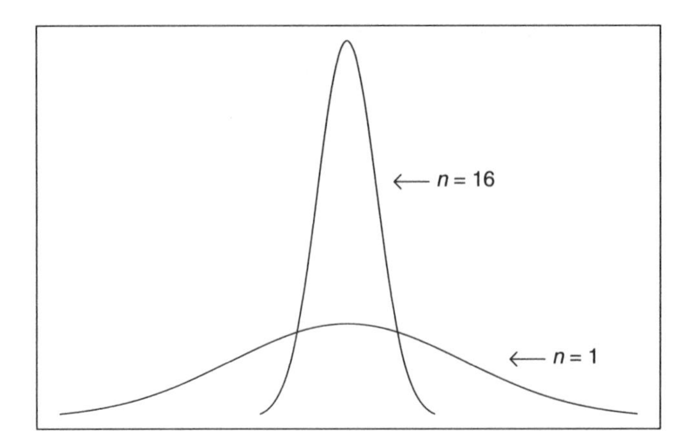

Figure 5.1 An illustration of how the sampling distribution changes with the sample size under normality.

we might get $\bar{X} = 9.92$. If we repeat this study, again with $n = 16$, now we might get $\bar{X} = 10.76$. If we continue to repeat this study many times, the average of the sample means will be 10. And if we were to compute the variance of these sample means we would get $\sigma^2/n = 1^2/16 = .0625$, and the standard error of the sample mean is $\sigma/\sqrt{n} = 1/4 = .25$.

Figure 5.1 shows the normal distribution used in the last example, which is indicated by $n = 1$. Recall that σ^2 reflects how closely a *single* observation tends to be to the population mean. Also shown is the sampling distribution of the sample mean, $\bar{X}$, which is the distribution indicated by $n = 16$. For example, the area under this curve and to left of the value 1, say, corresponds to the probability that a sample, based on 16 observations, will have a value less than or equal to 1, assuming normality. Because its variance is σ^2/n, which is smaller than the variance based on a single observation, the sample mean tends to be closer to the population mean, on average. Indeed, as we increase the sample size, σ^2/n decreases. Intuition suggests that the larger the sample size, the more accurate will be the sample mean in terms of estimating the population mean. The result just given helps quantify just how much more accurate it will be.

Definition An *estimator* is some expression, based on the observations made, intended to estimate some feature of the population under study. The sample mean, $\bar{X}$ is an estimator of the population mean, and its observed value is called an *estimate*. The sample variance, s^2 is an estimator, and its observed value is said to be an estimate of the population variance.

Definition An estimator is *unbiased* if its average value over millions of studies (and in theory infinitely many studies) is equal to the quantity it is trying to estimate. The sample mean is an unbiased estimate of the population mean because it can be shown that $E(\bar{X}) = \mu$. For the binomial, $E(\hat{p}) = p$, so $\hat{p}$ is an unbiased estimate of the true probability of success, p. It can be shown that on average, the sample variance, s^2, is equal to σ^2. That is, s^2 is an unbiased estimate of the population variance. In symbols, $E(s^2) = \sigma^2$.

Determining probabilities associated with the sample mean

The results just described make it possible to address the following type of problem. If we are told that data are randomly sampled from a normal distribution with a specified sample size, population mean and population variance, what is the probability that the sample mean is less than 10, less than 20, or less than c for any constant c we might pick? In symbols, we want to know $P(\bar{X} \leq c)$. Recall from chapter 4, that if X has a normal distribution, we can determine $P(X \leq c)$ by standardizing X. That is, if we subtract the population mean and divide by the population standard deviation, X is transformed to a standard normal random variable. In symbols, we used the fact that

$$P(X \leq c) = P\left(Z \leq \frac{c - \mu}{\sigma}\right),$$

where Z has a standard normal distribution. Here, the same strategy is used when working with the sample mean, the main difference being that the sample mean has standard deviation $\sigma/\sqrt{n}$ rather than σ. In symbols, under random sampling from a normal distribution,

$$Z = \frac{\bar{X} - \mu}{\sigma/\sqrt{n}}$$

has a standard normal distribution. This means that

$$P(\bar{X} \leq c) = P\left(Z \leq \frac{c - \mu}{\sigma/\sqrt{n}}\right), \tag{5.1}$$

which can be determined by referring to table 1 in appendix B. In addition, the probability that the sample mean is greater than or equal to c is

$$P(\bar{X} \geq c) = 1 - P\left(Z \leq \frac{c - \mu}{\sigma/\sqrt{n}}\right), \tag{5.2}$$

and the probability that the sample mean is between the constants a and b is

$$P(a \leq \bar{X} \leq b) = P\left(Z < \frac{b - \mu}{\sigma/\sqrt{n}}\right) - P\left(Z \leq \frac{a - \mu}{\sigma/\sqrt{n}}\right). \tag{5.3}$$

Example 3

If 25 observations are randomly sampled from a normal distribution with mean 50 and standard deviation 10, what is the probability that the sample mean will be less than 45? We have that $n = 25$, $\mu = 50$, $\sigma = 10$ and $c = 45$, so

$$\begin{aligned} P(\bar{X} \leq 45) &= P\left(Z \leq \frac{45 - 50}{10/\sqrt{25}}\right) \\ &= P(Z \leq -2.5) \\ &= .0062. \end{aligned}$$

Example 4

A company claims that after years of experience, students who take their training program typically increase their SAT scores by an average of 30 points.

They further claim that the increase in scores has a normal distribution with standard deviation 12. As a check on their claim, you randomly sample 16 students and find that the average increase is 21 points. The company argues that this does not refute their claim because getting a sample mean of 21 or less is not that unlikely. To determine whether their claim has merit, you compute

$$\begin{aligned}P(\bar{X} \leq 21) &= P\left(Z \leq \frac{21-30}{12/\sqrt{16}}\right)\\ &= P(Z \leq -3)\\ &= .0013,\end{aligned}$$

which indicates that getting a sample mean as small or smaller than 21 is a relatively unlikely event. That is, there is empirical evidence that the claim made by the company is probably incorrect.

Example 5

A researcher claims that for college students taking a particular test of spatial ability, the scores have a normal distribution with mean 27 and variance 49. If this claim is correct, and you randomly sample 36 students, what is the probability that the sample mean will be greater than 28? First compute

$$\frac{c-\mu}{\sigma/\sqrt{n}} = \frac{28-27}{\sqrt{49/36}} = .857.$$

Because $P(Z \leq .857) = .20$, equation (5.2) says that $P(\bar{X} > 28) = 1 - P(Z \leq .857) = 1 - .20 = .80$. This says that if we randomly sample $n = 25$ students, and the claims of the researcher are true, the probability of getting a sample mean greater than 28 is .8.

Example 6

Thirty-six observations are randomly sampled from a normal distribution with $\mu = 5$ and $\sigma = 3$. What is the probability that the sample mean will be between 4 and 6? So in the notation used here, $a = 4$ and $b = 6$. To find out, compute

$$\frac{b-\mu}{\sigma/\sqrt{n}} = \frac{6-5}{3/\sqrt{36}} = 2.$$

Referring to table 1 in appendix B, $P(\bar{X} \leq 4) = P(Z \leq 2) = .9772$. Similarly,

$$\frac{a-\mu}{\sigma/\sqrt{n}} = \frac{4-5}{3/\sqrt{36}} = -2,$$

and $P(Z \leq -2) = .0228$. So, according to equation (5.3),

$$P(2 \leq \bar{X} \leq 6) = .9772 - .0228 = .9544.$$

Problems

8. Suppose $n = 16$, $\sigma = 2$, and $\mu = 30$. Assume normality and determine (a) $P(\bar{X} \leq 29)$, (b) $P(\bar{X} > 30.5)$, (c) $P(29 \leq \bar{X} \leq 31)$.

9. Suppose $n = 25$, $\sigma = 5$, and $\mu = 5$. Assume normality and determine (a) $P(\bar{X} \leq 4)$, (b) $P(\bar{X} > 7)$, (c) $P(3 \leq \bar{X} \leq 7)$.

10. Someone claims that within a certain neighborhood, the average cost of a house is $\mu =$\$100,000 with a standard deviation of $\sigma =$ \$10,000. Suppose that based on $n = 16$ homes, you find that the average cost of a house is $\bar{X} =$ \$95,000. Assuming normality, what is the probability of getting a sample mean this low or lower if the claims about the mean and standard deviation are true?

11. In the previous problem, what is the probability of getting a sample mean between \$97,500 and \$102,500?

12. A company claims that the premiums paid by its clients for auto insurance has a normal distribution with mean $\mu = 750$ dollars and standard deviation $\sigma = 100$ dollars. Assuming normality, what is the probability that for $n = 9$ randomly sampled clients, the sample mean will a value between 700 and 800 dollars?

13. Imagine you are a health professional interested in the effects of medication on the diastolic blood pressure of adult women. For a particular drug being investigated, you find that for $n = 9$ women, the sample mean is $\bar{X} = 85$ and the sample variance is $s^2 = 160.78$. Estimate the standard error of the sample mean assuming random sampling.

5.3 Non-normality and the sampling distribution of the sample mean

During the early years of the nineteenth century, thanks to efforts made by Gauss, it was realized that assuming data have a normal distribution is highly convenient from both a theoretical and computational point of view. But this left open an issue of fundamental importance: How might one justify the use of the normal distribution beyond mere mathematical convenience? In particular, when approximating the sampling distribution of the sample mean, under what circumstances is it reasonable to assume that the normal distribution can be used as described and illustrated in the previous section? Gauss worked on this problem over a number of years, but it is a result derived by Laplace that is routinely used today. Announced in the year 1810, Laplace called his result the central limit theorem, where the word central is intended to mean fundamental.

Roughly, the *central limit theorem* says that under random sampling, as the sample size gets large, the sampling distribution of the sample mean approaches a normal distribution with mean μ and variance σ^2/n. Put another way, if the sample size is sufficiently large, we can assume that the sample mean has a normal distribution. This means that with a 'sufficiently large' sample size, it can be assumed that

$$Z = \frac{\bar{X} - \mu}{\sigma/\sqrt{n}}$$

has a standard normal distribution.

An important aspect of the central limit theorem, particularly in light of some modern insights, is the phrase 'sufficiently large'. This is rather vague. Just how large must the sample size be in order justify the assumption that the sample mean has a normal distribution? For reasons described in this chapter, currently, a common claim is that $n = 40$ generally suffices. But in subsequent chapters it will become evident that two key components of this issue were overlooked. In particular, general situations will be described where a much larger sample size is required when attention is restricted to the mean. There are many recently derived methods that provide strategies for dealing with small sample sizes, and a glimpse of some of these techniques will be provided.

Approximating the binomial distribution

We have seen that when using the binomial, we estimate the probability of success with $\hat{p}$, which is just a sample mean based on n variables having the value 0 or 1. Consequently, under random sampling, the central limit theorem says that if the sample size is sufficiently large, $\hat{p}$ will have, approximately, a normal distribution with mean p (the true probability of success) and variance $p(1-p)/n$. This means that if we standardize $\hat{p}$ by subtracting its mean and dividing by its standard error, the result will be a variable having, approximately, a standard normal distribution. In symbols,

$$Z = \frac{\hat{p} - p}{\sqrt{p(1-p)/n}}$$

will have, approximately, a standard normal distribution. This implies that for any constant c, if n is sufficiently large, it will be approximately true that

$$P(\hat{p} \le c) = P\left(Z \le \frac{c-p}{\sqrt{p(1-p)/n}}\right), \tag{5.4}$$

where Z is a standard normal random variable. That is, this probability can be determined with table 1 in appendix B. And for any constants a and b,

$$P(a \le \hat{p} \le b) = P\left(Z \le \frac{b-p}{\sqrt{p(1-p)/n}}\right) - P\left(Z \le \frac{a-p}{\sqrt{p(1-p)/n}}\right). \tag{5.5}$$

The accuracy of these approximations depends on both n and p. The approximation performs best when $p = .5$. When p is close to 0 or 1, much larger sample sizes are needed to get a good approximation. A commonly used rule is that if both np and $n(1-p)$ are greater than 15, the normal approximation will perform reasonably well. The left panel of figure 5.2 shows a plot of the probability function for $\hat{p}$ when $n = 10$ and $p = .5$, and the left panel is when $n = 100$, again with $p = .5$. Based on these plots, it does not seem too surprising that the normal distribution gives a good approximation of the sampling distribution of $\hat{p}$ when the sample size is not too small.

Example 1

Consider a binomial distribution with $p = .5$ and $n = 10$, and imagine we want to determine the probability that $\hat{p}$ will have a value less than or equal to 7/10. That is, the goal is to determine $P(\hat{p} \le .7)$. Using methods already described, the exact probability is 0.945. Using the approximation given by equation (5.4),

$$P(\hat{p} \le .7) = P\left(Z \le \frac{.7 - .5}{\sqrt{.5(1-.5)/10}}\right) = P(Z \le 1.264911).$$

Referring to table 1 in appendix B, $P(Z <= 1.264911) = .897$, which differs from the exact value by $.945 - .897 = .048$.

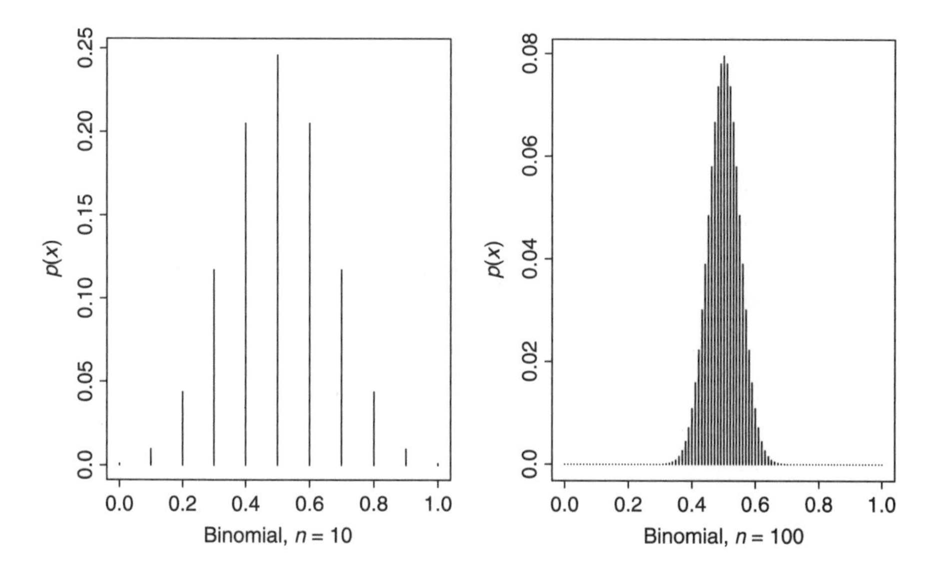

Figure 5.2 The sampling distribution of $\hat{p}$, the proportion of successes when working with the binomial distribution having probality of success $p = .5$.

Example 2

We repeat the previous example, only now we approximate $P(.3 \le \hat{p} \le .7)$. So referring to equation (5.5), $b = .7$, $(b - p)/\sqrt{p(1-p)/n} = (.7 - .5)/\sqrt{.5(1-.5)/10} = 1.264911$ $(a-p)/\sqrt{p(1-p)/n} = (.3 - .5)/\sqrt{.5(1-.5)/10} = -1.264911$, so the approximation of $P(.3 \le \hat{p} \le .7)$ is $P(Z \le 1.264911) - P(Z \le -1.264911) = 0.8970484 - 0.1029516 = 0.7940968$. The exact value is 0.890625.[1]

Example 3

Consider a binomial distribution with $p = .5$ and $n = 100$, and imagine we want to determine the probability that $\hat{p}$ will have a value less than or equal to .55. The exact value is 0.8643735. The normal approximation is

$$P(\hat{p} \le .55) = P(Z \le 1) = 0.8413447.$$

So compared to the case $n = 10$, we see that we get a better approximation here. This is to be expected based on the central limit theorem, which says that the approximation will improve as the sample size gets large.

Example 4

We repeat the last example, only now we compute the probability that $\hat{p}$ will have a value between .45 and .55. Referring to equation (5.5), the approximate value of this probability is

$$P(Z \le 1) - P(Z \le -1) = 0.6826895.$$

1. There are many ways of improving the approximation when n is small, some of which are mentioned in later chapters.

The exact value is 0.6802727. So again we see that the approximation is performing better compared to the situation where $n = 10$.

Example 5

Again consider the case $n = 100$, only now $p = .05$ and we want to determine the probability that $\hat{p}$ will be less than or equal to .03. So now

$$(c-p)/\sqrt{p(1-p)/n} = (.03 - .05)/\sqrt{.05(1-.05)/100} = -.917663,$$

and based on the central limit theorem, $P(\hat{p} \leq .03)$ is approximately equal to $P(Z \leq -.9176629) = 0.1794$. The exact value is 0.2578387, and so the approximation is less accurate than when $p = .5$.

Approximating the sampling distribution of the sample mean: The general case

Now we discuss approximating the distribution of the sample mean, via the central limit theorem, for the more general case where X is virtually any variable. In particular, we no longer assume that X has a normal distribution. With the aid of a computer,it is a fairly simple matter to illustrate how well a normal distribution approximates the sampling distribution of the mean. The binomial with $n = 100$ and $p = .5$ is an example of a relatively light-tailed distribution where, based on the boxplot rule, few outliers tend to occur. The immediate goal is to consider what happens when observations are sampled from a symmetric distribution where outliers tend to be common.

Imagine we randomly sample 10 observations from the contaminated normal distribution in figure 4.5 and compute the sample mean. To get some idea of what the distribution of the sample mean looks like, we can repeat this process 4,000 times yielding 4,000 sample means. A plot of the resulting sample means is shown in the left panel of figure 5.3. Also shown is the normal approximation of the sample mean stemming from the central limit theorem. The right panel of figure 5.3 shows the distribution of the sample means when the sample size is increased to $n = 40$ plus the normal approximation of the sample mean stemming from the central limit theorem. The approximation is fairly good with $n = 10$ and for $n = 40$ it is quite accurate.

Next we consider the sampling distribution of the sample mean when sampling from the skewed distributions shown in figure 5.4. The distribution shown in the left panel of figure 5.4 is relatively light-tailed, roughly meaning that a random sample tends to contain a relatively small proportion of outliers. The distribution in the right panel is heavier-tailed, meaning that outliers are more common.

The top left panel of figure 5.5 shows a plot of 4,000 sample means when sampling from the light-tailed distribution in figure 5.4, with $n = 10$, and the the right panel is the sampling distribution when $n = 40$. The bottom two panels show plots of 4,000 sample means when sampling from the heavy-tailed distribution instead, again with $n = 10$ and 40. We see that with a skewed, light-tailed distribution, the sampling distribution of the mean is approximately normal with $n = 40$. When sampling from the heavy-tailed distribution with $n = 10$, the normal approximation is noticeably worse versus the case when sampling from a light-tailed distribution instead. But even now,

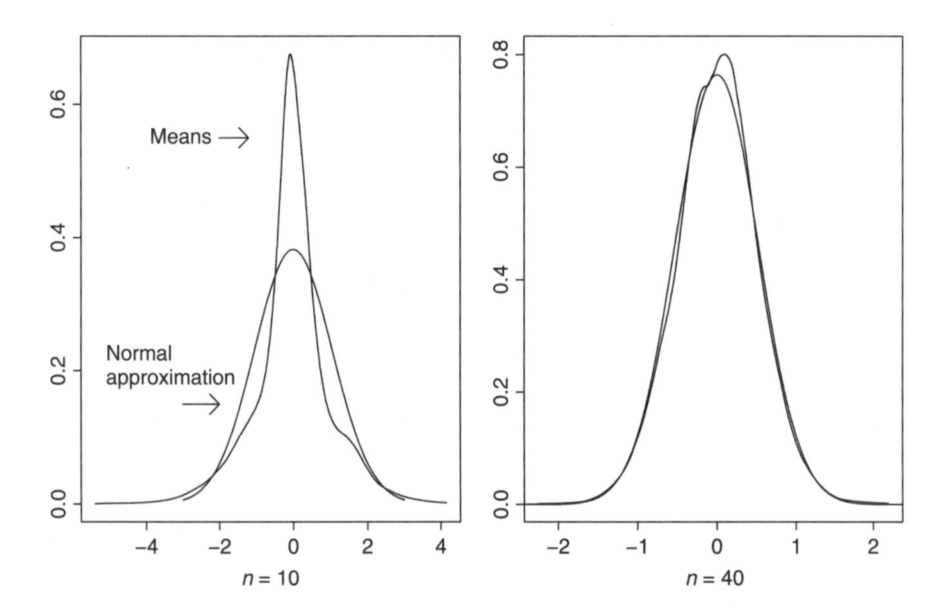

Figure 5.3 As the sample size gets large, the sampling distribution of the mean will approach a normal distribution under random sampling.

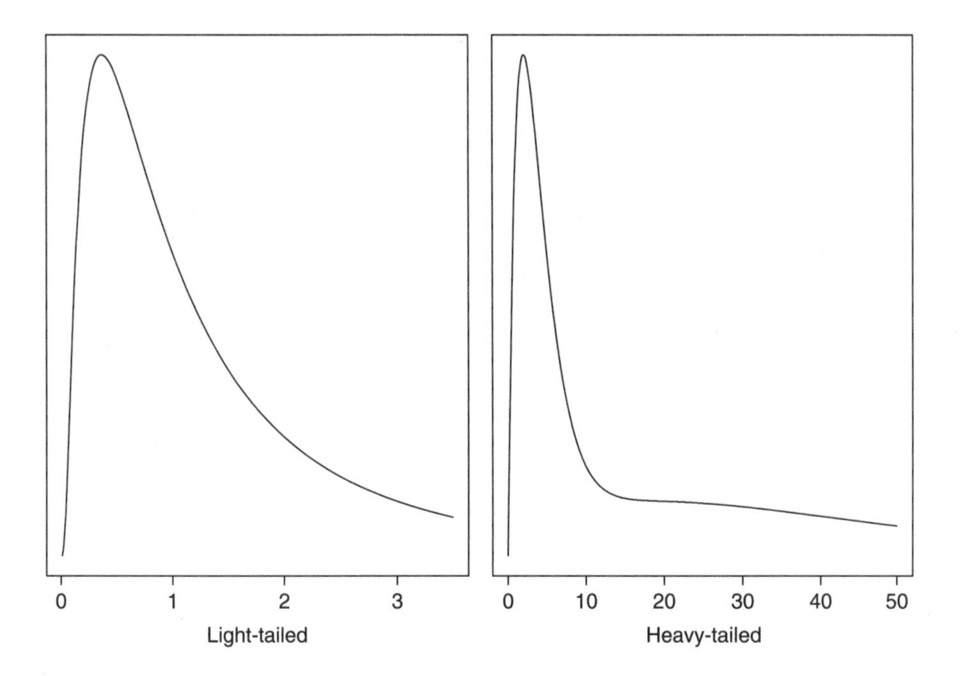

Figure 5.4 Examples of skewed distributions having light and heavy tails.

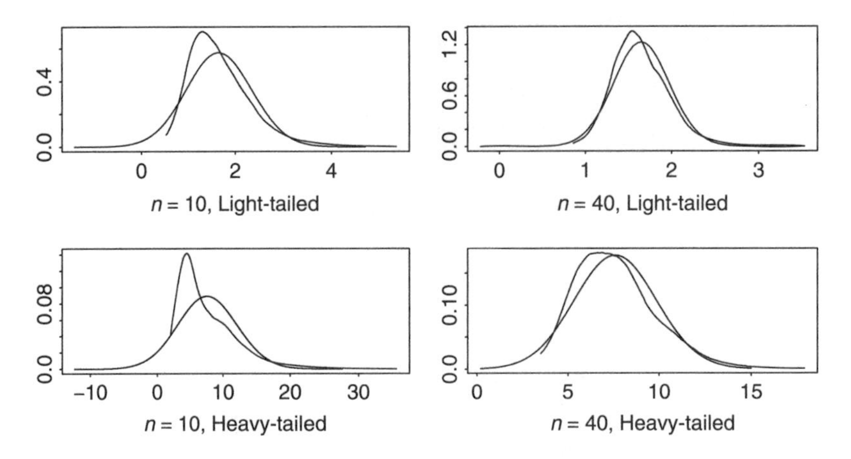

Figure 5.5 With skewed, light-tailed distributions, smaller sample sizes are needed to assume that the sample mean has a normal distribution versus situations where sampling is from a heavy-tailed distribution.

with $n = 40$, the approximation based on the normal distribution performs tolerably well.

Some key points

Figures 5.3 and 5.5 illustrate why it is commonly assumed that with a sample size of 40 or more, generally, normality can be assumed. Historically, the classic illustrations of the central limit theorem were based on two specific distributions. One is called a uniform distribution, which is symmetric and extremely light-tailed, and the other is called an exponential distribution, which is skewed and light-tailed as well. Both of these distributions look nothing like a normal distribution, and again we find that with $n = 40$, the sampling distribution of the sample mean is well approximated by a normal distribution. These findings have had a tremendous influence regarding views about how large the sample size must be to justify normality. But there is a very important point that cannot be stressed too strongly. Subsequent chapters will describe classic methods for comparing groups based on means, which are routinely taught and used. If the sampling distribution of the sample mean has, approximately, a normal distribution, does this necessarily imply that these methods will perform well? There are circumstances where the answer is yes, but under general conditions, the answer is no. Fortunately, many modern methods have been derived for dealing with known problems, some of which will be described.

Although an argument can be made that in general, the sampling distribution of the mean is approximately normal with $n = 40$, it should be noted that there are circumstances where this rule breaks down. One such situation is the binomial where the probability of success, p, is close to 0 or 1. Another situation is where extreme outliers occur.

Example 6

Imagine we resample, with replacement, 40 values from the data on sexual attitudes that are given in table 2.3. If we repeat this process 4,000 times, each time computing the sample mean, a plot of the means is as shown in the left

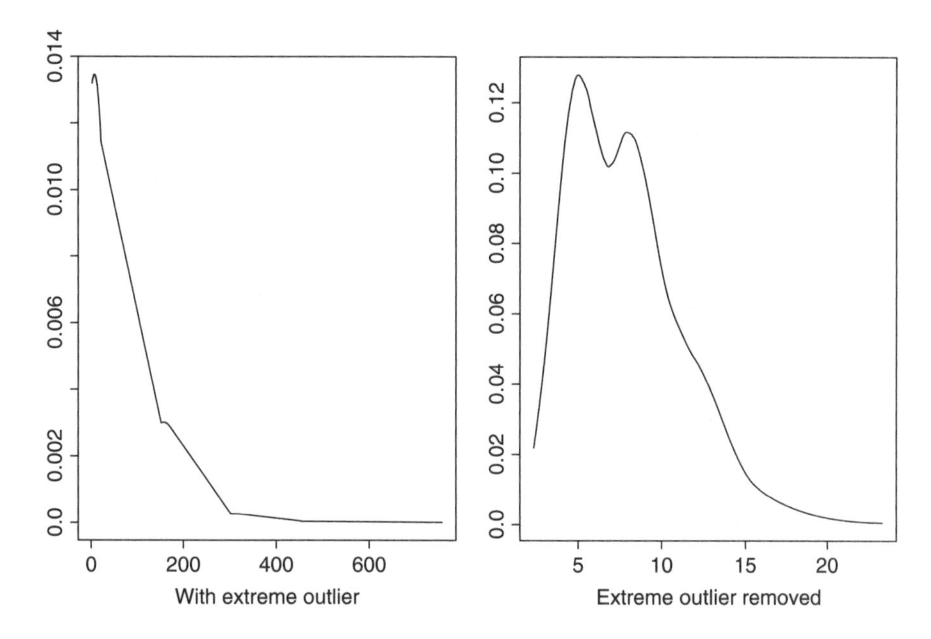

Figure 5.6 Although it is often the case that with a sample size of 40, the sampling distribution of the sample mean will be approximately normal, exceptions arise as illustrated here.

panel of figure 5.6. As is evident, now the sampling distribution of the mean looks nothing like a normal distribution. The problem here is that there is a single extreme outlier with the value 6,000. If we remove this outlier, the resulting plot of the means is shown in the right panel of figure 5.6. Even now, the plot of the means differs from a normal distribution in an obvious way.

Problems

14. You randomly sample 16 observations from a discrete distribution with mean $\mu = 36$ and variance $\sigma^2 = 25$. Use the central limit theorem to determine (a) $P(\bar{X} < 34)$, (b) $P(\bar{X} < 37)$, (c) $P(\bar{X} > 33)$, (d) $P(34 < \bar{X} < 37)$.
15. You sample 25 observations from a non-normal distribution with mean $\mu = 25$ and variance $\sigma^2 = 9$. Use the central limit theorem to determine (a) $P(\bar{X} < 24)$, (b) $P(\bar{X} < 26)$, (c) $P(\bar{X} > 24)$, (d) $P(24 < \bar{X} < 26)$.
16. Referring to the previous problem, describe a situation where reliance on the central limit theorem to determine $P(\bar{X} < 24)$ might be unsatisfactory.
17. Describe situations where a normal distribution provides a good approximation of the sampling distribution of the mean.

5.4 Sampling distribution of the median

All estimators, such as the median, 20% trimmed mean, the sample variance, and the interquartile range, have sampling distributions. Again we imagine repeating a study millions of times, in which case a plot of the estimates would indicate what the sampling

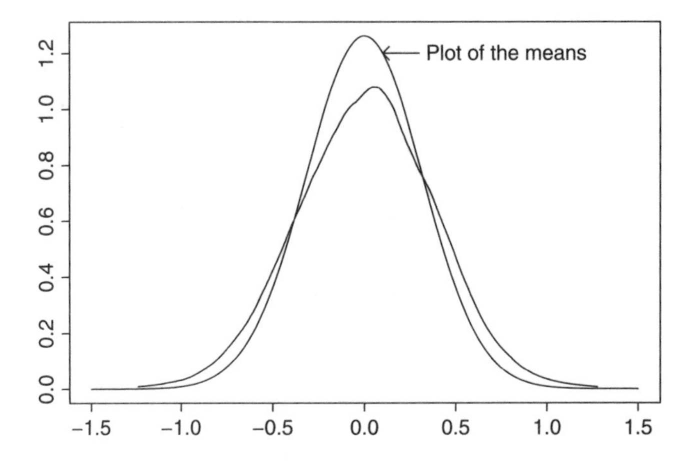

Figure 5.7 Plots of 4,000 means and medians, each based on a random sample of a size 10, taken from a normal distribution. Note that the sample means tend to be closer to the population, 0.

distribution is like. One positive feature of sampling distributions is that they provide a useful perspective on how estimators compare.

The focus here is on the sampling distribution of the median. The median is of interest in its own right, and it helps provide perspective when trying to understand the relative merits of the mean. Recall that when distributions are symmetric about a central value, the population mean and median are equal. This means that the sample mean and median are attempting to estimate the same quantity. Is there any practical advantage to using one estimator over the other in terms of getting a more accurate estimate? The notion of a sampling distribution helps address this problem.

Example 1

First consider the situation where observations are randomly sampled from a standard normal distribution. So both the mean, $\bar{X}$, and median, M, are attempting to estimate the same value: 0. For illustrative purposes, the sample size is taken to be $n = 10$. Will the sample mean tend to be more accurate than the median? To find out, we can use a computer to generate observations from a standard normal distribution, compute both the mean and median, and then we repeat this process 4,000 times yielding 4,000 sample means and medians. figure 5.7 shows a plot of the results. Notice that the sample means are more tightly clustered around 0. This indicates that on average, they are more likely to give a more accurate estimate than the median. The improvement of the mean over the median might not seem that striking, but in other contexts to be described, the mean offers a distinct advantage.

It can be shown that under normality, the sample mean has the smallest standard error of any location estimator we might choose. In particular, it performs better than the median or 20% trimmed mean. But under non-normality, there are situations where it performs poorly.

Example 2

To illustrate that the mean can perform more poorly than the median, we repeat the last example, only now observations are sampled from the symmetric,

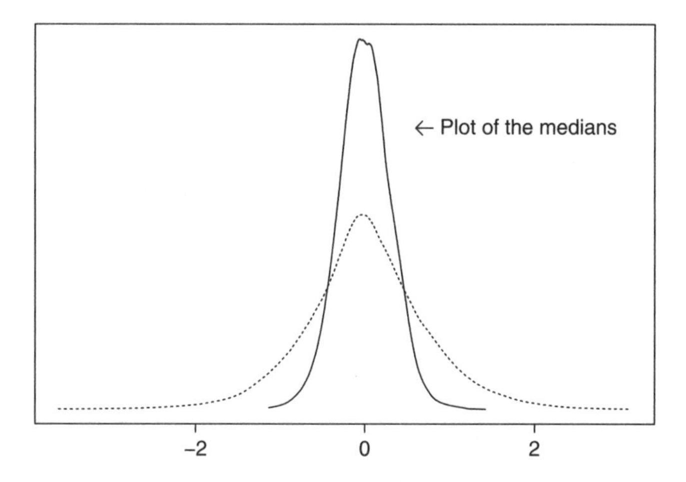

Figure 5.8 Plots of 4,000 means and medians, each based on a random sample of a size 10, taken from a symmetric heavy-tailed distribution. In this case, the median is much more accurate than the mean, on average.

heavy-tailed distribution shown in figure 4.5. Figure 5.8 shows a plot of the results. As indicated, now the sample medians are much more tightly clustered around the central value indicating that on average, the median provides a much more accurate estimate.

Example 2 illustrates a general result of some practical importance. When sampling from a distribution where outliers are relatively rare, the mean will have a smaller standard error than the median. But as we move toward situations where the number of outliers tends to be high, eventually the median will have a smaller standard error than the mean.

Estimating the standard error of the median

Like the squared standard error of the mean, the squared standard error of the median refers to the variation of the median over many studies. Estimating the standard error of the sample mean is straightforward: use $s/\sqrt{n}$. But when dealing with the median, no single estimator is routinely recommended. Here, a simple estimator is used that has practical value in situations to be described. Caution must be exercised because we will see that there are conditions where it performs poorly. (And many alternatives for estimating the standard error of the median suffer from a similar problem.)

The estimate used here was derived by McKean and Schrader (1984). To apply it, first put the observations in ascending order, which is denoted by $X_{(1)} \leq X_{(2)} \leq \cdots \leq X_{(n)}$. Next, compute

$$\frac{n+1}{2} - 2.5758\sqrt{\frac{n}{4}},$$

round this value to the nearest integer, and call it k. The McKean-Schrader estimate of the squared standard error of the median is

$$s_M^2 = \left(\frac{X_{(n-k+1)} - X_{(k)}}{5.1517}\right)^2.$$

Example 3

The values

$$2.2, -11.0, -7.6, 7.3, -12.5, 7.5, -3.2, -14.9, -15.0, 1.1$$

are used to illustrate how to estimate the standard error of the median. Putting these values in ascending order yields

$$-15.0, -14.9, -12.5, -11.0, -7.6, -3.2, 1.1, 2.2, 7.3, 7.5.$$

So $X_{(1)} = -15$, $X_{(2)} = -14.9$ and $X_{(10)} = 7.5$. The sample size is $n = 10$ and

$$\frac{10+1}{2} - 2.5758\sqrt{\frac{10}{4}} = 1.4.$$

Rounding 1.4 to the nearest integer yields $k = 1$. Because $n - k + 1 = 10 - 1 + 1 = 10$, the squared standard error of the sample median is estimated to be

$$\left(\frac{X_{(10)} - X_{(1)}}{5.1517}\right)^2 = \left(\frac{7.5 - (-15)}{5.1517}\right)^2 = 19.075.$$

Consequently, the standard error is estimated to be $\sqrt{19.075} = 4.4$.

Example 4

There are realistic situations where the estimated standard error of the median can be substantially smaller than the estimated standard error of the mean. For example, Harrison and Rubinfeld (1978) conducted a study dealing generally with the cost of homes in regions near Boston, Mass. One of the variables of interest dealt with crime rates. Based on a sample size of 504, the estimated standard error of the median was 0.035 versus 0.382 for the mean, which is more than 10 times as large as the standard error of the median.

For continuous variables, where tied (duplicated) values never occur, the estimate of the standard error of the median, just described and illustrated, performs reasonably well. But when dealing with discrete random variables, where tied (duplicated) values occur, the estimate can perform rather poorly.

Example 5

Imagine that a training program for investing in stocks is rated on a scale between 0 and 10. So there are only 11 possible outcomes. Further assume that for the population of individuals who have taken the training program, the probability function is as follows:

x	$p(x)$	x	$p(x)$
0	.028247524	6	.0367569090
1	.121060821	7	.0090016920
2	.233474441	8	.0014467005
3	.266827932	9	.0001377810
4	.200120949	10	.0000059049
5	.102919345		

If 100 observations are sampled from this distribution, it can be shown that the standard error of the median is approximately .098. But over 75% of the time, the McKean–Schrader estimate exceeds .19. That is, typically, the estimate is about twice as large as it should be, which will be seen to be a source of practical concern. Roughly, the problem here is that because the only possible values are the integers from 0 to 10, with $n = 100$, tied values will occur.[2]

The central limit theorem and the median

A version of the central limit theorem applies to the sample median, meaning that there are general conditions where the sampling distribution of the median approaches a normal distribution as the sample size gets large. When sampling from continuous distributions, or distributions where duplicate values occur infrequently if at all, typically the sample size does not have to be very large for the sampling distribution of the median to be approximately normal. Letting θ (a lower case Greek theta) represent the population median this means that if we are given the value of the standard error of the median, say σ_M, then

$$Z = \frac{M - \theta}{\sigma_M}$$

will have, approximately, a standard normal distribution. Consequently, for any constant c, it will be approximately true that

$$P(M \leq c) = P\left(Z \leq \frac{c - \theta}{\sigma_M}\right).$$

In words, this probability can be determined by computing $(c - \theta)/\sigma_M$ and using table 1 in appendix B. In practice, σ_M is not known, but it can be estimated with s_M, suggesting that we use the approximation

$$P(M \leq c) = P\left(Z \leq \frac{c - \theta}{s_M}\right), \tag{5.6}$$

and it turns out that this approximation can be very useful. But for discrete distributions, where only a few possible values might be observed, a normal approximation of the sampling distribution can be quite unsatisfactory. That is, when tied (duplicated) values occur, the approximation given by equation (5.6) can perform poorly. For instance, in example 5, approximating the sampling distribution of the median with a normal distribution is unsatisfactory with $n = 100$.

Example 6

For a continuous distribution where the standard error of the median is .5, someone claims that the population median is 10. In symbols, the claim is that $\theta = 10$. If you collect data and find that the median is $M = 9$, and if the claim is correct, is it unusual to get a sample median this small or smaller? To find out, we determine $P(M \leq 9)$. We see that $(c - \theta)/\sigma_M = (9 - 10)/.5 = -2$, and from table 1 in appendix B, $P(Z \leq -2) = .023$, suggesting that the claim might be incorrect.

2. Many other proposed methods for estimating the standard error of the median also perform poorly for this same situation.

Example 7

Based on 40 values sampled from a continuous distribution, a researcher computes the McKean-Schrader estimate of the standard error of M and gets $s_M = 2$. Of interest is the probability of getting a sample median less than or equal to 24 if the population median is 26. To approximate this probability, compute $(c - \theta)/S_m = (24 - 26)/2 = -1$. So the approximate probability is $P(Z \leq -1) = .1587$.

Example 8

A study is conducted yielding the following values: 1, 3, 4, 3, 4, 2, 3, 1, 2, 3, 4, 1, 2, 1, 2. In this case, it would not be advisable to assume that the sampling distribution of the median is normal because tied (duplicated) values are common. In particular, the McKean–Schrader estimate of the standard error might perform poorly in this case.

Problems

18. For the values 4, 8, 23, 43, 12, 11, 32, 15, 6, 29, verify that the McKean–Schrader estimate of the standard error of the median is 7.57.
19. In the previous example, how would you argue that the method used to estimate the standard error of the median is a reasonable approach?
20. For the values 5, 7, 2, 3, 4, 5, 2, 6, 7, 3, 4, 6, 1, 7, 4, verify that the McKean–Schrader estimate of the standard error of the median is .97.
21. In the previous example, how would you argue that the method used to estimate the standard error of the median might be highly inaccurate?
22. In problem 20, would it be advisable to approximate $P(M \leq 4)$ using equation (5.6)?
23. For the values 2, 3, 5, 6, 8, 12, 14, 18, 19, 22, 201, why would you suspect that the McKean–Schrader estimate of the standard error of the median will be smaller than the standard error of the mean? (*Hint*: Consider features of data that can have a large impact on s, the sample standard deviation, and recall that the standard error of the mean is $s/\sqrt{n}$.)Verify that this speculation is correct.
24. Summarize when it would and would not be reasonable to assume that the sampling distribution of M is normal.

5.5 Modern advances and insights

A fundamental goal is finding an estimator with a relatively small standard error. When a distribution is reasonably symmetric, for example, the mean and median are estimating, approximately, the same quantity, so it is desirable to use the estimator that on average is most accurate. That is, we would use the estimator that has the smallest standard error. Yet one more important issue is being able to estimate the standard error based on the observations available.

We have seen that the standard error of the mean can be relatively large compared to the standard error of the median. In general, the standard error of the mean might be relatively large when one or more outliers are present. Although the median can have a substantially smaller standard error than the mean, it has several practical problems that were described in this chapter. In some situations it provides an important and useful alternative to the mean, but there are general circumstances where it is less than satisfactory.

One strategy is to use means if no outliers are found, but we will see in chapters 6 and 7 that even when outliers rarely occur, serious practical concerns about the mean remain. And later in this section we will see that simply discarding outliers and using the mean of the remaining data creates technical problems when trying to estimate the standard error.

The immediate goal is to describe just one of several alternatives to the mean and median that has been found to have practical value in a wide range of situations. (chapter 13 will summarize the relative merits of a variety of methods aimed at dealing with non-normality.) It is motivated by the realization that practical problems with the median arise roughly because it uses an extreme amount of trimming—it trims all but one or two of the values. So one possibility is to trim fewer values. But how much trimming should be used? One approach is to choose an amount of trimming that maintains, to a reasonable degree, the positive features of the mean, but which eliminates the practical problems associated with the median. Although no specific amount of trimming is always optimal, it has been found that a 20% trimmed mean, already described in chapter 2, is often a good choice. One advantage of 20% trimming is that in terms of achieving a small standard error, it competes fairly well with the mean when sampling from a normal distribution, but unlike the mean, the standard error remains relatively small when outliers are common. Another advantage of trimming is that normality can be assumed in situations where normality should not be assumed when using the mean. (Details are given in chapter 6.) In addition, a good approximation of the sampling distribution can be obtained when sampling from a discrete distribution that has relatively few values. That is, trimming 20% deals with the problem of tied values when using the median, which was illustrated by example 5 in section 5.4. A related problem with the median, when sampling from a discrete distribution that has relatively few values, is getting a good estimate of the standard error. Again, by trimming only 20%, this problem becomes negligible.

There is, however, a negative feature associated with a 20% trimmed mean. If the number of outliers is sufficiently large, the median could have a substantially smaller standard error than the 20% trimmed mean.

Example 1

It is informative to repeat example 6 in section 5.3 where observations are sampled with replacement from the data in table 2.3. Recall that these data are skewed with an extreme outlier. figure 5.6 illustrated that a plot of 4,000 sample means, with $n = 40$, looks nothing like a normal distribution. The right panel of figure 5.9 shows a plot of 4,000 20% trimmed means, and the left panel shows the medians. The point is that the plot of the 20% trimmed means looks more like a normal distribution than a plot of the means or medians. Not only does the plot of the medians not resemble a normal curve, the sampling distribution of the median is discrete with only a few values for the sample median occurring. Also, the standard error of the mean is approximately 87 versus 1.02 and 0.9 for the median and 20% trimmed mean, respectively.

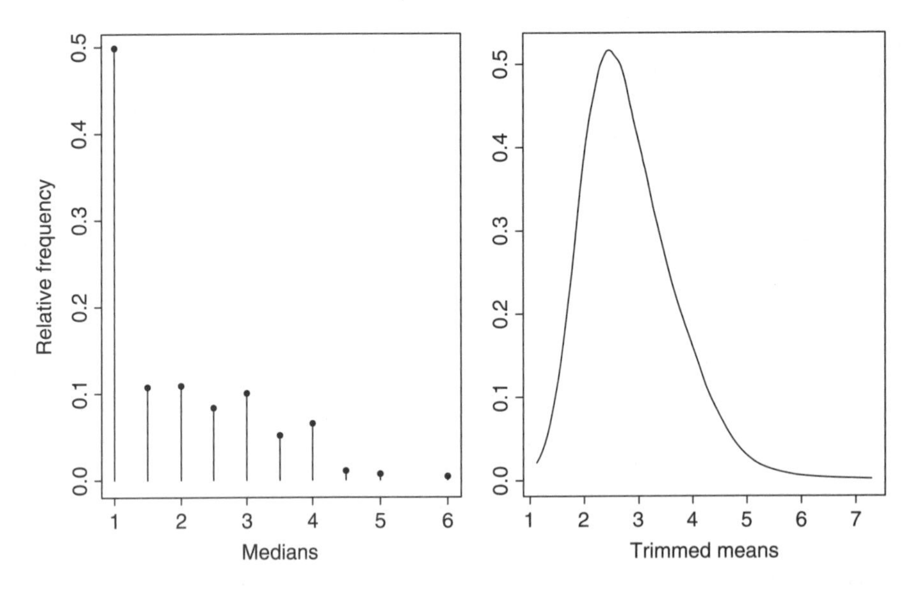

Figure 5.9 When tied values occur, situations are encountered where the sampling distribution of the sample median is highly discrete and non-normal, as illustrated in the left panel. For the same situation, the sampling distribution of the 20% trimmed mean is reasonably continuous and has a more normal shape, as indicated in the right panel.

The last example illustrated that the 20% trimmed mean can have an approximately normal distribution in situations where the sampling distribution of the mean and median is clearly non-normal. A criticism of this illustration might be that, although outliers appear to be common, extreme outliers such as the one in table 2.3, are rare. But we will see situations in chapters 6 and 7 where the choice among these three estimators is of considerable practical importance in situations that are much more common. It should be stressed, however, that despite any advantages enjoyed by both the 20% trimmed mean and median, for skewed distributions, they are not estimating the population mean. That is, if there is specific interest in estimating the population mean, the median and 20% trimmed mean can be highly unsatisfactory when a distribution is skewed. But if the goal is to avoid a measure of location that that lies in the tails of a distribution, the median and 20% trimmed mean can be much more satisfactory than the mean.

Estimating the standard error of a trimmed mean

There is a simple method for estimating the squared standard error of a 20% trimmed mean:

$$s_t^2 = \frac{s_w^2}{.6^2 n}, \tag{5.7}$$

where s_w^2 is the Winsorized sample variance introduced in chapter 2. (The .6 in the denominator is related to the amount of trimming, which is assumed to be 20%. If 10% trimming is used instead, the .6 is replaced by .8.) The standard error of the trimmed mean is estimated with the (positive) square root of this last equation:

$$s_t = \frac{s_w}{.6\sqrt{n}}. \tag{5.8}$$

In practice, situations are encountered where the standard error of the mean is about as small or slightly smaller than the standard error of the 20% trimmed mean. As will be illustrated, the standard error of the 20% trimmed mean can be substantially smaller than the standard error of the mean, but the reverse rarely, if ever, occurs.

Example 2

An earlier problem dealt with a study where rats were subjected to a drug that might affect aggression. The measures of aggression were

$$5, 12, 23, 24, 18, 9, 18, 11, 36, 15.$$

The 20% Winsorized variance can be computed as described in chapter 4 and is equal to 27.1667. Because $n = 10$, the squared standard error of the 20% trimmed mean is

$$\frac{27.1667}{.36(10)} = 7.546.$$

So the standard error is estimated to be $\sqrt{7.546} = 2.75$. As for the mean and median, the estimated standard error is 2.83 and 6.02, respectively. So the 20% trimmed mean has the smallest estimated standard error, but in this case, the improvement over the mean is not that striking.

Example 3

The data in table 5.3 are from a study on self-awareness and reflect how long a participant could keep a portion of an apparatus in contact with a specified target. The trimmed mean is $\bar{X}_t = 283$ and its estimated standard error is 56.1. In contrast, the standard error of the sample mean is $s/\sqrt{n} = 136$, a value approximately 2.4 times larger than the sample standard error of the trimmed mean. So in contrast to the previous example, the trimmed mean has a much smaller standard error than the mean. The sample median has an estimated standard error of 77.8.

Discarding outliers and using the mean

Another seemingly natural strategy is to check for outliers, remove any that are found, and compute the mean with the remaining data. There are situations where this approach has practical value, but there are serious technical issues that are not immediately evident. Methods for dealing with these technical issues are available, but generally they are not obvious based on methods covered in an introductory course. Although, these issues are not discussed in detail here, one of these technical issues is worth mentioning.

Imagine that for a random sample of n observations, we check for outliers, remove them, and now m values remain. How should the squared standard error be estimated? It might seem that the problem is trivial: compute the sample variance using the remaining m values and divide the result by m. This would be valid if $n - m$ values were randomly removed. But this was not done only extreme values were removed. It can be shown that when extreme values are removed, the remaining m values are no longer independent.[3] And the dependence among the remaining values requires special techniques when estimating the standard error of an estimator. Notice that when using the 20% trimmed mean, the sample variance of the values left after trimming was not used when estimating the squared standard error. Rather, the 20% Winsorized variance was used, and this value was divided by $.36n$, not m. It can be shown that this distinction is not trivial, but no details are given here. Suffice it to say that technically correct methods for estimating the standard depend in part of how outliers or extreme values are treated.

Example 4

For the data in table 2.3, the estimated standard error of the 20% trimmed mean is .532 using the technically correct estimate of the standard error based on the Winsorized variance and given by equation (5.8). There are 63 values left after trimming. Imagine that instead of using equation (5.8) we simply use the method for the sample mean using these 63 values only. That is, compute s using these 63 values and then compute $s/\sqrt{63}$. This yields 0.28, which is less than half of the value based on the equation (5.8). The discrepancy between these two values will be seen to be substantial.

A Summary of Some Key Points

- The sampling distribution of the sample mean reflects the likelihood of getting particular values for $\bar{X}$. In particular, it provides information about the likelihood that $\bar{X}$ will be close to the population mean, μ, when a study is repeated many times. That is, the sampling distribution of the sample mean provides some sense of how well $\bar{X}$ estimates the population mean μ. The accuracy of $\bar{X}$ is reflected by its standard error, $\sigma/\sqrt{n}$.
- All estimators, such as the sample median, the 20% trimmed mean, and the sample variance have a sampling distribution. If a study could be repeated millions of times, each time yielding a sample variance, we would know, for example, the sampling distribution of s^2 to a high degree of accuracy. That is, we could determine $P(s^2 \le c)$ for any constant c.
- Under normality, the sampling distribution of $\bar{X}$ is also normal with mean μ and variance σ^2/n.
- The standard errors corresponding to the mean, the 20% trimmed mean, and the median, can differ substantially. For the special case where sampling is from a perfectly symmetric distribution, this means that that are conditions where the median and 20% trimmed mean can provide a much more accurate estimate of the population mean relative to the sample mean, $\bar{X}$.
- Estimating the standard error of the mean and 20% trimmed mean is relatively simple. The same is true when using the median provided tied values never occur. But with tied values, obtaining a reasonably accurate estimate of the standard error of the median can be difficult even with very large sample sizes.

3. A relatively nontechnical explanation can be found in Wilcox, 2003.

- The central limit theorem indicates that with a sufficiently large sample size, the sample mean will have, approximately, a normal distribution. When the population standard deviation is known, it appears that under fairly general conditions, normality can be assumed with a sample size of $n \geq 40$. But we saw that exceptions occur, and subsequent chapters will make clear that when the population standard deviation σ is not known, assuming normality can be highly unsatisfactory.
- A version of the central limit theorem also indicates that the sampling distribution of the 20% trimmed mean will be approximately normal provided the sample size is not too small. Currently, it seems that tied values have little or no impact on this result. But when dealing with medians, assuming normality can be highly unsatisfactory when tied values can occur.

Problems

25. For the values

$$59, 106, 174, 207, 219, 237, 313, 365, 458, 497, 515,$$
$$529, 557, 615, 625, 645, 973, 1065, 3215,$$

estimate the standard error of the 20% trimmed mean.

26. For the data in problem 25, why would you suspect that the standard error of the sample mean will be larger than the standard error of the 20% trimmed mean? Verify that this speculation is correct.

27. The ideal estimator of location would have a smaller standard error than any other estimator we might use. Explain why such an estimator does not exist.

28. Under normality, the sample mean has a smaller standard error than the 20% trimmed mean or median. If observations are sampled from a distribution that appears to be normal, does this suggest that the mean should be preferred over the trimmed mean and median?

29. If the sample mean and 20% trimmed mean are nearly identical, it might be thought that for future studies, it will make little difference which measure of location is used. Comment on why this is not necessarily the case.

30. Imagine that you are able to generate 25 observations on a computer from the distribution shown in the left panel of figure 5.4. Outline how you would determine the sampling distribution of the sample median, M. In particular, how would you determine the probability that M will have a value less than 1.5?

6

ESTIMATION

Chapter 4 described what are called population parameters. They are unknown quantities that characterize the population of things or individuals under study. An example is the population mean, μ, which represents the average of all individuals if only they could be measured. Another example is p, the probability of success associated with a binomial distribution. A fundamental goal is making inferences about these unknown parameters based on data that are available, with the available data representing only a subset of all the individuals of interest. As in chapter 2, we might have n observations and compute the sample mean $\bar{X}$, which is an example of an estimator; it estimates the population mean μ. In a similar manner, if 12 of 100 persons develop negative side effects when taking a drug, then

$$\hat{p} = \frac{12}{100}$$

provides an estimate of the probability of negative side effects among all individuals who might take the drug. That is, $\hat{p}$ estimates p, the probability of success associated with a binomial distribution. In chapter 5, it was noted that $\bar{X}$ is a reasonable estimate of μ in the sense that if we imagine conducting millions of studies, the average of the resulting sample means will be equal to μ. In more formal terms, assuming random sampling, $E(\bar{X}) = \mu$. That is, $\bar{X}$ is an *unbiased* estimate of μ. In a similar manner $E(\hat{p}) = p$ when working with a binomial distribution. Nevertheless, for any given study, it is generally the case that $\bar{X}$ is not equal to the population mean μ, for the simple reason that not all individuals or things of interest have been measured. Similarly, in general, $\hat{p}$ is not equal to p, the true probability of success. This raises a fundamental question: Given some data, and if, for example, $\bar{X} = 12$, can certain values for the population mean be ruled out as being highly unlikely? Is there some way, for example, to conclude that the unknown value for μ is at least 2 and does not exceed 20? Put another way, given some data, what range of values for μ is reasonable? In a similar manner, when working with the binomial, if we observed 15 successes among 20 trials, is it reasonable to conclude that the unknown probability of success, p, is at least .5?

The classic strategy for addressing these problems, one that is routinely used today, was derived by Laplace about two centuries ago. The basic idea is to take advantage of results related to sampling distributions covered in chapter 5. There are conditions where this classic strategy performs very well, but today it is realized that there are general conditions where it can fail miserably, for reasons outlined at the end of this chapter. A brief indication of how modern methods deal with this problem is provided.

6.1 Confidence interval for the mean: Known variance

A *confidence interval* for the population mean μ is an interval, based on the observed data, that contains the unknown population mean with some specified probability. Generally, confidence intervals will vary over studies. For instance, one study might suggest that the interval (2, 14) contains μ, and another study might suggest that the interval (5, 10) contains μ. If, for example, the method for computing a confidence interval contains μ with probability .95, then the resulting confidence interval from any one study is said to be a .95 confidence interval for μ.

To elaborate on what this means and how confidence intervals are computed, we begin with the simplest case where sampling is from a normal distribution and the population variance, σ^2, is known. This is unrealistic in the sense that σ^2 is rarely if ever known, but this simplifies the reasoning and helps make clear the principles underlying the method. Once the basic principle is described, more realistic situations are covered where σ^2 is unknown and sampling is from non-normal distributions.

Example 1

Ten patients were given a drug to increase the number of hours they sleep. Table 6.1 shows some hypothetical data and, as indicated, the sample mean is $\bar{X} = 1.58$. Momentarily assume that among all patients we might measure, the increased amount of sleep has a normal distribution with variance $\sigma^2 = 1.664$. Then from chapter 5, the sampling distribution of the sample mean has a normal distribution with variance $\sigma^2/n = 1.664/10 = 0.1664$. So the standard error of the sample mean is $\sqrt{.1664} = .408$. From chapter 5, it also follows that, regardless of what the true value of the population mean happens to be,

$$Z = \frac{\bar{X} - \mu}{0.408}$$

Table 6.1 Additional hours sleep gained by using an experimental drug

Patient	Increase
1	1.2
2	2.4
3	1.3
4	1.3
5	0.0
6	1.0
7	1.8
8	0.8
9	4.6
10	1.4
$\bar{X} = 1.58$	

has a standard normal distribution. But from chapter 4, we know that with probability .95, a standard normal random variable will have a value between −1.96 and 1.96. In symbols,

$$P\left(-1.96 \le \frac{\bar{X}-\mu}{0.408} \le 1.96\right) = .95.$$

Rearranging terms in this last equation, we see that

$$P(\bar{X}-0.8 \le \mu \le \bar{X}+0.8) = .95. \quad (6.1)$$

In words, if we were to repeat this study millions of times (and in theory, infinitely many times), 95% of the intervals $(\bar{X}-0.8, \bar{X}+0.8)$ would contain the unknown population mean, μ. In our example, $\bar{X}=1.58$, and

$$(\bar{X}-0.8, \bar{X}+0.8) = (0.78, 2.38) \quad (6.2)$$

is a .95 confidence interval meaning that the interval was constructed so that with probability .95, it will contain the unknown population mean.

Rather than compute a .95 confidence interval, you might want to compute a .99 or .90 confidence interval instead. From table 1 in appendix B, or as pointed out in chapter 4, it can be seen that the probability of a standard normal random variable having a value between −2.58 and 2.58 is .99. That is,

$$P(-2.58 \le Z \le 2.58) = .99.$$

Proceeding as was done in the last paragraph, this implies that

$$\left(\bar{X}-2.58\frac{\sigma}{\sqrt{n}}, \bar{X}+2.58\frac{\sigma}{\sqrt{n}}\right)$$

is a .99 confidence interval for μ. In the example, $\sigma/\sqrt{10} = .408$, so now the confidence interval for μ is (0.53, 2.63).

> **Notation** A common notation for the probability that a confidence interval does *not* contain the population mean, μ, is α, where α is a lower case Greek alpha. When computing a .95 confidence interval, $\alpha = 1-.95 = .05$. For a .99 confidence interval, $\alpha = .01$. The quantity α is the probability of making a mistake. That is, if we perform an experiment with the goal of computing a .95 confidence interval, there is a .95 probability that the resulting interval contains the mean, but there is a $\alpha = 1-.95 = .05$ probability that it does not.

The method can be described in a slightly more general context as follows, still assuming that sampling is from a normal distribution. Imagine the goal is to compute a $1-\alpha$ confidence interval for some value for $1-\alpha$ you have chosen. The first step is to determine c such that the probability of a standard normal random variable being between $-c$ and c is $1-\alpha$. In symbols, determine c such that

$$P(-c \le Z \le c) = 1-\alpha.$$

Table 6.2 Common choices for $1-\alpha$ and c

$1-\alpha$	c
.90	1.645
.95	1.96
.99	2.58

From chapter 4, this means that you determine c such that

$$P(Z \leq c) = \frac{1+(1-\alpha)}{2}$$
$$= 1 - \frac{\alpha}{2}.$$

Put another way, c is the $1-\alpha/2$ quantile of a standard normal distribution. For example, if you want to compute a $1-\alpha = .95$ confidence interval, then

$$\frac{1+(1-\alpha)}{2} = \frac{1+.95}{2} = .975,$$

and from table 1 in appendix B we know that

$$P(Z \leq 1.96) = .975,$$

so $c = 1.96$. For convenience, table 6.2 lists the value of c for three common choices for $1-\alpha$.

Once c is determined, a $1-\alpha$ confidence interval for μ is

$$\left(\bar{X} - c\frac{\sigma}{\sqrt{n}}, \bar{X} + c\frac{\sigma}{\sqrt{n}}\right). \tag{6.3}$$

Definition The *probability coverage* of a confidence interval is the probability that the interval contains the unknown parameter being estimated. Roughly, if we repeat a study millions of times, the probability coverage refers to the proportion of the resulting intervals that contain the unknown parameter. The desired probability coverage is typically represented by $1-\alpha$, with $1-\alpha = .95$ or .99 being common choices. The value $1-\alpha$ is called the *confidence level* or *confidence coefficient* associated with the confidence interval.

Example 2

Imagine that a training program for improving SAT scores has been used for years and that among the thousands of students who have enrolled in the program, the average increase in their scores is 48. For illustrative purposes, we imagine that the number of students is so large that for all practical purposes, we know that the population mean is $\mu = 48$. Now imagine that we estimate the effectiveness of the new method by trying it on $n = 25$ students and computing the sample mean yielding $\bar{X} = 54$. This means that based on our experiment, the average effectiveness of the new method is estimated to be 54, which is larger than the average increase using the standard method, suggesting that the new method is better for the typical student. But we know that the sample mean is probably not equal to the population mean, so there is uncertainty about whether the experimental method would be better, on average, than the standard method if all students were to attend the new training program. For illustrative purposes, assume that the population standard deviation is known and equal to 9. That is, $\sigma = 9$. Then a .95 confidence interval for the unknown mean associated with the experimental method is

$$\left(54 - 1.96\frac{9}{\sqrt{25}},\ 54 + 1.96\frac{9}{\sqrt{25}}\right) = (50.5,\ 57.5).$$

So based on the 25 students available to us, their SAT scores indicates that μ is somewhere between 50.5 and 57.5, and by design, the probability coverage (or confidence level) will be .95, assuming normality. Observe that this interval does not contain the value 48 suggesting that the experimental method is better, on average, than the standard training technique.

Example 3

We repeat the last example, only this time we compute a .99 confidence interval instead. The .99 confidence interval for μ is

$$\left(54 - 2.58\frac{9}{\sqrt{25}},\ 54 + 2.58\frac{9}{\sqrt{25}}\right) = (49.4,\ 58.6).$$

Among the millions of times we might repeat this experiment with $n = 25$ students, there is a .99 probability that we have computed a confidence interval that contains μ. Note that again this interval does not contain 48, again suggesting that the experimental method is better on average.

Example 4

Suppose that for $n = 16$ observations randomly sampled from a normal distribution, $\bar{X} = 32$ and $\sigma = 4$. Compute a .9 confidence interval. Here, $1-\alpha = .9$, so we can determine c simply by referring to table 6.2, and we see that $c = 1.645$. Without table 6.2, we proceed by first noting that $(1+0.9)/2 = 0.95$. Referring to table 1 in appendix B, $P(Z \leq 1.645) = .95$, so again $c = 1.645$. Therefore, a .9 confidence interval for μ is

$$\left(32 - 1.645\frac{4}{\sqrt{16}},\ 32 + 1.645\frac{4}{\sqrt{16}}\right) = (30.355, 33.645),$$

and there is a .9 probability that this interval contains μ. Although $\bar{X}$ is not, in general, equal to μ, note that the length of the confidence interval provides some sense of how well $\bar{X}$ estimates the population mean. Here the length is $33.645 - 30.355 = 3.29$.

Example 5

A college president claims that IQ scores at her institution are normally distributed with a mean of $\mu = 123$ and a standard deviation of $\sigma = 12$. Suppose you randomly sample $n = 20$ students and find that $\bar{X} = 110$. Does the $1 - \alpha = .95$ confidence interval for the mean support the claim that the average of all IQ scores at the college is $\mu = 123$? Because $1-\alpha = .95$, $c = 1.96$ as just explained, so the .95 confidence interval is

$$\left(110 - 1.96\frac{14}{\sqrt{20}}, 110 + 1.96\frac{14}{\sqrt{20}}\right) = (103.9, 116.1).$$

The interval (103.9, 116.1) does not contain the value 123 suggesting that the president's claim might be false. Note that there is a .05 probability that the confidence interval will not contain the true population mean, so there is some possibility that the president's claim is correct.

Important conceptual point

In this last example, suppose that the .95 confidence interval is (119, 125). Would it be reasonable to conclude that the president's claim (that $\mu = 123$) is correct? The answer is no. It could be that the population mean μ is 120 or 124 for example; these values would not be ruled out based on the confidence interval. Confidence intervals can provide empirical evidence that certain values for the parameter being estimated (in this case the population mean) can probably be ruled out. But proving that the population mean is exactly 120 is virtually impossible without measuring the entire population of students under study. We can, however, design studies to improve the precision of our estimates simply by increasing the sample size, n. Looking at equation (6.3) we see that the length of a confidence interval is

$$\left(\bar{X} + c\frac{\sigma}{\sqrt{n}}\right) - \left(\bar{X} - c\frac{\sigma}{\sqrt{n}}\right) = 2c\frac{\sigma}{\sqrt{n}}.$$

So, by increasing the sample size n, the length of the confidence interval decreases and reflects the extent to which the sample mean gives an improved estimate of the population mean.

Use caution when interpreting confidence intervals

Care must be taken not to read more into a confidence interval than is warranted. For the situation at hand, the probability coverage of a confidence interval reflects the likelihood, over many studies, that the confidence interval will contain the unknown population mean, μ. But there are several ways in which a confidence interval can be interpreted incorrectly, which are illustrated with the last example where .95 confidence interval for the average IQ score at some university was found to be (103.9, 116.1).

- 95% of all students have an IQ between 103.9 and 116.1. The error here is interpreting the ends of the confidence intervals as quantiles. That is, if among all students, the .025 quantile is 100.5 and the .975 quantile is 120.2, this means that 95% of all students have an IQ between 100.5 and 120.2. But confidence intervals for the population mean tell us nothing about the quantiles associated with all students at this university.
- There is a .95 probability that a randomly sampled student will have an IQ between 100.5 and 120.2. This erroneous interpretation is similar to the one just described. Again, confidence intervals do not indicate the likelihood of observing a particular IQ, but rather indicate a range of values that are likely to include μ.

- All sample means among future studies will have a value between 103.9 and 116.1 with probability .95. This statement is incorrect because it is about the sample mean, not the population mean. (For more details about this particular misinterpretation, see Cumming and Maillardet, 2006.)

Problems

1. Explain the meaning of a .95 confidence interval.
2. If you want to compute a .80, or .92, or a .98 confidence interval for μ when σ is known, and sampling is from a normal distribution, what values for c should you use in equation (6.3)?
3. Assuming random sampling is from a normal distribution with standard deviation $\sigma = 5$, if you get a sample mean of $\bar{X} = 45$ based on $n = 25$ subjects, what is the .95 confidence interval for μ?
4. Repeat the previous example, only compute a .99 confidence interval instead.
5. A manufacturer claims that their light bulbs have an average life span that follows a normal distribution with $\mu = 1{,}200$ hours and a standard deviation of $\sigma = 25$. If you randomly test 36 light bulbs and find that their average life span is $\bar{X} = 1{,}150$, does a .95 confidence interval for μ suggest that the claim $\mu = 1{,}200$ is reasonable?
6. For the following situations, (a) $n = 12$, $\sigma = 22$, $\bar{X} = 65$, (b) $n = 22$, $\sigma = 10$, $\bar{X} = 185$, (c) $n = 50$, $\sigma = 30$, $\bar{X} = 19$, compute a .95 confidence interval for the mean assuming normality.
7. What happens to the length of a confidence interval for the mean of a normal distribution when the sample size is doubled? What happens if it is quadrupled?
8. The length of a bolt made by a machine parts company is a normal random variable with standard deviation σ equal to .01 mm. The lengths of four randomly selected bolts are: 20.01, 19.88, 20.00, 19.99. (a) Compute a .95 confidence interval of the mean. (b). Specifications require a mean length μ of 20.00 mm for the population of bolts. Do the data indicate that this specification is not being met? (c) Given that the .95 confidence interval contains the value 20, why might it be inappropriate to conclude that the specification is being met?
9. The weight of trout sold at a trout farm has a standard deviation of .25. Based on a sample of 10 trout, the average weight is 2.10 lb. Compute a .99 confidence interval for the population mean, assuming normality.
10. A machine to measure the bounce of a ball is used on 45 randomly selected tennis balls. Experience has shown that the standard deviation of the bounce is .30. If $\bar{X} = 1.70$, and assuming normality, what is a .90 confidence interval for the average bounce?

6.2 Confidence intervals for the mean: σ not known

There are two practical concerns with the method for computing confidence intervals described in the previous section. The first is that the population variance σ^2 is, in general, not known. The other is that distributions are rarely, if ever, exactly normal.

Assuming normality is convenient from a technical point view, but to what extent does it provide a reasonably accurate approximation? If, for example, we claim that a .95 confidence interval for the population mean has been computed, is it possible that the probability coverage is only .9 or even .8? Under what conditions is the actual probability coverage reasonably close to .95? Here, we first focus on how to proceed when the standard deviation σ is not known, still assuming normality. Then we will discuss the effects of non-normality and how they might be addressed.

Example 1

Imagine you are a health professional interested in the effects of medication on the diastolic blood pressure of adult women. For a particular drug being investigated, you find that for nine women, the sample mean is $\bar{X} = 85$ and the sample variance is $s^2 = 160.78$. So although we do not know the population variance σ^2, we have an estimate of it, namely $s^2 = 160.78$, as noted in chapter 5. If we momentarily assume that this estimate is reasonably accurate, then a natural strategy is to compute confidence intervals as described in the previous section and given by equation (6.3). Because $n = 9$, we see that $s/\sqrt{n} = 4.2$, so a .95 confidence interval for the mean would be

$$(85 - 1.96(4.2),\ 85 + 1.96(4.2)) = (76.8,\ 93.2).$$

Prior to the year 1900, the process just illustrated was the strategy used to compute confidence intervals, and it turns out that this approach is reasonable if the sample size is sufficiently large, assuming random sampling. That is, a form of the central limit theorem tells us that provided the sample size is reasonably large,

$$T = \frac{\bar{X} - \mu}{s/\sqrt{n}} \tag{6.4}$$

will have, approximately, a standard normal distribution. Moreover, the accuracy of this approximation improves as the sample size increases.

However, William Sealy Gosset (1876–1937) noted that the sample standard deviation s is an erratic estimator of σ when the sample size, n, is small. Even when sampling from a normal distribution, concerns arise. Gosset worked for Arthur Guinness and Son, a Dublin brewery. His applied work dealt with quality control issues relevant to making beer, typically he was forced to make inferences based on small sample sizes, and so he set out to find a method that would take into account the fact that s might be a rather unsatisfactory estimate of σ. He published his results in a now famous 1908 paper, "The Probable Error of a Mean." In essence, Gosset determined the sampling distribution of T, assuming that observations are randomly sampled from a normal distribution. But initially Guinness did not allow Gosset to publish his results. Eventually, however, he was allowed to publish his results provided that he use a pseudonym chosen by the managing director of Guinness, C. D. La Touche. The pseudonym chosen by La Touche was 'Student.' For this reason, the distribution of T is called *Student's t-distribution.* The main point is that under normality, we can determine the probability that T is less than 1, less than 2, or less than c for any constant c we might choose. It turns out that the distribution depends on the sample size, n. By convention, the quantiles of the distribution are reported in terms of *degrees of freedom*: $\nu = n - 1$, where ν is a lower case Greek nu. Figure 6.1 shows Student's t-distribution with $\nu = 4$ degrees of freedom.

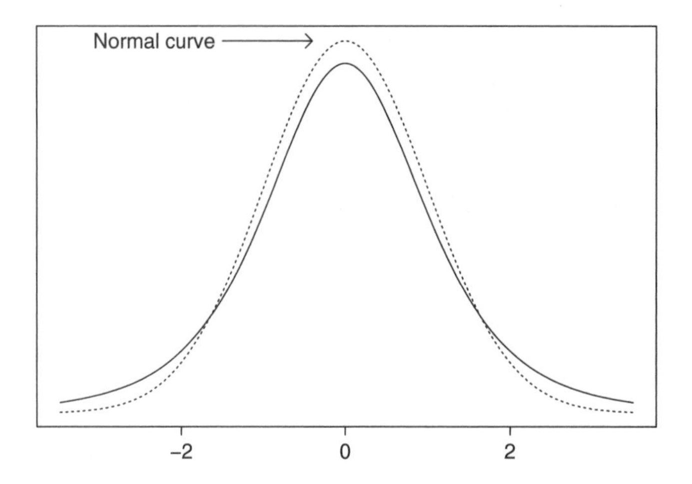

Figure 6.1 Shown is a standard normal curve and a Student's t-distribution with four degrees of freedom. Student's t-distributions are symmetric about zero, but they have thicker or heavier tails.

Note that the distribution is similar to a standard normal. In particular, it is symmetric about zero. With infinite degrees of freedom, Student's t and the standard normal are identical. But otherwise, Student's t-distribution does not belong to the class of normal distributions, even though it has a symmetric, bell shape. That is, a distribution is called normal if the equation for the curve has the form indicated by equation (4.6). Generally, the equation for Student's t does not have this form. Put another way, all normal distributions are symmetric and bell-shaped, but there are infinitely many symmetric and bell-shaped distributions that are not normal.

Table 4 in appendix B reports some quantiles of Student's t-distribution. The first column gives the degrees of freedom. The next column, headed by $t_{.9}$, reports the .9 quantiles. For example, with $\nu = 1$, we see 3.078 under the column headed by $t_{.9}$. This means that $P(T \leq 3.078) = .9$. That is, if we randomly sample two observations from a normal distribution, in which case $\nu = n - 1 = 1$, there is a .9 probability that the resulting value for T is less than 3.078. Similarly, if $\nu = 24$, $P(T \leq 1.318) = .9$. The column headed by $t_{.99}$ lists the .99 quantiles. For example, if $\nu = 3$, we see 4.541 under the column headed $t_{.99}$, so the probability that T is less than 4.541 is .99. If $\nu = 40$, Table 2 indicates that $P(T \leq 2.423) = .99$. Many software packages, such as Minitab, R and S-PLUS, contain functions that compute Student's t-distribution for any $\nu \geq 1$.

Similar to the situation when working with normal distributions,

$$P(T \geq c) = 1 - P(T \leq c), \qquad (6.5)$$

where c is any constant that might be of interest. For example, with $\nu = 4$, $P(T \leq 2.132) = .95$, as previously indicated, so $P(T \geq 2.132) = 1 - P(T \leq 2.132) = .05$.

Example 2

Imagine you are involved in a study on the effects of a cold medication on reaction times. Assuming normality, you randomly sample 13 observations and compute the sample mean and variance. What is the probability that T is less than 2.179? The degrees of freedom are $\nu = n - 1 = 13 - 1 = 12$. From table 4 in appendix B, looking at the row with $\nu = 12$, we see 2.179 in the column headed by .975, so $P(T < 2.179) = .975$. That is, we do not know the value of the population mean, μ, but regardless of what its true value happens to be, the probability that T will be less than 2.179 is .975. The practical implication of this result is that we can compute a confidence interval for μ, as illustrated momentarily.

Example 3

If $\nu = 30$ and $P(T \geq c) = .005$, what is c? Because table 4 gives the probability that T is less, than or equal to some constant, we must convert the present problem into one where table 4 can be used. Based on equation (6.5), if $P(T \geq c) = .005$, then $P(T \leq c) = 1 - P(T \geq c) = 1 - .005 = .995$. Looking at the column headed by $t_{.995}$ in table 4, we see that with $\nu = 30$, $P(T \leq 2.75) = .995$, so $c = 2.75$.

With Student's t-distribution, we can now compute a confidence interval for μ when σ is not known, assuming that observations are randomly sampled from a normal distribution. Recall that when σ is known, the $1 - \alpha$ confidence interval for μ is

$$\bar{X} \pm c\frac{\sigma}{\sqrt{n}} = \left(\bar{X} - c\frac{\sigma}{\sqrt{n}}, \bar{X} + c\frac{\sigma}{\sqrt{n}}\right),$$

where c is the $1 - \alpha/2$ quantile of a standard normal distribution and read from table 1 in appendix B. When σ is not known, this last equation becomes

$$\bar{X} \pm c\frac{s}{\sqrt{n}} = \left(\bar{X} - c\frac{s}{\sqrt{n}}, \bar{X} + c\frac{s}{\sqrt{n}}\right), \tag{6.6}$$

where now c is the $1 - \alpha/2$ quantile of Student's t-distribution with $n - 1$ degrees of freedom and read from table 4 of appendix B. If observations are randomly sampled from a normal distribution, then the probability coverage is exactly $1 - \alpha$.

Example 1 (*Continued*)

In the first example of this section we computed a .95 confidence interval for μ assuming that $\sigma^2 = s^2$. The result was

$$(85 - 1.96(4.2),\ 85 + 1.96(4.2)) = (76.8,\ 93.2).$$

But based on Student's t, with $\nu = 8$ degrees of freedom, the value for c should be 2.306 rather than 1.96. So now the .95 confidence interval is

$$(85 - 2.306(4.2),\ 85 + 2.306(4.2)) = (75.3,\ 94.7).$$

Example 4

Doksum and Sievers (1976) report data on weight gain among rats. One group was the control and the other was exposed to an ozone environment. For illustrative purposes, attention is focused on the control group and the goal is to determine a range of possible values for the population mean that are reasonable based on the data available. Here, we compute a .95 confidence interval. Because there are $n = 22$ rats, the degrees of freedom are $n - 1 = 22 - 1 = 21$. Because $1 - \alpha = .95$, $\alpha = .05$, so $\alpha/2 = .025$, and $1 - \alpha/2 = .975$. Referring to table 4 in appendix B, we see that the .975 quantile of Student's t-distribution with 21 degrees of freedom is approximately $c = 2.08$. Because $\bar{X} = 11$ and $s = 19$, a .95 confidence interval is

$$11 \pm 2.08 \frac{19}{\sqrt{22}} = (2.6, 19.4).$$

That is, although both the population mean and variance are not known, we can be reasonably certain that the population mean, μ, is between 2.6 and 19.4, if the assumption of sampling from a normal distribution is true.

Example 5

Imagine you are interested in the reading abilities of fourth graders. A new method for enhancing reading is being considered, you try the new method on 11 students and then administer a reading test yielding the scores

$$12, 20, 34, 45, 34, 36, 37, 50, 11, 32, 29.$$

For illustrative purposes, imagine that after years of using a standard method for teaching reading, the average scores on the reading test has been found to be $\mu = 25$. Someone claims that if the new teaching method is used, the population mean will remain 25. The goal here is to determine whether this claim is consistent with the .99 confidence interval for μ. That is, does the .99 confidence interval contain the value 25? It can be seen that the sample mean is $\bar{X} = 30.9$ and $s/\sqrt{11} = 3.7$. There are $n = 11$ observations, so the degrees of freedom are $\nu = 11 - 1 = 10$. Because $1 - \alpha = .99$, it can be seen that $1 - \alpha/2 = .995$, so from table 4 in appendix B, $c = 3.169$. Consequently, the .995 confidence interval is

$$30.9 \pm 3.169(3.7) = (19.2, 42.6).$$

This interval contains the value 25, so the claim that $\mu = 25$ cannot be refuted based on the available data. Note, however, that the confidence interval also contains 35 and even 40. Although we cannot rule out the possibility that the mean is 25, there is some possibility that the new teaching method enhances reading by a substantial amount, but with only 11 participants, the confidence interval is too long to resolve how effective the new method happens to be.[1]

A conceptual issue

Given the goal of computing a confidence interval, there is an important conceptual issue that is worth stressing. Notice that the key to computing a confidence interval when the standard deviation σ is known is being able to determine the quantiles of the distribution of

$$Z = \frac{\bar{X} - \mu}{\sigma/\sqrt{n}}.$$

Under normality, Z has a standard normal distribution and these quantiles are given in table 1 in appendix B, which in turn yield a confidence interval for the population mean.

1. There are methods for determining how many more observations are needed so that confidence interval will have some specified length (e.g., Wilcox, 2003), but the details go beyond the scope of this book.

When the standard deviation σ is not known, the key to computing a confidence interval is to determine the quantiles of the distribution of

$$T = \frac{\bar{X} - \mu}{s/\sqrt{n}}$$

and again this can be done assuming normality. During the last two centuries, the strategy has been to appeal to the central limit theorem when dealing with non-normal distributions. That is, hope that the sample size is large enough so that even under non-normality, T has approximately a Student's t-distribution. Since about 1990, however, it has become clear that there are general conditions under which the sample size must be much larger than previously thought. A description and illustration of why problems arise are provided in section 6.5. (And another fundamental concern, described in chapter 7, has been evident since at least 1960.) Fortunately, modern methods are now available that deal with this problem in a very effective manner.

Problems

11. Assuming the degrees of freedom are 20, find the value c for which

    (a) $P(T > c) = .025$,
    (b) $P(T \leq c) = .995$
    (c) $P(-c \leq T \leq c) = .90$.

12. Compute a .95 confidence interval if

    (a) $n = 10$, $\bar{X} = 26$, $s = 9$,
    (b) $n = 18$, $\bar{X} = 132$, $s = 20$,
    (c) $n = 25$, $\bar{X} = 52$, $s = 12$.

13. Repeat the previous exercise, but compute a .99 confidence interval instead.
14. For a study on self-awareness, the observed values for one of the groups were

    $$77, 87, 88, 114, 151, 210, 219, 246, 253, 262, 296,$$
    $$299, 306, 376, 428, 515, 666, 1{,}310, 2{,}611.$$

    Compute .95 confidence interval for the mean assuming normality.
15. Rats are subjected to a drug that might affect aggression. Suppose that for a random sample of rats, measures of aggression are found to be

    $$5, 12, 23, 24, 18, 9, 18, 11, 36, 15.$$

    Compute a .95 confidence interval for the mean assuming the scores are from a normal distribution.

6.3 Confidence intervals for the population median

This section describes and illustrates three methods for computing a confidence interval for the population median, θ. (Again, θ is a lower case Greek theta.) The first method is included partly to illustrate a general principle used by many conventional methods for computing confidence intervals. But the other two have certain practical advantages to be described.

Recall from chapter 5 that like the sample mean, the sample median has a sampling distribution. A rough but useful way of conceptualizing it is to think of millions of studies, each resulting in a sample median based on n observations. The variance of these millions of sample medians is the squared standard error of the sample median. As noted in chapter 5, one way of estimating this squared standard error is with the McKean–Schrader estimator, S_M^2. The (positive) square root of this last value, S_M, estimates the standard error.

A version of the central limit theorem says that under random sampling,

$$Z_M = \frac{M - \theta}{S_M}$$

will have, approximately, a standard normal distribution if the sample size is sufficiently large. Note the similarity between this last equation and the methods for computing confidence intervals for the population mean. In practical terms, a confidence interval for the population median can be computed by assuming that Z_M has a standard normal distribution. So an approximate .95 confidence interval for the population median is

$$(M - 1.96S_M, M + 1.96S_M) \tag{6.7}$$

and more generally, if c is the $1 - \alpha/2$ quantile of a standard normal distribution,

$$(M - cS_M, M + cS_M) \tag{6.8}$$

is an approximate $1 - \alpha$ confidence interval.

Example 1

Consider again the sleep data in table 6.1. The sample median is $M = 1.3$, the McKean-Schrader estimate of the standard error is $S_M = 0.8929$ and the .95 confidence interval for the population median is

$$(1.3 - 1.96(0.8929), 1.3 + 1.96(0.8929)) = (-0.45, 3.05).$$

This is in contrast to the .95 confidence interval for the mean, based on Student's t: (0.7001, 2.4599). Note that the length of the confidence interval based on the median is $3.05 - (-0.45) = 3.5$ versus a length of 1.76 using the mean.

Important conceptual issue

This method for computing a confidence interval for the median illustrates a general strategy for computing confidence intervals that currently dominates applied research. It is important to be aware of this strategy and to develop some sense about its relative merits compared to more modern techniques. Consider any unknown parameter of interest, such as the population median, mean or the population standard deviation. When computing a confidence interval, the most basic form of this strategy stems from Laplace and consists of standardizing the estimator being used. That is, subtract the parameter being estimated, divide by the standard error of the estimator and then assume that this standardized variable has, approximately, a standard normal distribution. Under general conditions, the central limit theorem says that this approximation will perform reasonably well if the sample size is sufficiently large. When working with means, $\bar{X}$ estimates μ, an estimate of the standard error of $\bar{X}$ is $s/\sqrt{n}$, so this standardization takes the form

$$T = \frac{\bar{X} - \mu}{s/\sqrt{n}},$$

and an improvement on Laplace's approach was to approximate the distribution of T with Student's t-distribution. When the goal is to make inferences about the population median, the sample median M estimates the population median θ, S_M estimates the standard error of M, so this same strategy suggests assuming that

$$Z_M = \frac{M - \theta}{S_M}$$

has a standard normal distribution, which in turn means that a confidence interval for the median can be computed. This general strategy has proven to be very useful, but it is important to be aware that under general conditions, the confidence interval for the median given by equation (6.8) performs poorly, namely situations where duplicate or *tied* values tend to occur. For example, if we observe the values 4, 7, 8, 19, 34, 1, all values occur only once and so it is said that there are no tied values. But if we observe 23, 11, 9, 8, 23, 6, 7, 1, 11, 19, 11, then tied values occur because the value 11 occurred three times and the value 23 occurred twice. In chapter 5, it was pointed out that when tied values tend to occur, S_M can be a highly unsatisfactory estimate of the standard error of the median.

There are at least two ways of improving upon the above method for computing a confidence interval for the median. One of these is a very general technique called a percentile bootstrap method, the details of which are outlined in the final section of this chapter. (It is general in the sense that it can applied to a wide range of problems.) Another is the method given in box 6.1.

Example 2

To illustrate the method in box 6.1, imagine a study yielding the values

$$4, 12, 14, 19, 23, 26, 28, 32, 39, 43.$$

Note that the values are written in ascending order so for the notation used in box 6.1, the $X_{(1)} = 4$, $X_{(2)} = 12$ and $X_{(10)} = 43$. Also note that the sample size is $n = 10$. The method in box 6.1 uses confidence intervals having the form

$$(X_{(k)}, X_{(n-k+1)}),$$

where k is some integer between 1 and $n/2$. The problem is, given some choice for k, what is the probability coverage? For illustrative purposes, first consider $k = 1$ in which case $n - k + 1 = 10$. Consider a binomial random variable Y with probability of success $p = .5$ and $n = 10$. As indicated in chapter 4, we can use table 2 in appendix B to determine that $P(k \leq Y \leq n - k) = P(1 \leq Y \leq 9) = .998$. This says that $(X_{(1)}, X_{(10)}) = (4, 43)$ is a .998 confidence interval for the population median. For $k = 2$, the confidence interval becomes (12, 39). Box 6.1 indicates that the probability that this interval contains the population median can be determined from table 2 in appendix B and is given by $P(k \leq Y \leq n - k) = P(2 \leq Y \leq 8) = .978$.

BOX 6.1

Goal: Compute a $1-\alpha$ confidence interval for the median based on the observations $X_1, \ldots, X_n$.

Consider a binomial distribution with probability of success $p = .5$. Given the sample size n, consider any integer k greater than 0 but less than or equal to $n/2$. Let P_k be the probability that the number of successes is between k and $n-k$. In symbols, if Y is a binomial with probability of success .5, $P_k = P(k \le Y \le n-k)$. Illustrations are covered in chapter 4. For example, if $n = 15$ and $k = 3$, then $n-k = 12$ and from table 2 in appendix B, $P_k = .996 - .004 = .992$. Put the observations $X_1, \ldots, X_n$ in ascending order yielding $X_{(1)} \le \cdots \le X_{(n)}$. Then

$$(X_{(k)}, X_{(n-k+1)})$$

is a confidence interval for the population median having probability coverage P_k, assuming random sampling only.

(The software, Minitab, has a built-in function that performs the calculations, called sint, and an S-PLUS or R version is available as well, which is stored in the library of functions Rallfunv1-v7 mentioned in chapter 1. These functions use a refinement of the method just described and allow you to choose the α value.)

Given k, we can determine the probability coverage based on the method in box 6.1, as just illustrated. But if, for example, we want to compute a .95 confidence interval, there might not be any value of k that accomplishes this goal exactly. There is an extension of the method aimed at dealing with this problem; the details are not covered here, but it can be applied with the Minitab and R function sint mentioned in box 6.1.

Example 3

It is illustrated that with small sample sizes, the choice between the more accurate method used by the R function sint, versus equation (6.8), can make a practical difference. (The method leading to equation 6.8 is included primarily to illustrate the conceptual issue that was just described.) Consider again the data in table 6.1. Using the R function sint, a .95 confidence interval for the (population) median is (0.93, 2.01). Note that this interval differs substantially from the confidence interval obtained in example 1. It is also noted that the .95 confidence interval for the mean, based on Student's t, is considerably longer. The length is 1.7 versus $2.01 - .03 = 1.08$ using the function sint.

Problems

16. Suppose $M = 34$ and the McKean–Schrader estimate of the standard error of M is $S_M = 3$. Compute a .95 confidence interval for the population median.
17. For the data in problem 14 of section 6.2, the McKean–Schrader estimate of the standard error of M is $S_M = 77.8$ and the sample median is 262. Compute a .99 confidence interval for the population median.

18. If $n = 10$ and you compute a confidence interval for the median using $(X_{(k)}, X_{(n-k)})$, as described in box 6.1, what is the probability the probability that this interval contains the population median if $k = 2$?

19. Repeat the previous problem, only with $n = 15$ and $k = 4$.

20. For the data in problem 14 of section 6.2, if we use (88, 515) as a confidence interval for the median, what is the probability that this interval contains the population median?

6.4 The binomial: Confidence interval for the probability of success

Consider the binomial distribution introduced in chapter 4. A common goal is to make inferences about the unknown probability of success, p, based on observations we make where the only two possible outcomes are 1, for success, and 0 for failure. In formal terms, we observe $X_1, \ldots, X_n$, where now $X_1 = 1$ or 0, $X_2 = 1$, or 0, and so on. The sample mean of these n values corresponds to the proportion of successes among the n observations made. But rather than label the average of these values $\bar{X}$, the more common notation is to use $\hat{p}$. That is

$$\hat{p} = \frac{1}{n} \sum X_i \tag{6.9}$$

is the observed proportion of successes and estimates the unknown probability of success, p. In fact, $\hat{p}$ can be shown to be an unbiased estimate of p (meaning that $E(\hat{p}) = p$) and its squared standard error or variance is

$$\mathrm{VAR}(\hat{p}) = \frac{p(1-p)}{n}.$$

So its standard error is

$$\mathrm{SE}(\hat{p}) = \sqrt{\frac{p(1-p)}{n}}.$$

As noted in chapter 5, the central limit theorem applies and says that if the sample size is sufficiently large,

$$\frac{\hat{p} - p}{\sqrt{p(1-p)/n}}$$

will have, approximately, a standard normal distribution. In the present context, this means that Laplace's method is readily applied when computing a confidence interval for p. It is given by

$$\hat{p} \pm c\sqrt{\frac{p(1-p)}{n}},$$

where c is the $1 - \alpha/2$ quantile of a standard normal distribution. We do not know the value of the quantity under the radical, but it can be estimated with $\hat{p}$, in which case a simple $1 - \alpha$ confidence interval for p is

$$\hat{p} \pm c\sqrt{\frac{\hat{p}(1-\hat{p})}{n}}. \tag{6.10}$$

Example 1

Among all registered voters, let p be the proportion who approve of how the president of the United States is handling his job. Suppose 10 people are asked to indicate whether they approve of the president's performance. Further imagine that the responses are

$$1, 1, 1, 0, 0, 1, 0, 1, 1, 0$$

and you want a .95 confidence interval for p. So, as before, $c = 1.96$, which is read from table 1 in appendix B and corresponds to a .975 quantile of a standard normal distribution. In terms of the above notation, $X_1 = 1$ (the first individual responded that she approves), $X_2 = 1$ and $X_4 = 0$, and the proportion who approve is $\hat{p} = 6/10$. So the estimated standard error associated with $\hat{p}$ is $\sqrt{.6(1 - .6)/10} = .155$ and a .95 confidence interval is

$$.6 \pm 1.96(.155) = (.296, .904).$$

In words, we do not know p, but based on the available information, if we assume random sampling and that $\hat{p}$ has a normal distribution, then we can be reasonably certain that p has a value between .296 and .904.

Example 2

A sample of 100 voters in a town contained 64 persons who favored a bond issue. To assess the proportion of all voters who favor the bond issue, we compute a .95 confidence interval for p. Here, $n = 100$, $X = 64$, $\hat{p} = .64$, $c = 1.96$, and

$$\sqrt{\frac{\hat{p}(1-\hat{p})}{n}} = \sqrt{\frac{.64 \times .36}{100}} = .048.$$

So the .95 confidence interval is

$$(.64 - .094,\ .64 + .094) = (.546,\ .734).$$

Agresti-Coull method

The accuracy of the confidence interval just described and illustrated depends on both the sample size n and the unknown probability of success, p. Generally, the closer p happens to be to 0 or 1, the larger the sample size must be to get an accurate confidence interval. As is probably evident, a practical problem is that we do not know p, this raises concerns, and many improvements have been proposed. Brown et al. (2002) compared several methods and found that what they call the Agresti-Coull method, which is a simple generalization of method derived by Agresti and Coull (1998), performs relatively well. The Agresti-Coull method is applied as follows.[2]

Let X represent the total number of successes among n observations, in which case

$$\hat{p} = \frac{X}{n},$$

2. Blyth's (1986) comparisons of various methods suggest using Pratt's (1968) approximate confidence interval, which can be applied with the S-PLUS or R software mentioned in chapter 1. The function binomci performs the calculations. It appears to be unknown how Pratt's method compares to the Agresti-Coull method.

the proportion of successes among the n observations.[3] As before, let c be the $1-\alpha/2$ quantile of a standard normal distribution. Compute

$$\tilde{n} = n + c^2,$$

$$\tilde{X} = X + \frac{c^2}{2},$$

and

$$\tilde{p} = \frac{\tilde{X}}{\tilde{n}}.$$

Then the $1-\alpha$ confidence interval for p is

$$\tilde{p} \pm c\sqrt{\frac{\tilde{p}(1-\tilde{p})}{\tilde{n}}}.$$

Blyth's method

For the special cases where the number of successes is $X = 0$, 1, $n-1$ and n, Blyth (1986) suggests computing a $1-\alpha$ confidence interval for p as follows:

- If $X = 0$, use

$$(0,\ 1-\alpha^{1/n}).$$

- If $X = 1$, use

$$\left(1-(1-\frac{\alpha}{2})^{1/n},\ 1-(\frac{\alpha}{2})^{1/n}\right).$$

- If $X = n-1$, use

$$\left((\frac{\alpha}{2})^{1/n},\ (1-\frac{\alpha}{2})^{1/n}\right).$$

- If $X = n$, use

$$(\alpha^{1/n},\ 1).$$

Example 3

Some years ago, a television news program covered a story about a problem that sometimes arose among people undergoing surgery: Some patients would wake up and become aware of what was happening to them and they would have nightmares later about their experience. Some surgeons decided to monitor brain function in order to be alerted if someone was regaining consciousness. In the event this happened, they would give the patient more medication to keep them under. Among 200,000 surgeries, zero patients woke up under this experimental method, but hospital administrators, worried about the added cost of monitoring brain function, argued that with only 200,000 surgeries, it is impossible to be very certain about the actual probability of someone waking up. Note, however, that we can compute a .95 confidence interval using the method just described. The number of times a patient woke up is $X = 0$, the

3. In the notation used at the beginning of this section, $X = \sum X_i$.

number of observations is $n = 200,000$, so a .95 confidence interval for the probability of waking up is

$$(0, 1 - .05^{1/200000}) = (0, .000015).$$

Example 4

Example 1 is repeated, only now we compute the Agresti-Coull .95 confidence interval. As before, $X = 6$, $n = 10$ and $c = 1.96$. So $\tilde{n} = 10 + 1.96^2 = 13.84$, $\tilde{X} = 6 + 1.96^2/2 = 7.92$ and $\tilde{p} = 7.92/13.84 = .5723$. And the .95 confidence interval is

$$.5723 \pm 1.96\sqrt{\frac{.5723(1 - .5723)}{13.84}} = (.31, .83).$$

In contrast, using equation (6.10), the .95 confidence interval is (.296, .904). Notice that the upper end of this confidence interval, .904, differs substantially from the Agresti-Coull upper end, .83.

Problems

21. You observe the following successes and failures: 1, 1, 1, 1, 1, 0, 0, 0, 0, 0, 0, 0, 0, 0, 0. Compute a .95 confidence interval for p using equation (6.10) as well as the Agresti-Coull method.
22. Given the following results for a sample from a binomial distribution, compute the squared standard error of $\hat{p}$ when (a) $n = 25$, $X = 5$, (b) $n = 48$, $X = 12$, (c) $n = 100$, $X = 80$, (d) $n = 300$, $X = 160$.
23. Among 100 randomly sampled adults, 10 were found to be unemployed. Give a .95 confidence interval for the percentage of adults unemployed using equation (6.10). Compare this result to the Agresti-Coull confidence interval.
24. A sample of 1,000 fish was obtained from a lake. It was found that 290 were members of the bass family. Give a .95 confidence interval for the percentage of bass in the lake using equation (6.10).
25. Among a random sample of 1,000 adults, 60 reported never having any legal problems. Give a .95 confidence interval for the percentage of adults who never had legal problems using equation (6.10).
26. Among a sample of 600 items, only one was found to be defective. Explain why using equation (6.10) might be unsatisfactory when computing a confidence interval for the probability that an item is defective.
27. In the previous problem, compute a .90 confidence interval for the probability that an item is defective.
28. One-fourth of 300 persons in a large city stated that they are opposed to a certain political issue favored by the mayor. Calculate a 99% confidence interval for the fraction of people of individuals opposed to this issue using equation (6.10).
29. A test to detect a certain type of cancer has been developed and it is of interest to know the probability of a false-negative indication, meaning the test fails to detect cancer when it is present. The test is given to 250 patients known to have cancer

and five tests fail to show its presence. Determine a .95 confidence interval for the probability of a false-negative indication using equation (6.10).

30. In the previous problem, imagine that 0 false-negative indications were found. Determine a .99 confidence interval for the probability of a false-negative indication.

31. A cosmetic company found that 180 of 1,000 randomly selected women in New York city have seen the company's latest television advert. Compute a .95 confidence interval for the percentage of women in New York city who have seen the advert using equation (6.10).

6.5 Modern advances and insights

This chapter has covered the basic strategies used to compute confidence intervals, strategies that are routinely taught and used. It would be remiss, however, not to mention that there have been many advances and insights related to the methods covered in this chapter, a few of which are outlined here.

Student's t and non-normality

Computing confidence intervals using Student's t technique is a classic method that is routinely used and therefore essential to know. Also, Student's t represents a major breakthrough in our ability to assess the information in data. At one time, there were reasons to suspect that it continues to perform reasonably well when sampling from non-normal distributions, but modern insights make it clear that this is not always the case. As seems evident, getting the most accurate information possible from data requires a good understanding of when Student's t gives accurate results when computing confidence intervals as well as when it fails.

To address this issue, we first stress a connection between sampling distributions, covered in chapter 5, and Student's t. Momentarily, suppose we know the population mean μ, and imagine that we sample 21 observations and compute the sample mean $\bar{X}$ and the standard deviation, s. Then, of course, we can compute

$$T = \frac{\bar{X} - \mu}{s/\sqrt{n}}.$$

Further imagine we repeat this process thousands and perhaps even millions of times. Then a plot of the resulting T values would provide a very accurate approximation of the actual distribution of T. If, for example, the sample size were $n = 21$ and we found that the proportion of T values less than 2.086, say, is .975, and if the proportion of T values less than -2.086 is .025, then we would know how to compute a .95 confidence interval for the population mean: simply use $c = 2.086$ in equation (6.6). And indeed, the exact value under normality is $c = 2.086$.

Now we consider what happens when we sample from non-normal distributions. If we repeat the process just described, again with a sample size of $n = 21$, only we sample from a non-normal distribution, will the proportion of T values less than 2.086 be approximately .975? For many years it was conjectured that the answer is yes. This conjecture was not based on wild speculations, but today it is known that a poor approximation might result and we understand why problems were missed for so many years.

Example 1

Imagine that 25 observations are randomly sampled from the skewed distribution shown in figure 6.2. This distribution has a population mean $\mu = 1.6487$. So once we compute the mean $\bar{X}$ and sample standard deviation, s, we can compute $T = \sqrt{n}(\bar{X} - \mu)/s$. The method for computing a confidence interval for the mean, covered in section 6.2, assumes that if we repeat this process many times, a plot of the resulting T values will be approximately the same as a plot of Student's t-distribution with 24 degrees of freedom. Here we repeat this process 4000 times resulting in 4000 T values. Figure 6.3 shows a plot

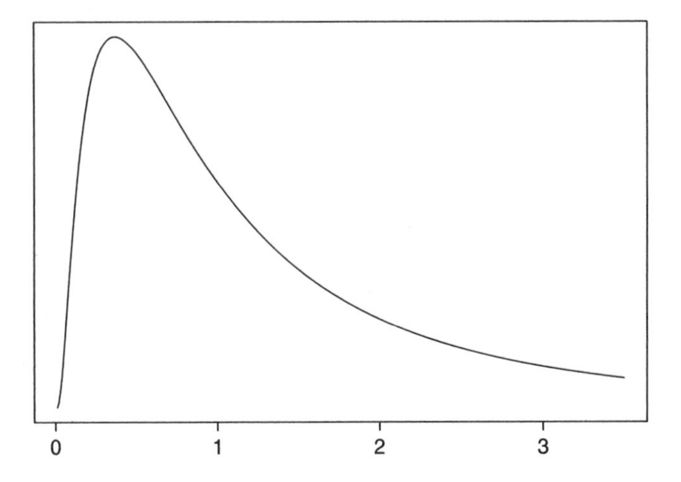

Figure 6.2 A skewed, light-tailed distribution used to illustrate the effects of non-normality when using Student's t.

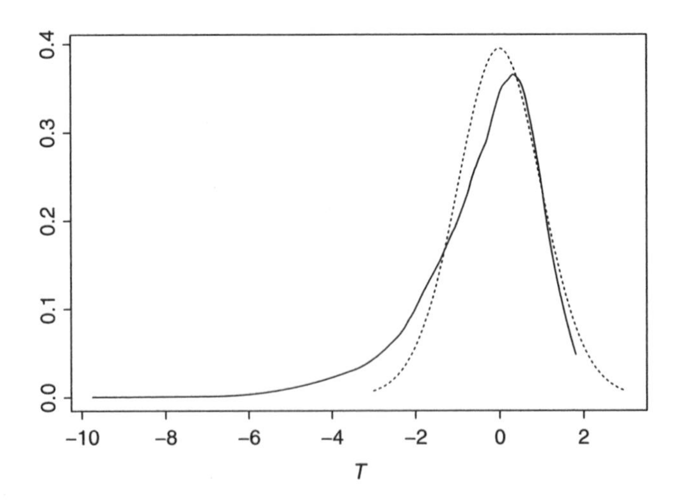

Figure 6.3 Shown is the distribution of T when sampling from the distribution in figure 6.2 and the distribution of T when sampling from a normal distribution.

of these T values plus a plot of Student's t-distribution with 24 degrees of freedom. As is evident, the actual distribution of T is poorly approximated by a Student's t-distribution with 24 degrees of freedom. One obvious concern is that the actual distribution of T is not even symmetric. Under normality, the probability that T is less than or equal to -2.086 is .025. But when sampling from the skewed distribution used here, the probability is approximately .12. And the probability that T is greater than 2.086 is only .001.

The process used in this last example, to study the properties of Student's t, is an example of what is called a *simulation study*. Roughly, using a computer, we sample observations from a distribution we pick. For example, we could sample data from a normal distribution having mean 0 and variance 1.[4] In the example, because we know the mean, we are able to assess whether a confidence interval based on Student's t will contain the population mean. We then repeat this process many times and for illustrative purposes, imagine we repeat it 10,000 times. So when computing a .95 confidence interval, for example, we have 10,000 confidence intervals. If a method for computing a confidence interval is performing well, then it should be the case that approximately 95% of all confidence intervals contain the population mean.

It was mentioned in chapter 5 that roughly, when dealing with skewed distributions, as the likelihood of observing outliers increases, the larger the sample size we might need when trying to approximate the sampling distribution of the sample mean with a normal curve. When dealing with Student's t, this remains true. That is, large sample sizes might be needed so that the actual distribution of the T values will be approximately the same as the distribution under normality. Indeed, realistic situations are known where a sample size greater than 300 might be required to get accurate results with Student's t.

Example 2

Table 2.3 reports the responses of 105 undergraduate males regarding how many sexual partners they desired over the next 30 years. Here the extreme outlier is removed leaving 104 observations. The median response was one. Imagine that we resample, with replacement, 104 observations from these values. (Sampling with replacement means that when an observation is sampled, it is put back, and so has the possibility of being chosen again.) In effect, we are sampling from a distribution with a known mean, which is the sample mean of our observed data. Consequently, we can compute T. Repeating this process 1,000 times yields an approximation of the distribution of T, which is shown in figure 6.4. The smooth symmetric curve is the distribution of T assuming normality. As is evident, the two curves differ substantially. (In essence, this is a bootstrap-t technique, which is discussed later in this chapter.)

If the sample size is large enough so that the sampling distribution of the sample mean is approximately normal, does it follow that it is safe to use Student's t to compute a confidence interval? The answer is, not necessarily, as illustrated by the following example.

4. Using the software R, this can be done with the function rnorm.

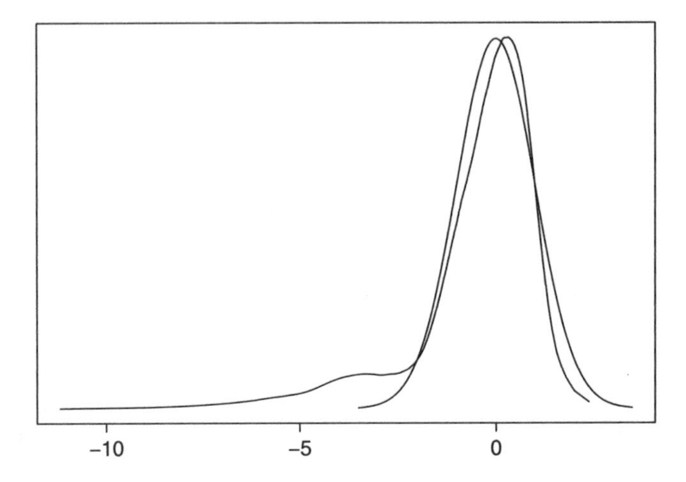

Figure 6.4 When dealing with data from actual studies, we sometimes encounter situations where the actual distribution of T differs even more from the distribution of T under normality than is indicated in figure 6.3. Shown is the distribution of T when sampling from the data in table 2.3, with the extreme outlier removed. The sample size is $n = 104$. With the extreme outlier included, the distribution of T becomes even more skewed to the left.

Example 3

Chapter 5 provided an example where a plot of 1,000 sample means, each sample mean based on 40 observations, is approximately normal even though sampling is from a distribution that is clearly non-normal. For convenience, the results are duplicated in figure 6.5. The left panel shows the distribution from which observations are sampled, and the right panel shows a plot of the resulting sample means. The dashed curve is the normal curve used to approximate the plot of the sample means via the central limit theorem. Now look at figure 6.6. The solid line is a plot of 1,000 T values and the dashed line is the distribution of T assuming normality. The left tail of the two distributions are reasonably similar, but the right tails are not. When using T, assuming normality implies that the .025 and .975 quantiles are −2.02 and 2.02, respectively. That is, when computing a confidence interval, it is assumed that $c = 2.02$ should be used. But the actual .975 quantile of Student's t-distribution, when observations follow the distribution in the left panel of figure 6.4, is approximately 1.37, which is not reasonably close to the value under normality, namely, 2.02.

In summary, Student's t might provide an accurate confidence interval for the population mean, but there are realistic situations where this is not the case. Modern methods, briefly outlined in the final section of this chapter, provide various strategies for dealing with this problem.

Comments about Bell-shaped distributions

We have just seen that when sampling from skewed distributions, Student's t might be unsatisfactory when computing a confidence interval for the population mean. A positive feature of Student's t is that when observations follow a symmetric distribution, all indications are that under fairly general conditions, the actual probability that the

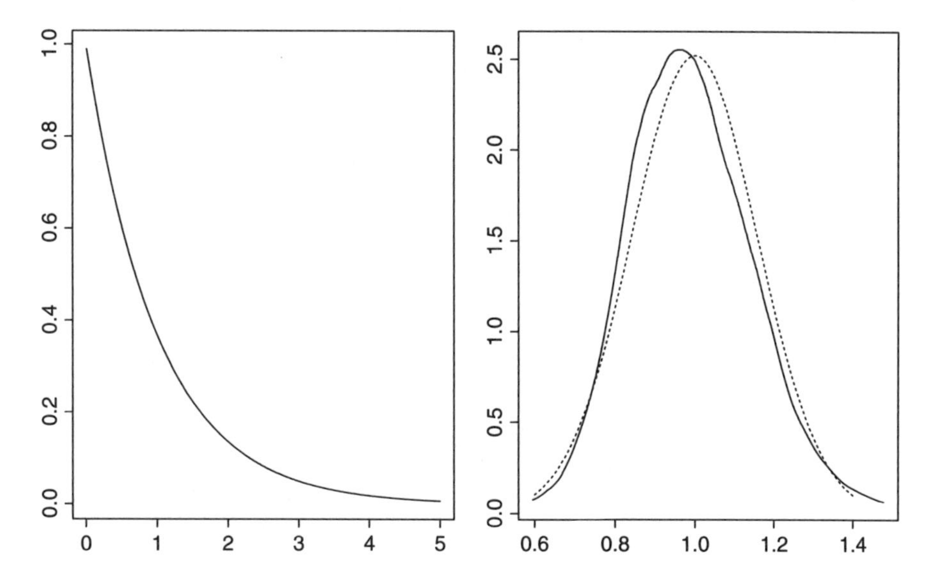

Figure 6.5 When sampling from the distribution in the left panel, with $n = 40$, the sampling distribution of the sample mean is approximately normal. But compare this to the sampling distribution of T shown in figure 6.6.

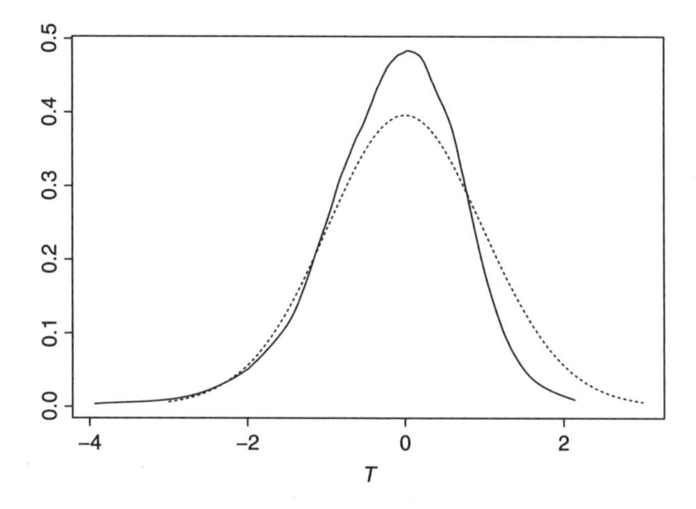

Figure 6.6 When sampling from the distribution in the left panel, with $n = 40$, even though the sampling distribution of the sample mean is approximately normal, the sampling distribution of T can differs substantially from a Student's t-distribution, as illustrated here.

confidence interval will contain the population mean is at least $1 - \alpha$. So if we compute a .95 confidence interval, we can be reasonably certain that the actual probability coverage is at least .95 if the distribution is approximately symmetric. There is, however, another concern: Under very small departures from normality, the length of the confidence interval might be relatively higher than the length we would get using some other measure of location such as the median.

Chapter 4 described what is called a mixed normal distribution. As was indicated, this distribution is very similar to a standard normal distribution, yet its variance is 10.9

versus 1 for the standard normal. Also recall that because the distribution is symmetric about zero, the population mean and median are identical. Here, this means that the sample mean and median are attempting to estimate the same unknown parameter. Chapter 5 illustrated that when sampling from a mixed normal distribution, the standard error of the median is strikingly smaller than the standard error of the mean. This suggests that in the present context, confidence intervals based on the median will, in general, be substantially shorter, and this turns out to be the case.

Example 4

To illustrate the point just made, 30 observations were sampled from a contaminated normal distribution. The .95 confidence interval based on Student's t was $(-3.62, 1.71)$, versus $(-0.49, 0.34)$ using the median. Both confidence intervals contain zero (the true value of the population mean and median), as they should, but the confidence interval based on the median gives us a much more precise indication of the value of the population mean and median.

We can summarize the performance of Student's t as follows:

- When sampling from a symmetric distribution where outliers tend to be rare, such as any normal distribution, Student's t performs well in terms of both probability coverage and the length of the resulting confidence interval relative other methods you might use.
- When sampling from an asymmetric distribution where outliers tend to be rare, a sample size of 200 or more might be needed to get a reasonably accurate confidence interval. A positive feature is that the length of the confidence interval continues to compete well with other methods. Also, even when the distribution of the sample means is approximately normal, it is possible that Student's t is providing inaccurate confidence intervals unless the sample size is fairly large.
- When sampling from a symmetric distribution where outliers tend to be common, the length of the confidence interval based on Student's t might be very high relative to other methods that might be used.
- When sampling from an asymmetric distribution where outliers tend to be common, a sample size of 300 or more might be needed to get accurate probability coverage, and the length of the confidence interval will tend to be larger than that obtained by alternative techniques.

Dealing with practical problems

When sampling from a skewed distribution, what can be done to improve probability coverage? One suggestion might be to check for outliers, and if none are found, assume Student's t gives accurate results. Currently, it is unclear how large the sample size must be in order to recommend this approach. And even if it were used, what do we do when there is good evidence that sampling is from a distribution where outliers are common?

One of the many alternative strategies is to use a *bootstrap-t method.* To provide a brief indication of how the method is applied, consider again the simulation studies illustrated by examples 1 and 2. Note that using a computer, simulations could be used to determine the distribution of Student's t under normality. That is, we could generate observations from a normal distribution resulting in a value for T, and by repeating this process many times, we would have an accurate approximation of the distribution of T. For example, based on 1,000 T values, we might find that 97.5% of the values are less than 2.03, in which case we would estimate that the probability of getting a T value less than 2.03 is .975. The bootstrap-t is based on the same strategy, only rather than sample observations from a normal distribution, you randomly sample observations, with replacement, from the data at hand. That is, we perform a simulation study based on the observed data rather a hypothetical (normal) distribution. (Gosset used a similar strategy as an empirical check of the validity of his Student's t-test. Without a computer, the calculations to him over a year to complete.)

When the goal is to make inferences about the population mean, the bootstrap-t method for computing confidence intervals performs well relative to other methods that have been proposed, but even it can fail to provide accurate probability coverage. In figure 6.4, for example, published papers suggest that the bootstrap-t method underestimates the negative consequences of sampling from a non-normal distribution.[5] However, despite the fact that a bootstrap-t method fails to correct all problems problems with Student's t, it has considerable practical value in situations not covered here.

A common suggestion is to transform the data by taking logarithms, for example, and more complicated transformations have been proposed. Here it is merely noted that this approach can be relatively unsatisfactory for reasons outlined in chapter 4.

Recall from chapter 2 that the median belongs to the class of trimmed means: it trims all but the middle one or two values after putting the observations in ascending order. Generally, the more we trim, the less effect skewness has on accurate probability coverage. But a negative consequence of using a median (the maximum amount of trimming) is that when dealing with distributions where outliers are relatively rare, such as normal distributions, the length of the confidence interval might be relatively long compared to using Student's t. (And there is the added problem of dealing with tied values, as previously noted.) Also note that when dealing with skewed distributions, confidence intervals based on the median generally tells us little or nothing about the population mean for the simple reason that the population mean and median differ, and they might differ substantially.

One of several alternative possibilities is to simply trim less so as to get reasonably short confidence intervals when dealing with normal distributions, and simultaneously get accurate probability coverage when dealing with skewed distributions. As in previous chapters, a good choice for general use is 20% trimming. To compute a $1-\alpha$ confidence interval for the population 20% trimmed mean, let h be the number of observations left after trimming. Let c be the $1-\alpha/2$ quantile of the Student's t-distribution with $h-1$ degrees of freedom and let s_w be the Winsorized sample standard deviation. The $1-\alpha$ confidence interval for the trimmed is

$$\left(\bar{X}_t - c\frac{s_w}{.6\sqrt{n}}, \bar{X}_t + c\frac{s_w}{.6\sqrt{n}}\right).$$

5. For more information about the relative merits of the bootstrap-t, see Wilcox (2003, 2005). These books describe easy-to-use R and S-PLUS software.

Table 6.3 Self-awareness data

77	87	88	114	151	210	219	246	253	262
296	299	306	376	428	515	666	1310	2611	

Example 5

The data in table 6.3 are from a study on self-awareness and reflect how long a participant could keep a portion of an apparatus in contact with a specified target. (These data were used in problem 14 and are reproduced here for convenience.) The trimmed mean is $\bar{X}_t = 283$ and its estimated standard error is $s_w/(.6\sqrt{n}) = 146.804/(.6\sqrt{(19)}) = 56.1$. In contrast, the standard error of the sample mean is $s/\sqrt{n} = 136$, a value approximately 2.4 times larger than the estimated standard error of the 20% trimmed mean. The number of observations trimmed is 6, so the number of observations left after trimming is $h = 19 - 6 = 13$ and the degrees of freedom are $\nu = h - 1 = 12$. When computing a .95 confidence interval, $1 - \alpha/2 = .975$, and referring to table 4 in appendix B, $c = 2.18$. So the .95 confidence interval for the population trimmed mean is

$$283 - 2.18(56.1),\ 283 + 2.18(56.1) = (160.7,\ 405.3).$$

The percentile bootstrap method

We conclude this chapter with a brief outline of a fairly modern method that offers an alternative to Laplace's strategy for computing a confidence interval. Imagine we have n observations, say $X_1, \ldots, X_n$. A *bootstrap sample* of size n is obtained by randomly sampling, with replacement, n observations from $X_1, \ldots, X_n$ yielding say $X_1^*, \ldots, X_n^*$.

Example 6

Suppose we observe the 10 values

$$1, 4, 2, 19, 4, 12, 29, 4, 9, 16.$$

If we randomly sample a single observation from among these 10 values, we might get the value 2. If we randomly sample another observation, we might again get the value 2, or we might get the value 9 or any of the other values listed. If we perform this process 10 times, we might get the values

$$2, 9, 16, 2, 4, 12, 4, 29, 16, 19.$$

This is an example of a bootstrap sample of size 10. If we were to get another bootstrap sample of size 10, this time we might get

$$29, 16, 29, 19, 2, 16, 2, 9, 4, 29.$$

Now suppose we create many bootstrap samples in the manner just described and for each bootstrap sample we compute the median. For illustrative purposes, imagine we repeat this process 1,000 times yielding 1,000 (bootstrap) sample medians, which can be done very quickly on modern computers. If we put these 1,000 sample medians is ascending order, then the middle 95% form an approximate .95 confidence interval for the population median. In terms of achieving accurate probability coverage, this method performs very well when using a median or 20% trimmed mean, but it does not perform well when using the mean; a bootstrap-t method is preferable. When sample sizes are sufficiently large, the percentile bootstrap is not necessary when working with a 20% trimmed mean, but for small sample sizes it is preferable and in fact works very well, even in situations where all methods based on means perform poorly. Although the percentile bootstrap method is not necessary when computing a confidence interval for a single median, when working with more complicated problems covered in subsequent chapters, it currently seems to be the best method to use.[6]

A Summary of Some Key Points

- Confidence intervals provide a fundamental method for assessing how well an estimator, such as the sample mean, $\bar{X}$, estimates its population analog, such as the population μ. Probability coverage of a confidence interval refers to the probability, over many studies, that the resulting confidence intervals will contain the population parameter being estimated.
- The classic and routinely used method for computing a confidence interval for μ is based on Student's t. For symmetric, light-tailed distributions, its actual probability coverage tends to be reasonably close to nominal level. For symmetric, heavy-tailed distributions, the actual probability coverage tends to be higher than the stated level. But for skewed distributions, probability coverage can be poor, and when outliers are common, practical problems are exacerbated. Problems can occur even with $n = 300$.
- In terms of achieving accurate probability coverage, methods based on 20% trimmed means perform better than methods based on means for a wide range of situations. For symmetric, heavy-tailed distributions, using a 20% trimmed mean can result in a substantially shorter confidence interval for the population mean than methods based on the sample mean. For skewed distributions, methods based on 20% trimmed means do not provide a satisfactory confidence interval for μ, but they do provide a satisfactory confidence interval for the population trimmed mean that might better reflect what is typical.
- Fairly accurate confidence intervals for the population median can be computed using equation (6.8) when tied values never occur. But with tied values, a percentile bootstrap method should be used.
- Equation (6.10) is the basic method for computing a confidence interval for p, the probability of success associated with a binomial distribution, that is typically covered in an introductory course. But all indications are that the Agresti-Coull method is preferable in practice. For the special cases where the number of successes is 0, 1, $n - 1$ or n, use Blyth's method.
- Bootstrap methods can provide more accurate confidence intervals. There are conditions where they provide little improvement, but there are conditions where they have practical value. With a large enough sample size, bootstrap methods are not needed, except possibly when dealing with the median, but it is unclear how large the sample size must be for this to be the case.

6. For easy-to-use software, see the discussion of S-PLUS and R software in chapter 1.

Problems

32. Describe in general terms how non-normality can affect Student's t-distribution.

33. Chapter 5 illustrated that when (randomly) sampling observations from a skewed distribution where outliers are rare, it is generally reasonable to assume that plots of sample means over many studies has, approximately, a normal distribution. This means that reasonably accurate confidence intervals for the mean can be computed when the standard deviation, σ, is known. Based on information summarized in this section, how does this contrast with the situation where the variance is not known and confidence intervals are computed using Student's t-distribution?

34. Listed here are the average LSAT scores for the 1973 entering classes of 15 American law schools.

    545 555 558 572 575 576 578 580
    594 605 635 651 653 661 666

    (LSAT is a national test for prospective lawyers.) The .95 confidence interval for the population μ is $(577.1, 623.4)$. (a) Use a boxplot to verify that there are no outliers. (b) A boxplot suggests that observations might have been sampled from a skewed distribution. Argue that as a result, the confidence interval for the mean, based on Student's t, might be inaccurate.

35. Compute a .95 confidence for the trimmed mean if (a) $n = 24$, $s_w^2 = 12$, $\bar{X}_t = 52$, (b) $n = 36$, $s_w^2 = 30$, $\bar{X}_t = 10$, (c) $n = 12$, $s_w^2 = 9$, $\bar{X}_t = 16$.

36. Repeat the previous exercise, but compute a .99 confidence interval instead.

37. Problem 14 (in this chapter) used data from a study of self-awareness. In another portion of the study, a group of participants had the following values.

    $$59, 106, 174, 207, 219, 237, 313, 365, 458, 497, 515,$$

    $$529, 557, 615, 625, 645, 973, 1{,}065, 3{,}215.$$

    Compute a .95 confidence interval for both the population mean and 20% trimmed mean.

38. The ideal estimator of location would have a smaller standard error than any other estimator we might use. Explain why such an estimator does not exist.

39. For the values

    $$5, 60, 43, 56, 32, 43, 47, 79, 39, 41,$$

    compute a .95 confidence interval for the trimmed mean and compare the results to the .95 confidence interval for the mean.

40. In the previous exercise, the confidence interval for the 20% trimmed mean is shorter than the confidence interval for the mean. Explain why this is not surprising.

7

HYPOTHESIS TESTING

The previous chapter described confidence intervals, which provide a fundamental strategy for making inferences about population measures of location such as the population mean μ and the population median. About a century after Laplace's ground breaking work on sampling distributions, a new set of tools was developed for making inferences about parameters that adds perspective and which are routinely used when attempting to understand data. One of the main architects of this new perspective was Jerzy Neyman. Forced to leave Poland due to the war between Poland and Russia, Neyman eventually moved to London in 1924 where he met Egon Pearson (son of Karl Pearson, whom we met in chapter 1). Their joint efforts led to the Neyman–Pearson framework for testing hypotheses that is the subject of this chapter.

7.1 Testing hypotheses about the mean of a normal distribution, σ known

The essence of hypothesis testing methods is quite simple: A researcher formulates some speculation about the population under study. Typically the speculation has to do with the value of some unknown population parameter, such as the population mean. Then data are analyzed with the goal of determining whether the stated speculation is unreasonable. That is, is there empirical evidence indicating that the speculation is probably incorrect? The speculation about the value of a parameter is called a *null hypothesis*.

Example 1

A manufacturer of a high definition television claims that the average life of the bulb used in their rear projection model lasts an average of at least 48 months. Imagine that you want to collect data with the goal of determining whether this claim is correct. Then the null hypothesis is that the population mean for all televisions is at least 48. This is written as

$$H_0 : \mu \geq 48, \tag{7.1}$$

where the notation H_0 is read "H nought." The alternative to the null hypothesis is written as

$$H_a : \mu < 48. \tag{7.2}$$

In this last example, imagine that for 10 televisions, you determine how long the bulb lasts and find that the average life is $\bar{X} = 50$ months. That is, you estimate that μ is equal to 50, which supports the claim that the population mean is at least 48. So in particular, you would not reject the stated hypothesis. But suppose you get a sample mean of $\bar{X} = 46$. Now the estimate is that the population mean is less than 48, but the manufacturer might claim that if the population mean is 48, a sample mean of 46 is not all that unusual. The issue is, how low does the sample mean have to be in order to argue that the null hypothesis is unlikely to be true?

For the moment, we are making three fundamental assumptions:

- Random sampling
- Normality
- σ known.

As noted in chapter 6, assuming that σ is known is generally unrealistic, it is rarely known exactly, but it provides a convenient framework for describing the basic principles and concepts underlying hypothesis testing methods. For convenience, we denote some hypothesized value for μ by μ_0. In the last example, $\mu_0 = 48$.

The logic of hypothesis testing

There are three types of null hypotheses that are of general interest:

1. $H_0 : \mu \geq \mu_0$.
2. $H_0 : \mu \leq \mu_0$.
3. $H_0 : \mu = \mu_0$.

Consider the first hypothesis. The basic strategy is to momentarily assume that $\mu = \mu_0$ and to consider the probabilistic implications associated with the observed value of the sample mean. So in example 1 dealing with how long a bulb lasts, we give the manufacturer the benefit of the doubt and momentarily assume that $\mu = 48$. The idea is that if we find empirical evidence strongly suggesting that the mean is less than 48, then we would conclude that the hypothesis H_0: $\mu \geq 48$ should be rejected.

If $\mu = \mu_0$ and the three assumptions made are true, then from chapter 6,

$$Z = \frac{\bar{X} - \mu_0}{\sigma/\sqrt{n}} \tag{7.3}$$

has a standard normal distribution. Consider the rule: reject the null hypothesis if $Z \leq c$, where c is the .05 quantile of a standard normal distribution. So from table 1 in appendix B, $c = -1.645$. Then the probability of erroneously rejecting, when in fact $\mu = \mu_0$, is .05.

It might help to note that the rule just given is tantamount to rejecting if the sample mean is sufficiently smaller than the hypothesized value. More precisely, rejecting if $Z \leq c$

is the same as rejecting if

$$\bar{X} \leq \mu_0 + c\frac{\sigma}{\sqrt{n}}.$$

In the last example, $c = -1.645$, so the rule would be to reject if $\bar{X} \leq \mu_0 - 1.645\sigma/\sqrt{n}$. But for convenience, the rule is usually stated in terms of Z.

Definition Rejecting a null hypothesis, when in fact it is true, is called a *Type I error*. For the situation just described, assuming the underlying assumptions are true, the probability of a Type I error is .05 when $\mu = \mu_0$.

We can summarize a decision rule for H_0: $\mu \geq \mu_0$ in more general terms as follows. Let c be the α quantile of a standard normal distribution and suppose this null hypothesis is rejected if $Z \leq c$. Then, if the three underlying assumptions are true, the probability of a Type I error is α.

Example 2

A researcher claims that on a test of open mindedness, the population mean for adult men is at least 50 with a standard deviation of $\sigma = 12$. So in the notation used in this chapter, $\mu_0 = 50$ and the null hypothesis is H_0: $\mu \geq 50$. Based on 10 adult males, would the researcher's claim be rejected if the sample mean is $\bar{X} = 48$ and the Type I error probability is to be .025? To find out, first note that from table 1 in appendix B, the .025 quantile of a standard normal distribution is $c = -1.96$. Because $n = 10$, we have that

$$Z = \frac{\bar{X} - \mu_0}{\sigma/\sqrt{n}} = \frac{48 - 50}{12/\sqrt{10}} = -0.53.$$

Because -0.53 is greater than -1.96, the null hypothesis is not rejected.

Example 3

We repeat the last example, only now suppose the sample mean is $\bar{X} = 40$. Then $Z = -2.635$, this is less than -1.96, so reject the null hypothesis. That is, there is empirical evidence that the claim is unlikely to be true.

Now consider the second type of null hypothesis, H_0: $\mu \leq \mu_0$. For this case, we would not reject if the sample mean is less than the hypothesized value μ_0, for the simple reason that if the null hypothesis is true, this is the type of result we would expect. Rather, we would reject if the sample mean is sufficiently larger than the null value, which means that we would reject if Z is sufficiently large. More formally, if we want the probability of a Type I error to be α when in fact $\mu = \mu_0$, let c be the $1 - \alpha$ quantile of a standard normal distribution and reject null hypothesis if $Z \geq c$.

Example 4

Imagine that the goal is to test H_0: $\mu \leq 15$ assuming normality, $\sigma = 4$, and that the Type I error probability is to be .01. Then $1 - \alpha = .99$ and from table 1 in appendix B, $c = 2.33$. If based on a sample of size 25, the sample mean is $\bar{X} = 18$, we have that

$$Z = \frac{18 - 15}{4/\sqrt{25}} = 3.75.$$

Because 3.75 is greater than 2.33, reject the null hypothesis.

Finally, consider the third type of null hypothesis, H_0: $\mu = \mu_0$. Now we would reject if the sample mean is sufficiently smaller or larger than the hypothesized value, μ_0. In symbols, if the goal is to have the probability of a Type I error equal to α, reject if Z is less than or equal to the $\alpha/2$ quantile of a standard normal distribution or if Z is greater than or equal to the $1-\alpha/2$ quantile. Put another way, reject if $|Z| \geq c$, where now c is the $1-\alpha/2$ quantile of a standard normal distribution.

Example 5

Imagine you want to test H_0: $\mu = \mu_0$ with $\alpha = .05$. Then $\alpha/2 = .025$, so $1-\alpha/2 = .975$, meaning that you would reject if $|Z| \geq 1.96$, the .975 quantile of a standard normal distribution. Figure 7.1 indicates for which values of Z we would reject. The shaded region indicates the *critical region*, meaning the collection of Z values for which the null hypothesis would be rejected. Here, the critical region consists of all values less than or equal to -1.96 as well as all values greater than or equal to 1.96. The constant c is called a *critical value.*

Example 6

We repeat example 4, only now we test H_0: $\mu = 15$, again assuming normality, $\sigma = 4$, and that the Type I error probability is to be .01. So $\alpha/2 = .005$, $1-\alpha/2 = .995$, and from table 1 in appendix B, $c = 2.58$. Because $|Z| = 3.75 > 2.58$, reject the null hypothesis.

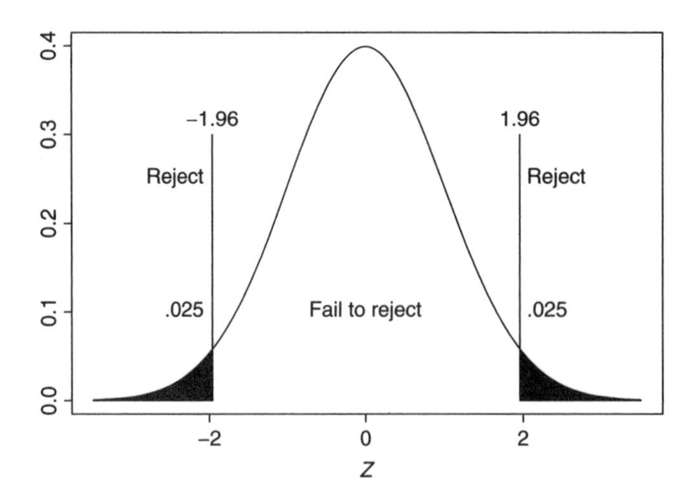

Figure 7.1 A graphical depiction of the rejection rule when using Z and $\alpha = .05$. The shaded portions indicate the rejection regions.

Summary

The details of the three hypothesis testing methods are summarized here for convenience, assuming that the probability of a Type I error is to be α. Once the sample mean has been determined, compute

$$Z = \frac{\bar{X} - \mu_0}{\sigma/\sqrt{n}}.$$

Case 1. H_0: $\mu \geq \mu_0$. Reject H_0 if $Z \leq c$, the α quantile of a standard normal distribution.

Case 2. H_0: $\mu \leq \mu_0$. Reject H_0 if $Z \geq c$, the $1 - \alpha$ quantile of a standard normal distribution.

Case 3. H_0: $\mu = \mu_0$. Reject H_0 if $Z \geq c$ or if $Z \leq -c$, where now c is the $1 - \frac{\alpha}{2}$ quantile of a standard normal distribution. Equivalently, reject if $|Z| \geq c$.

The hypotheses $H_0 : \mu \geq \mu_0$ and $H_0 : \mu \leq \mu_0$ are called *one-sided hypotheses*. In contrast, $H_0 : \mu = \mu_0$ is called a *two-sided hypothesis*. The constant c is called a *critical value*.

p-Values

Notice that if you are told that the null hypothesis is rejected with $\alpha = .05$, and nothing else, it is unknown whether you would also reject with $\alpha = .025$, $\alpha = .01$ or $\alpha = .001$. Generally, what is the smallest α value that would result in rejecting the null hypothesis? This smallest α value is called a *p-value* and can be determined by using the observed value of Z as a critical value. Generally, p-values have values between 0 and 1. The closer the p-value is to zero, the stronger the evidence that the null hypothesis should be rejected.

Example 7

Imagine that the goal is to test H_0: $\mu \geq 6$ and that the value of the test statistic is $Z = -1.8$. If you wanted the probability of a Type I error to be $\alpha = .05$, then the critical value would be $c = -1.645$. And because the value of Z is less than the critical value, you would reject. But suppose you had used a critical value equal to the observed value of Z, namely, -1.8. Then the corresponding probability of a Type I error is called the p-value and is equal to $P(Z \leq= -1.8)$. From table 1 in appendix B, the p-value is equal to 0.036. So in particular, you would reject if the desired Type I error probability were .05, .04 or even .036, but not for $\alpha = .03$.

For the three types of hypotheses that arise, here is a more detailed description of how to compute the p-value given $\bar{X}$, μ_0, n and σ, which will help make clear a certain conceptual issue to be described.

Case 1. H_0: $\mu \geq \mu_0$. The p-value is

$$p = P\left(Z \leq \frac{\bar{X} - \mu_0}{\sigma/\sqrt{n}}\right), \tag{7.4}$$

which can be determined using table 1 in appendix B.

Case 2. H_0: $\mu \le \mu_0$. The p-value is

$$p = P\left(Z \ge \frac{\bar{X} - \mu_0}{\sigma/\sqrt{n}}\right). \tag{7.5}$$

Case 3. H_0: $\mu = \mu_0$. The p-value is

$$p = 2\left(1 - P\left(Z \le \frac{|\bar{X} - \mu_0|}{\sigma/\sqrt{n}}\right)\right). \tag{7.6}$$

Example 8

Imagine that $n = 36, \sigma = 5, \bar{X} = 44$ and you test H_0: $\mu \ge 45$, which corresponds to case 1. Then $Z = -1.5$, and the p-value is the probability that a standard normal random variable has a value less than or equal to -1.5, which is $P(Z \le -1.5) = 0.0668$. Because the p-value is greater than .05, you would not reject if you wanted the Type I error to be .05. If the p-value were .015, then in particular you would reject if the largest Type I error probability you wanted to allow were $\alpha = .05$ or .025, but not if $\alpha = .01$.

Example 9

We repeat the previous example, only with $\bar{X} = 46$. Then $Z = 1.5$ and the p-value is $P(Z \le 1.5) = .933$. So this is not even close to rejecting, which is not surprising because the hypothesis is that μ is greater than or equal to 45, and the sample mean is consistent with this speculation.

Example 10

A company claims that on average, their beef hotdogs have 150 calories. Assume normality and that $\sigma = 10$. Imagine that the average calories for 12 hotdogs is $\bar{X} = 145$ and the goal is to check the company's claim. In formal terms, the goal is to test H_0: $\mu = 150$, which corresponds to case 3. The test statistic is $Z = -1.73$. The corresponding p-value is

$$p = 2(1 - P(Z \le |-1.73|)) = 2(1 - 0.958) = .084.$$

So the null hypothesis would not be rejected if the desired Type I error probability were .05, but it would be rejected with a Type I error probability of .1.

Comments on interpreting *p*-values

Care must be taken not to read more into a p-value than is warranted. Notice that the expressions for computing a p-value make it clear that it is influenced by three factors:

1. The difference between the sample mean and the hypothesized value, $\bar{X} - \mu_0$,
2. the magnitude of the standard deviation, σ, and
3. the value of the sample size, n.

Said another way, a p-value is determined by the difference between the sample mean and the hypothesized population mean, $\bar{X} - \mu_0$, as well as the standard error of the sample mean, $\sigma/\sqrt{n}$. The point is that if a p-value is close to zero, and we are told nothing else, it is not necessarily the case that the estimate of the population mean, $\bar{X}$, differs substantially from the hypothesized value. The reason could be that the standard deviation is small relative to the sample size. In particular, a small p-value does *not* necessarily mean that the difference between the null value of the mean and its actual value, namely $\mu - \mu_0$, is large.

Here is another perspective that is helpful. A p-value reflects a property of the sampling distribution of the test statistic Z. Roughly, it reflects the likelihood that Z will have a value greater than 0. If the null hypothesis is true, and the assumptions made here are true, the probability that Z will be greater than 0 is .5. When the null hypothesis is false, this probability will be less than or greater than .5, depending on how μ differs from the hypothesized value, μ_0. It can be shown that on average, if the null hypothesis is true and sampling is from a normal distribution, the p-value will be .5. That is, the p-value will differ from one study to the next, but over many studies, its average value is .5 if H_0 is true. This is related to the fact that when the null hypothesis is true, Z has a standard normal distribution, meaning that with probability .5, its value will be greater than 0. Under non-normality, we will see that skewness and outliers also affect the magnitude of the p-value.

A common convention is to describe a test of some hypothesis as *significant* if the p-value is small, say less than or equal to .05. But the term significant is a statistical term meaning that there is strong evidence that the null hypothesis should be rejected. It does not necessarily mean that there is an important or large difference between the hypothesized value of the mean and its actual value. One way to asses the magnitude of $\mu - \mu_0$ is to compute a confidence interval for μ using the methods in chapter 6.

Power and type II errors

There are two fundamental types of errors when testing hypotheses. The first is a Type I error, which has already been discussed. The second is called a *Type II error*, meaning that the null hypothesis is false, but we failed to reject. The probability of a Type II error is usually labeled β, a lower case Greek beta. The probability of rejecting, when the null hypothesis is false, is called *power* and is labeled $1 - \beta$. That is, power is the probability of making a correct decision about the null hypothesis when the hypothesis is false. Table 7.1 summarizes the four possible outcomes when testing hypotheses. Power is of critical importance for reasons illustrated by the next example.

Example 11

Imagine that a new drug for treating hypertension is being studied, but there is an issue about whether the drug causes liver damage. For illustrative purposes, suppose that based on some appropriate measure, liver damage is of little concern if $\mu \leq 14$. If a pharmaceutical firm reports that they tested H_0: $\mu \leq 14$

Table 7.1 Four possible outcomes when testing hypotheses

	Reality	
Decision	H_0 true	H_0 false
H_0 true	Correct decision	Type II error (probability β)
H_0 false	Type I error (probability α)	Correct decision (power)

and failed to reject, is it reasonable to assume that the drug is safe? Based only on the information given, the answer is not known. Imagine, for example, that if $\mu = 16$, severe liver damage could result. The concern is that if in reality $\mu = 16$, by chance we might fail to reject. For example, if the pharmaceutical firm reveals that a sample size of only 10 was used, intuitively, one would be suspicious about any conclusions regarding the entire population of individuals who might take this drug. In statistical terms, the issue is how much power is afforded with a sample size of only 10. If power is only $1 - \beta = .2$, there are serious concerns because this means that there is only a 20% chance of discovering a practical problem with this drug.

For the situation at hand where sampling is from a normal distribution and σ is known, power depends on four quantities: σ, α, n, and the value of $\mu - \mu_0$, which is the difference between the true value and the hypothesized value of the population mean. Here is how power is computed for the three types of hypotheses previously described.

Case 1. H_0: $\mu < \mu_0$. Determine the critical value c as previously described. (The critical value is the $1 - \alpha$ quantile of a standard normal distribution.) Then power, the probability of rejecting the null hypothesis, is

$$1 - \beta = P\left(Z \geq c - \frac{\sqrt{n}(\mu - \mu_0)}{\sigma}\right).$$

In words, power is equal to the probability that a standard normal random variable is greater than or equal to

$$c - \frac{\sqrt{n}(\mu - \mu_0)}{\sigma}.$$

Case 2. H_0: $\mu > \mu_0$. Determine the critical value c, which is now the α quantile of a standard normal distribution. Then power is

$$1 - \beta = P\left(Z \leq c - \frac{\sqrt{n}(\mu - \mu_0)}{\sigma}\right).$$

Case 3. H_0: $\mu = \mu_0$. Now c is the $1 - \frac{\alpha}{2}$ quantile of a standard normal distribution. Power is

$$1 - \beta = P\left(Z \leq -c - \frac{\sqrt{n}(\mu - \mu_0)}{\sigma}\right) + P\left(Z \geq c - \frac{\sqrt{n}(\mu - \mu_0)}{\sigma}\right).$$

Example 12

After years of production, a manufacturer of batteries for automobiles finds that on average, their batteries last 42.3 months with a standard deviation of $\sigma = 4$. A new manufacturing process is being contemplated and one goal is to determine whether the batteries have a longer life on average based on 10 batteries produced by the new method. For illustrative purposes, assume that the standard deviation is again $\sigma = 4$ and that the manufacturer decides to test H_0: $\mu \leq 42.3$ with the idea that if they reject, there is evidence that the new manufacturing process is better on average. Moreover, if in reality $\mu = 44$, and the Type I error probability is set at $\alpha = .05$, the manufacturer wants a high probability of rejecting the null hypothesis and adopting the new method. So a practical issue is how much power there is when $\mu = 44$. With $\alpha = .05$, the critical value is $c = 1.645$. So power is

$$\begin{aligned} 1-\beta &= P\left(Z \geq c - \frac{\sqrt{n}(\mu - \mu_0)}{\sigma}\right) \\ &= P\left(Z \geq 1.645 - \frac{\sqrt{10}(44 - 42.3)}{4}\right) \\ &= P(Z \geq .30) \\ &= .38. \end{aligned}$$

Example 13

We repeat the last example, only now we determine power when the sample size is increased to 20. Now, power is

$$\begin{aligned} 1-\beta &= P\left(Z \geq c - \frac{\sqrt{n}(\mu - \mu_0)}{\sigma}\right) \\ &= P\left(Z \geq 1.645 - \frac{\sqrt{20}(44 - 42.3)}{4}\right) \\ &= P(Z \geq -0.256) \\ &= .60 \end{aligned}$$

Increasing n to 30, it can be seen that power is now $1-\beta = .75$, meaning that the probability of correctly identifying an improved manufacturing process, when $\mu = 44$, is now .75. A rough explanation of why this occurs is as follows. Recall that the accuracy of the sample mean, as an estimate of the population mean, is reflected by its squared standard error, σ^2/n. So as n gets large, its squared standard error decreases, meaning that it is more likely that the sample mean will be close to the true population mean, 44, as opposed to the hypothesized value, 42.3.

We have just illustrated that increasing the sample size can increase power, roughly because this lowers the squared standard error. Now it is illustrated that the larger σ happens to be, the lower the power will be given α, n and a value for $\mu - \mu_0$. This is because increasing σ increases the standard error.

Example 14

For the battery example, again consider $\alpha = .05$, $\mu - \mu_0 = 44 - 42.3 = 1.7$ and $n = 30$ with $\sigma = 4$. Then power is .75 as previously indicated. But if $\sigma = 8$, power is now .31. If $\sigma = 12$, power is only .19.

Finally, the smaller α happens to be, the lower will be the power, given values for n, σ and $\mu - \mu_0$. This is because, as we lower α, the critical value increases, meaning that the value for Z needed to reject increases as well.

Example 15

For the battery example with $n = 30$, consider three choices for α: .05, .025, and .01. For $\alpha = .05$ we have already seen that power is .75 when testing $H_0 : \mu < 42.3$ and $\mu = 44$. For $\alpha = .025$, the critical value is now $c = 1.96$, so power is

$$\begin{aligned} 1 - \beta &= P\left(Z \geq c - \frac{\sqrt{n}(\mu - \mu_0)}{\sigma}\right) \\ &= P\left(Z \geq 1.96 - \frac{\sqrt{20}(44 - 42.3)}{4}\right) \\ &= P(Z \geq .059) \\ &= .47. \end{aligned}$$

If instead you use $\alpha = .01$, the critical value is now $c = 2.33$ and power can be seen to be .33. This illustrates that if we adjust the critical value so that the probability of a Type I error goes down, power goes down as well. Put another way, the more careful you are not to commit a Type I error by choosing α close to zero, the more likely you are to commit a Type II error if the null hypothesis happens to be false.

The results just described on how n, α and σ are related to power and can be summarized as follows:

- As the sample size, n, gets large, power goes up, so the probability of a Type II error goes down.
- As α goes down, in which case the probability of a Type I error goes down, power goes down and the probability of a Type II error goes up.
- As the standard deviation, σ, goes up, with n, α and $\mu - \mu_0$ fixed, power goes down.

Confidence intervals versus testing hypotheses

There is a simple connection between confidence intervals covered in chapter 6 and tests of hypotheses. As an illustration, imagine that the goal is to test H_0: $\mu = 64$ with a Type I error probability of $\alpha = .05$. Rather than proceed as was done here, you could simply compute a .95 confidence interval for the population mean and reject if this interval does not contain the hypothesized value of 64. Under the assumptions made here, the probability of a Type I error will be $1 - .95 = .05$. More generally, if a $1 - \alpha$ confidence interval is computed, and you reject the null hypothesis if this interval does not contain the hypothesized value, the probability of a Type I error will be α. Both approaches have their own advantages. Confidence intervals not only tell us whether we should reject some hypothesis of interest, they also provide information about the accuracy of our estimate of μ. A related advantage is that they provide information about which values for μ appear to be reasonable and which do not. If we are told only that we reject the null hypothesis, or if we are only told the p-value, this information is not available to us. But hypothesis testing makes clear the notion of a Type I error. And it has the added importance of highlighting the notion of power. A criticism of p-values is that it tells us nothing about the magnitude of $\mu - \mu_0$, the difference between actual population mean and its hypothesized value. This difference is an example of what is called an *effect size*, which is a general term for measures aimed at quantifying the extent to which the null hypothesis is false.

Another important point is that p-values do not tell us the likelihood of replicating a decision about the null hypothesis. For example, if we get a p-value of .001, what is the probability of rejecting at the .05 level if we were to conduct the study again? This is a power issue. If unknown to us, power is .2 when testing with $\alpha = .05$, then the probability of rejecting again, if the study is replicated, is .2. If the null hypothesis happens to be true, and by chance we got a p-value of .001, the probability of rejecting again is .05 or whatever α value we happen to use.

Problems

1. Given that $\bar{X} = 78$, $\sigma^2 = 25$, $n = 10$ and $\alpha = .05$, test $H_0 : \mu > 80$, assuming observations are randomly sampled from a normal distribution. Also, draw the standard normal distribution indicating where Z and the critical value are located.
2. Repeat the previous problem but test $H_0 : \mu = 80$.
3. For problem 2, compute a .95 confidence interval and verify that this interval is consistent with your decision about whether to reject the null hypothesis.
4. For problem 1, determine the p-value.
5. For problem 2, determine the p-value.
6. Given that $\bar{X} = 120$, $\sigma = 5$, $n = 49$ and $\alpha = .05$, test $H_0 : \mu > 130$, assuming observations are randomly sampled from a normal distribution.
7. Repeat the previous problem but test $H_0 : \mu = 130$.
8. For the previous problem, compute a .95 confidence interval and compare the result with your decision about whether to reject H_0.
9. If $\bar{X} = 23$ and $\alpha = .025$, can you make a decision about whether to reject $H_0 : \mu < 25$ without knowing σ?
10. An electronics firm mass produces a component for which there is a standard measure of quality. Based on testing vast numbers of these components, the company has found that the average quality is $\mu = 232$ with $\sigma = 4$. However, in recent years the quality has not been checked, so management asks you to check their claim with the goal of being reasonably certain that an average quality of less than 232 can be ruled out. That is, assume the quality is poor and in fact less than 232 with the goal of empirically establishing that this assumption is unlikely. You get $\bar{X} = 240$ based on a sample $n = 25$ components, and you want the probability of a Type I error to be .01. State the null hypothesis and perform the appropriate test assuming
11. An antipollution device for cars is claimed to have an average effectiveness of exactly 546. Based on a test of 20 such devices you find that $\bar{X} = 565$.

Assuming normality and that $\sigma = 40$, would you rule out the claim with a Type I error probability of .05?

12. Comment on the relative merits of using a .95 confidence interval for addressing the effectiveness of the antipollution device in the previous problem.

13. For $n = 25$, $\alpha = .01$, $\sigma = 5$ and $H_0 : \mu \geq 60$, verify that power is .95 when $\mu = 56$.

14. For $n = 36$, $\alpha = .025$, $\sigma = 8$ and $H_0 : \mu \leq 100$, verify that power is .61 when $\mu = 103$.

15. For $n = 49$, $\alpha = .05$, $\sigma = 10$ and $H_0 : \mu = 50$, verify that power is approximately .56 when $\mu = 47$.

16. A manufacturer of medication for migraine headaches knows that their product can damage the stomach if taken too often. Imagine that by a standard measuring process, the average damage is $\mu = 48$. A modification of their product is being contemplated, and based on ten trials, it is found that $\bar{X} = 46$. Assuming $\sigma = 5$, they test $H_0 : \mu \geq 48$, the idea being that if they reject, there is convincing evidence that the average amount of damage is less than 48. Then

$$Z = \frac{46 - 48}{5/\sqrt{10}} = -1.3.$$

With $\alpha = .05$, the critical value is -1.645, so they do not reject because Z is not less than the critical value. What might be wrong with accepting H_0 and concluding that the modification results in an average amount of damage greater than or equal to 48?

17. For the previous problem, verify that power is .35 if $\mu = 46$.

18. The previous problem indicates that power is relatively low with only $n = 10$ observations. Imagine that you want power to be at least .8. One way of getting more power is to increase the sample size, n. Verify that for sample sizes of 20, 30, and 40, power is .56, .71 and .81, respectively.

19. For the previous problem, rather than increase the sample size, what else might you do to increase power? What is a negative consequence of using this strategy?

7.2 Testing hypotheses about the mean of a normal distribution, σ not known

When the variance is not known, one simply estimates it with the sample variance, in which case the test statistic

$$Z = \frac{\bar{X} - \mu_0}{\sigma/\sqrt{n}}$$

becomes

$$T = \frac{\bar{X} - \mu_0}{s/\sqrt{n}}. \tag{7.7}$$

When the null hypothesis is true, T has a Student's t-distribution with $\nu = n - 1$ degrees of freedom when sampling from a normal distribution. This means that the the critical value c is read from table 4 in appendix B. The details can be summarized as follows.

Assumptions:

- Random sampling
- normality

Decision Rules:

- For $H_0 : \mu \geq \mu_0$, reject if $T \leq c$, where c is the α quantile of Student's t-distribution with $\nu = n - 1$ degrees of freedom and T is given by equation (7.7).
- For $H_0 : \mu \leq \mu_0$, reject if $T \geq c$, where now c is the $1-\alpha$ quantile of Student's t-distribution with $\nu = n-1$ degrees of freedom.
- For $H_0 : \mu = \mu_0$, reject if $T \geq c$ or $T \leq -c$, where now c is the $1 - \frac{\alpha}{2}$ quantile of Student's t-distribution with $\nu = n-1$ degrees of freedom. Equivalently, reject if $|T| \geq c$.

Example 1

Imagine that for a general measure of anxiety, a researcher claims that the population of college students has a mean of at least 50. As a check on this claim, suppose that ten college students are randomly sampled with the goal of testing H_0: $\mu \geq 50$ with $\alpha = .05$. Further imagine that the sample standard deviation is $s = 11.4$ and the sample mean is $\bar{X} = 44.5$. Because $n = 10$, the degrees of freedom are $\nu = n - 1 = 9$ and

$$T = \frac{\bar{X} - \mu_0}{s/\sqrt{n}} = \frac{44.5 - 50}{11.4/\sqrt{10}} = -1.5.$$

Referring to table 4 in appendix B, $P(T \leq -1.83) = .05$, so the critical value is -1.83. This means that if we reject when T is less than or equal to -1.83, the probability of a Type I error will be .05, assuming normality. Because the observed value of T is -1.5, which is greater than the critical value, you fail to reject. In other words, the sample mean is not sufficiently smaller than 50 to be reasonably certain that the speculation $\mu \geq 50$ is false. As you can see, the steps you follow when σ is not known mirror the steps you use to test hypotheses when σ is known.

Example 2

Suppose you observe the values

$$12, 20, 34, 45, 34, 36, 37, 50, 11, 32, 29$$

and the goal is to test H_0: $\mu = 25$ such that the probability of a Type I error is $\alpha = .05$. Here, $n = 11$, $\mu_0 = 25$ and it can be seen that $\bar{X} = 33.24$, $s/\sqrt{11} = 3.7$, so

$$T = \frac{\bar{X} - \mu_0}{s/\sqrt{n}} = \frac{33.24 - 25}{3.7} = 2.23.$$

The null hypothesis is that the population mean is exactly equal to 25. So the critical value is the $1 - \frac{\alpha}{2} = .975$ quantile of Student's t-distribution with degrees of freedom $\nu = 11 - 1 = 10$. Table 4 in appendix B indicates that

$$P(T \leq 2.28) = .975,$$

so our decision rule is to reject H_0 if the value of T is greater than or equal to 2.28 or less than or equal to -2.28. Because the absolute value of T is less than 2.28, you fail to reject.

Comments on interpreting two-sided tests: Turkey's three decision rule

There are some issues regarding the goal of testing for exact equality that should be discussed. Some authorities criticize this commonly sought goal on the grounds that exact equality is rarely if ever true. The argument is that if we test H_0 $\mu = 89$, for example, we can be reasonably certain that the actual value of the population mean differs at some decimal place from 89. For example, it might be 89.00012. In more general terms, it can be argued that without any data, one can make the argument that the hypothesis of exact equality is false. A related criticism is that as a consequence, we should never accept the null hypothesis as being true, so why test this hypothesis in the first place?

Momentarily imagine that we accept the argument that the population mean is never exactly equal to the hypothesized value. There is a way of salvaging hypothesis testing by interpreting it in terms of what is called *Tukey's three decision rule.* (For more details, see Jones and Tukey, 2000.) This means that when assessing the meaning of the test statistic T, one of three conclusions are reached regarding how the population mean compares to the hypothesized value. Let c be the $1 - \alpha/2$ quantile of Student's t-distribution with $n - 1$ degrees of freedom. That is, we reject if $T \leq -c$ or if $T \geq c$. The three possible conclusions are as follows:

- If we reject because T is less than or equal to $-c$, the lower critical value, conclude that $\mu < \mu_0$.
- If we reject because T is greater than or equal to c, the upper critical value, conclude that $\mu > \mu_0$.
- Otherwise, make no decision about whether μ is greater than or less than the hypothesized value.

Note that the first two conclusions are subject to a possible error. You might erroneously conclude, for example, that the population mean is greater than the hypothesized value, and the reverse conclusion might be made erroneously as well. But if we make our decision based on whether T is less than the $\alpha/2$ quantile of Student's t-distribution, or if it is greater than the $1-\alpha/2$ quantile, then the probability of making an error is at most α, assuming normality. Put more simply, we apply Student's t test as already described and illustrated, but we interpret the results within the framework just outlined. We do not view the goal as testing for exact equality, but rather, the goal is to make a decision about whether μ is less than or greater than μ_0.

A related issue is making a decision about whether the difference between μ and μ_0 is small. Although it can be argued that perhaps μ is not exactly equal to μ_0, this leaves open the issue of whether there is little difference between these two values.

Example 3

Consider an expensive prescription medication, for which its average effectiveness is found to be 48, based on some relevant measure. A company wants to market an alternative drug that is less expensive, but there is the issue of how its effectiveness compares to the medication currently being used. One possibility is to compute a confidence interval for μ. If the length of the confidence interval is judged to be small, and it contains μ_0, conclude that there is little difference between the average effectiveness of the two drugs.[1]

Testing hypotheses about medians

One way of testing hypotheses about the median is to use a strategy similar to Student's t. That is, subtract the hypothesized value from the sample median and divide by an estimate of the standard error. In symbols, if θ_0 represents some hypothesized value for the population median, the test statistic is

$$Z = \frac{M - \theta_0}{S_M}, \tag{7.8}$$

where S_M is the McKean–Schrader estimate of the standard error described in chapter 6. As pointed out in chapter 6, if tied (meaning duplicated) values tend to be rare, Z has, approximately, a standard normal distribution, assuming that the null hypothesis is true. This means that hypotheses are tested in essentially the same manner as in section 7.1. For instance, when testing H_0: $\theta = \theta_0$, reject if $|Z| \geq c$, the $1 - \alpha/2$ quantile of a standard normal distribution.

Example 4

A researcher claims that in 4000 BC, the median height of adult male skulls was 132mm. As a check on this claim, consider the following values for 30 skulls:

121, 124, 129, 129, 130, 130, 131, 131, 132, 132, 132, 133, 133, 134, 134,
134, 134, 135, 135, 136, 136, 136, 136, 137, 137, 138, 138, 138, 140, 143.

To test H_0: $\theta = 132$ with a Type I error probability of .05, compute the sample median, yielding $M = 134$. The value of S_M can be seen to be 0.97056, so the test statistic is $Z = (134 - 132)/0.97056 = 2.06$. The critical value is the .975 quantile of a standard normal distribution, which is read from table 1 in appendix B and found to be 1.96. So reject and conclude that the population median is greater than 132. The p-value is

$$2(1 - P(Z \leq 2.06)) = .039.$$

A concern about the method just illustrated is that there are tied values, and as noted in chapter 7, this might mean that S_M is a poor estimate of the true standard error, which in turn means that control over the probability of a Type I error might be relatively poor. Having tied values does not necessarily mean disaster, but the reality is that serious practical problems can occur. One way of dealing with this issue is to compute a confidence interval for the median using the method in box 6.1 and then reject if this interval does not contain the hypothesized value. Another possibility is to use a percentile bootstrap method as described in the final section of this chapter. This latter approach is more interesting in the sense that it provides a method for comparing medians in more complex situations to be covered.

1. Other approaches have been proposed, but the details are not discussed here.

Problems

20. Given the following values for $\bar{X}$ and s: (a) $\bar{X} = 44$, $s = 10$, (b) $\bar{X} = 43$, $s = 10$, (c) $\bar{X} = 43$, $s = 2$, test the hypothesis H_0: $\mu = 42$ with $\alpha = .05$ and $n = 25$.

21. For part b of the last problem you fail to reject but you reject for the situation in part c. What does this illustrate about power?

22. Given the following values for $\bar{X}$ and s: (a) $\bar{X} = 44$, $s = 10$, (b) $\bar{X} = 43$, $s = 10$, (c) $\bar{X} = 43$, $s = 2$, test the hypothesis H_0: $\mu < 42$ with $\alpha = .05$ and $n = 16$.

23. Repeat the previous problem only test H_0: $\mu > 42$.

24. A company claims that on average, when exposed to their toothpaste, 45% of all bacteria related to gingivitis is killed. You run 10 tests and find that the percentages of bacteria killed among these tests are 38, 44, 62, 72, 43, 40, 43, 42, 39, 41. The mean and standard deviation of these values are $\bar{X} = 46.4$ and $s = 11.27$. Assuming normality, test the hypothesis that the average percentage is 45 with $\alpha = .05$.

25. A portion of a study by Wechsler (1958) reports that for 100 males taking the Wechsler Adult Intelligent Scale (WAIS), the sample mean and variance on picture completion are $\bar{X} = 9.79$ and $s = 2.72$. Test the hypothesis H_0: $\mu \geq 10.5$ with $\alpha = .025$.

26. Given that $n = 16$, $\bar{X} = 40$, and $s = 4$, test H_0: $\mu \leq 38$ with $\alpha = .01$.

27. Given that $n = 9$, $\bar{X} = 76$, and $s = 4$, test H_0: $\mu = 32$ with $\alpha = .05$.

28. An engineer believes it takes an average of 150 man-hours to assemble a portion of an automobile. As a check, the time to assemble 10 such parts was ascertained yielding $\bar{X} = 146$ and $s = 2.5$. Test the engineer's belief with $\alpha = .05$.

29. In a study of court administration, the following times to disposition were determined for twenty cases and found to be

$$42, 90, 84, 87, 116, 95, 86, 99, 93, 92$$

$$121, 71, 66, 98, 79, 102, 60, 112, 105, 98.$$

Test the hypothesis that the average time to disposition is less than or equal to 80, using $\alpha = .01$.

7.3 Modern advances and insights

Student's t and non-normality

An issue of fundamental importance is how Student's t performs under non-normality. Based on properties of t described in chapter 6, a natural guess is that practical problems might arise and this speculation turns out to be correct. Although these concerns have been known for years, currently, they are rarely described and illustrated in an introductory course.

As done in chapter 6, four types of non-normal distributions are considered. The first is a symmetric, light-tailed distribution where outliers tend to be rare. For this case, Student's t performs well in terms of both Type I errors and power, compared to alternative methods we might use. For example, the mean and median are estimating, approximately, the same unknown value, but there is little or no practical reason to prefer the median over the mean. The median offers no advantage in terms of power, roughly because its standard error tends to be larger than the standard error of the mean due to trimming nearly all of the data.

When sampling from a symmetric, heavy-tailed distribution, generally the actual probability of a Type I error, when using Student's t, is less than the nominal level. A crude explanation is that outliers tend to inflate the sample variance, a large sample variance lowers the value of T, which in turn makes it less likely to reject. For example, if you test the hypothesis of a zero mean when sampling from the contaminated normal shown in figure 4.5, with $n = 20$ and $\alpha = .05$, the actual probability of a Type I error is approximately .025. But in terms of power, Student's t can perform rather poorly. The essential reason is that its standard error can be relatively large. As an example we again test the hypothesis of a zero mean, only now we sample from the contaminated normal with a population mean of .5, in which case power is approximately .11. But if we test the hypothesis of a zero median instead, using equation (7.8), power is approximately .35.

Next, consider skewed, light-tailed distributions. We saw in chapter 6 that the distribution of t can be asymmetric, rather than symmetric. (When sampling from any symmetric distribution, t also has a symmetric distribution, but otherwise its distribution is generally asymmetric.) In addition, the actual quantiles can differ substantially from the quantiles of Student's t under normality resulting in poor control over the probability of a Type I error. Imagine, for example, we sample twenty observations from the distribution shown in figure 6.2 with the goal of having a Type I error probability of .05. The actual probability of a Type I error is .15. For sample sizes 40, 80 and 160, the actual probability of a Type I error is .149, .124 and .109, respectively. So control over the probability of a Type I error is improving as the sample size increases, in accordance with the central limit theorem, but if we want the probability of a Type I error to be under .075, a sample size of about 200 is needed.

As for skewed, heavy-tailed distributions, control over the probability of a Type I error deteriorates. Now a sample size of more than 300 might be needed to get adequate control over the probability of a Type I error. And an added concern is that with outliers being common, the standard error of the mean is relatively large, which could mean relatively poor power. Note that by implication, skewness and outliers affect p-values, which complicates their interpretation. Despite this, we will see situations in later chapters where p-values have practical value.

Brief summary of how Student's t performs under non-normality

- For symmetric, light-tailed distributions, Student's t performs relatively well in terms of both Type I errors and power.
- For symmetric, heavy-tailed distributions, it controls Type I errors reasonably well, but power can be relatively poor.
- For asymmetric, light-tailed distributions, good control over the probability of a Type I error might require a sample size of 200 or more. Poor power could be an issue.
- For asymmetric, heavy-tailed distributions, serious concerns about Type I errors and power arise, and a fairly large sample size might be needed to correct any practical problems.

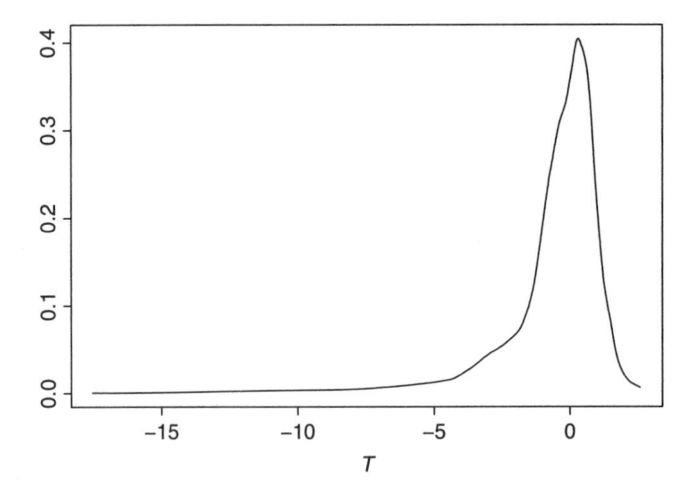

Figure 7.2 The distribution of t when sampling, with replacement, from data stemming from a study on hangover symptoms.

It was illustrated in chapter 6 that the actual quantiles of Student's t can differ substantially from the quantiles under normality. In the present context, this means that control over the probability of a Type I error can be poor and power can be affected as well. Here is another example, based on data from a study on hangover symptoms when drinking alcohol. (The data are from another portion of the study in example 2 of section 3.2.) The values from the study were

$$1, 0, 3, 0, 3, 0, 15, 0, 6, 10, 1, 1, 0, 2, 24, 42, 0, 0, 0, 2.$$

Here, we resample with replacement from the 20 values just given, compute T and repeat this process 1000 times. (In essence, a bootstrap-t method is being used.) A plot of the resulting T values is shown in figure 7.2.[2] In effect, data are being sampled from a distribution having mean $\mu = 5.5$. So for the data at hand, under the assumption of normality and with $\alpha = .05$, we would reject H_0: $\mu = 5.5$ if $T \leq -2.09$ or if $T \geq 2.09$. But based on the plot in figure 7.2, we should reject if $T \leq -5.9$ or if $T \geq 1.4$.

Strategies for dealing with non-normality

It would be extremely convenient if a single hypothesis testing method could be found that always performs relatively well in comparison to all other methods we might use. During the last half-century, it has become abundantly clear that, due to the effects of non-normality, such a method does not exist. To get the most accurate information from data, an understanding of the relative merits of various techniques is required.

Within the context of testing hypotheses, several strategies that have been considered when dealing with non-normality. Perhaps the most obvious strategy is to search for a method based on means that improves upon Student's t. Various methods have been proposed, in some cases they do give better results, but under

2. A possible criticism is that perhaps the bootstrap-t method is not giving an accurate approximation of the actual distribution of T. There is evidence that this is true—practical problems are probably worse than indicated by figure 7.2.

general conditions they can be unsatisfactory. Some of these methods are based on what are called bootstrap techniques, but the details go beyond the scope of this book. Another common suggestion is to transform the data. For example, use logarithms as noted in chapter 6. This might help, but as previously indicated, a distribution can remain skewed, making inferences about the mean of the original scores difficult, and problems with outliers are not necessarily eliminated. Put simply, do not assume that all practical problems have been addressed by transforming the data. One could use diagnostic tools, such as tests for normality, or one might use numerical quantities aimed at measuring skewness with the goal of determining whether a distribution is reasonably symmetric. At issue is not whether a distribution is non-normal, but rather, does it differ from normality to the point that practical problems arise. Tests for normality, and diagnostic tools in general, are only useful if they have enough power to detect situations where the normality assumption should be abandoned. Currently, it is unclear when such tests have adequate power. (In chapter 8 we will summarize published papers outlining situations where tests of assumptions are unsatisfactory as a diagnostic tool aimed at justifying Student's *t*.) Another common suggestion is to use what are called rank-based or nonparametric techniques, which are discussed in chapter 13. These methods have certain practical advantages and deserve serious consideration, but their relative merits must be postponed for now. Yet another possible approach is to use some measure of location other than the mean. The median has the potential of high power, relative to the mean, when dealing with distributions where outliers are common. Concerns are that power might be less satisfactory when sampling from normal or light-tailed distributions, and control over the probability of a Type I error, when using equation (7.8), can be poor when tied values occur. Tied values do not necessarily spell disaster when using the median, but the reality is that disaster can strike.

Handling tied values when using the median

One way of dealing with tied values, when testing hypotheses about the population median, is to use a percentile bootstrap technique, which was introduced at the end of chapter 6. Consider the goal of testing H_0: $\theta = \theta_0$, the hypothesis that the population median is equal to the specified value, θ_0. It turns out that the percentile bootstrap technique in chapter 6 can be used to compute a p-value. A rough explanation of the method is as follows. If the null hypothesis is true, then with probability .5, the sample median will be greater than the hypothesized value. (With small sample sizes this probability can differ from .5, but under random sampling, it will be very close to .5 even with fairly small sample sizes.) The closer this probability is to 0 or 1, the stronger the evidence that the null hypothesis should be rejected. The problem is that we do not know the sampling distribution of M, but it can be approximated using bootstrap samples.

As was done in chapter 6, imagine that we generate a bootstrap sample by resampling with replacement n observations from our observed data, and then we compute the median. For convenience, label this sample median M^* to distinguish it from M, the median based on the original data. We repeat this process B times, and here we use $B = 1000$. Let A be the number of M^* values that are less than the hypothesized value, θ_0, and let C be the number of times $M^* = \theta_0$. Let $Q = (A + .5C)/B$ and set P equal to Q or $1 - Q$, whichever is smaller. Then a p-value for testing H_0: $\theta = \theta_0$ is $p = 2P$.

Unlike the method based on the McKean-Schrader estimate of the standard error, this method performs well when there are tied values.

Example 1

Imagine that the goal is to test the hypothesis that the population median is 45. We generate a bootstrap sample, compute the median, repeat this 1000 times, and find that 900 of these bootstrap medians are less than 45 and 10 are equal to 45. So $A = 900$, $C = 10$, and $Q = (A + .5C)/B = (900 + .5(10))/1000 = .905$. Because $Q > .5$, $P = 1 - .905 = .095$, and the p-value is $2(.095) = .19$.

Example 2

Consider again the skull data used in example 4 in section 7.2. We saw that when testing the hypothesis that the median height is 132mm, the p-value is .039 when using the test statistic given by equation (7.8). Using the percentile bootstrap method instead, the p-value is .046. So in this case, despite the tied values, there is not that much difference between the p-values. But consider again the sexual attitude data in table 2.3, which also has tied values, and suppose the goal is to test H_0: $\theta = 0$. Then the p-value using equation (7.8) is .086, but when using the percentile bootstrap method, the p-value is less than .001, illustrating that the choice of method can make a substantial difference.

Trimming less: The Tukey-Mclaughlin method

As noted in chapter 6, practical problems with the median arise, roughly because it trims all but one or two values. One possibility is to trim less, say 20%, in which case the test statistic is

$$T = \frac{.6(\bar{X}_t - \mu_0)}{s_w/\sqrt{n}}, \tag{7.9}$$

where now μ_0 is some hypothesized value for the population trimmed mean, $\bar{X}_t$ is the 20% trimmed mean and s_w is the 20% Winsorized standard deviation. The hypothesis H_0: $\mu_t = \mu_0$ is rejected if $|T| \geq c$, where c is the $1 - \alpha/2$ quantile of a Student's t-distribution with $\nu = n - h - 1$ degrees of freedom, where h is the number of observations trimmed. This method was derived by Tukey and McLaughlin (1963).

Example 3

Doksum and Sievers (1976) report data on weight gain among rats. One group was the control and the other was exposed to an ozone environment. For illustrative purposes, attention is focused on the control group and we consider the claim that the typical weight gain is 26.4, as measured by the 20% trimmed mean. The sample size is $n = 23$, the 20% trimmed mean is $\bar{X}_t = 23.3$, $s_w = 3.9$, and for H_0: $\mu_t = 26.4$ we see that

$$T_t = \frac{.6(\bar{X}_t - \mu_0)}{s_w/\sqrt{n}} = \frac{.6(23.3 - 26.4)}{3.9/\sqrt{23}} = -2.3.$$

Because there are 23 rats, $h = 8$, so the degrees of freedom are $\nu = 23 - 8 - 1 = 14$, and the critical value (read from table 4 in appendix B) is $c = 2.14$. Because $|T| = |-2.3| = 2.3$ is greater than the critical value, reject the hypothesis that the trimmed mean is 26.4.

In the context of hypothesis testing, the Tukey-McLaughlin method has several advantages. It performs much better than Student's t in terms of controlling the probability of a Type I error, meaning that it performs well over a much broader range of situations. But even the Tukey-McLaughlin method can be unsatisfactory in terms of Type I errors when the sample size is small and the departure from normality is sufficiently severe. (This problem can be further reduced by using the percentile bootstrap method previously described. Using a percentile bootstrap method with a 20% appears to perform very well even with sample sizes of 10.) Another advantage is that under normality, its power is nearly comparable to Student's t and it has higher power than methods based on the median. Compared to Student's t, it maintains high power when sampling from symmetric distributions where outliers are common, but if the number of outliers is sufficiently high, using medians can have more power instead. When using equation (7.9), problems with tied values are of little or no concern, in contrast to comparing medians with equation (7.8). However, all indications are that problems due to tied values can be avoided when working with the median by using the percentile bootstrap method.

Another point to keep in mind is that when distributions are skewed, generally the population mean, median and 20% trimmed mean differ. So, for example, if the goal is to test the hypothesis that the population mean is 20, this hypothesis could be true even when the population median has some other value. In practical terms, if there is interest in making inferences about the mean when a distribution is skewed, methods based on a 20% trimmed mean or median can be highly unsatisfactory.

It should be stressed that currently, most applied researchers are familiar with the mean and median, but relatively few are familiar with a 20% trimmed mean. The 20% trimmed mean is used, but its practical advantages are rarely discussed in an introductory course.

One final comment might help. Based on the many negative features associated with Student's t, the utility of methods based on means might seem rather bleak, in the sense that there is concern about the possibility of making a Type I error or having relatively low power, unless perhaps the sample size is quite large. Despite known problems covered here, plus some additional concerns covered in subsequent chapters, methods based on means can be argued to have practical value, provided the results are interpreted in a manner that takes into account modern insights, some of which are yet to be described.

A Summary of Some Key Points

- When using Student's t test, the actual probability of a Type I error will be fairly close to the nominal level when sampling from symmetric distributions where outliers are rare. If outliers are common, the actual probability of a Type I error can drop well below the intended level.

Continued

A Summary of Some Key Points (*cont'd*)

- When using Student's t test, the control over the probability of a Type I error can be poor, with practical problems increasing as we move toward situations where outliers are common. In some cases, even with a sample size of $n = 300$, Student's t test can perform poorly.
- Skewness and outliers can adversely affect the power of Student's t.
- As a measure of effect size, meaning the extent to which the null hypothesis is false, p-values can be highly unsatisfactory.
- When using equation (7.8) to test hypotheses about the median, control over the probability of a Type I error is reasonably good when tied values never occur. But when tied values can occur, it should not be used; currently the percentile bootstrap is the best method.
- In terms of controlling the probability of a Type I error, the most effective (non-bootstrap) method covered in this chapter is the Tukey-McLaughlin method based on equation (7.9). That is, it performs reasonably well over a broader range of situations. But note that for skewed distributions, it is inappropriate in terms of testing hyotheses about the population mean.
- Generally, some type of bootstrap method has practical value in terms of controlling the probability of a Type I error.

Problems

30. Given the following values for $\bar{X}_t$ and s_w: (a) $\bar{X}_t = 44$, $s_w = 9$, (b) $\bar{X}_t = 43$, $s_w = 9$, (c) $\bar{X}_t = 43$, $s_w = 3$. Assuming 20% trimming, test the hypothesis H_0: $\mu_t = 42$ with $\alpha = .05$ and $n = 20$.

31. Repeat the previous problem, only test the hypothesis H_0: $\mu_t < 42$ with $\alpha = .05$ and $n = 16$.

32. For the data in problem 24, the trimmed mean is $\bar{X}_t = 42.17$ with a Winsorized standard deviation of $s_w = 1.73$. Test the hypothesis that the population trimmed mean is 45 with $\alpha = .05$.

33. A standard measure of aggression in 7-year-old children has been found to have a 20% trimmed mean of 4.8 based on years of experience. A psychologist wants to know whether the trimmed mean for children with divorced parents differs from 4.8. Suppose $\bar{X}_t = 5.1$ with $s_w = 7$ based on $n = 25$. Test the hypothesis that the population trimmed mean is exactly 4.8 with $\alpha = .01$.

34. Summarize the relative merits of using a percentile bootstrap method.

8

CORRELATION AND REGRESSION

One of the most common goals in applied research is trying to determine whether two variables are dependent, and if they are dependent, trying to understand the nature of the association. Is there an association between aggression in the home and the cognitive functioning of children living in the home? If yes, how might we describe it in an effective manner? Is there an association between breast cancer rates and exposure to solar radiation? What is the association between weight gain in infants and the amount of ozone in the air? This chapter describes the most commonly used tools for answering these type of questions. As usual, we provide a description of the mechanics and the underlying assumptions associated with conventional methods. Then we describe situations where these techniques perform well and where they can be highly inaccurate. Many modern tools have been designed for dealing with practical problems associated with the basic methods described here, but complete details go well beyond the scope of this book. However, a brief outline is provided of how a few modern methods deal with practical problems.

8.1 Least squares regression

One approach to studying the association between two variables is to collect data and then search for a straight line that summarizes the data in a reasonable manner. For example, imagine that X represents aggression in the home and that Y represents the cognitive functioning of children living in the home. Then roughly, assuming that a straight line summarizes the data in a reasonable manner means that if we are told X, the amount of aggression in the home, we get a reasonable estimate of Y, the cognitive functioning of children living in the home, using the equation

$$Y = \beta_0 + \beta_1 X, \tag{8.1}$$

where β_1 and β_0 are the unknown slope and intercept, respectively, that are to be determined based on observations to be made.[1] For instance, if $\beta_1 = .5$ and $\beta_0 = 10$, and we are told that the agression in the home is $X = 6$, then the estimated cognitive functioning of a child living in this home is $.5 \times 6 + 10 = 13$.

One problem is determining β_1, the slope, and β_0, the intercept based on the data available to us. There are many methods aimed at addressing this problem. Today, the most popular method is based on what is called the *least squares* principle, and therefore it is essential that students understand its relative merits. It has various advantages over competing methods, but it also has serious deficiencies, as we shall see. To provide some sense of when it performs well and when it fails, and why the method remains popular today, it helps to describe some of the reasons the method rose to prominence in applied research.

A simple example of some historical interest will help set the stage. Newton, in his *Principia*, argued that the earth should bulge at the equator due to its rotation. In contrast, based on empirical measures, the astronomer Cassini suspected that the earth bulges at the poles, but because of possible measurement errors, there was uncertainty about whether Cassini's data could be used to make valid inferences. In an attempt to resolve the issue, the French government funded expeditions to measure the linear length of a degree of latitude at several places on the earth. Newton's arguments suggested that there would be an approximately linear association between a certain transformation of the latitude and measures of arc length. For convenience, let X represent the transformed latitude and let Y represent the arc length. According to Newton, it should be the case that $\beta_1/3\beta_0$ is approximately 1/230.

During the second half of the eighteenth century, Roger Boscovich attempted to resolve the shape of the earth using the data in table 8.1. Recall that any two distinct points determine a line. If we denote the two points by (X_1, Y_1) and (X_2, Y_2), then the slope, for example, is given by

$$b_1 = \frac{Y_2 - Y_1}{X_2 - X_1},$$

the difference between the Y values divided by the difference between the X values. The intercept is given by

$$b_0 = Y_1 - b_1 X_1.$$

So, for the first and last points in table 8.1, the slope is

$$b_1 = \frac{57{,}422 - 56{,}751}{.8{,}386 - 0} = 800.14,$$

and the intercept is given by

$$b_0 = 57{,}422 - 800.14(57{,}422) = 56{,}751.$$

But Boscovich has a problem: he has five points and each pair of points gives a different value for the slope and intercept. Differences were to be expected because the measurements of arc length and latitude cannot be done precisely. That is, there is

1. The use of the Greek letter beta used in regression should not be confused with the use of β in chapter 7. In chapter 7, where β has no subscript, it represents the probability of a Type II error. When dealing with regression, β_1 and β_0 are typically used to represent the unknown slope and intercept of a line, and the goal is to estimate β_1 and β_0 based on observations we make.

Table 8.1 Boscovich's data on meridian arcs

Place	Transformed latitude	Arc length
Quito	0.0000	56,751
Cape of Good Hope	0.2987	57,037
Rome	0.4648	56,979
Paris	0.5762	57,074
Lapland	0.8386	57,422

measurement error due to the instruments being used and variations due to the individuals taking the measurements. So an issue is whether the different slopes can be combined in some manner to produce a reasonable estimate of the true slope and intercept (β_1 and β_0) if no measurement errors were made.

For the data at hand, there are 10 pairs of points, and the corresponding estimates of the slopes, written in ascending order, are

$$-349.19, 133.33, 490.53, 560.57, 713.09, 800.14, 852.79, 957.48, 1{,}185.13, 1{,}326.22.$$

One possibility is to simply average these 10 values yielding a single estimate of the unknown slope, an idea that dates back to at least the year 1750. This yields 667. Although averaging the slopes is reasonable, from a practical point of view it has two serious difficulties. The first is that in general, as the number of points increases, the number of pairs of points that must be averaged quickly becomes impractical without a computer. For example, if Boscovich had 50 points rather than only five, he would need to average 1,225 slopes. With a 100 points, he would need to average 4,950 slopes. The second problem is that it is not immediately clear how to assess the precision of the estimate. That is, how might we compute a confidence interval for the true slope? It might seem that we could somehow extend the methods in chapters 5 and 6 to get a solution, but there are technical problems that are very difficult to address in a reasonably simple fashion.[2]

Before continuing, note that another strategy is to take the median of the slopes rather than their average. This reflects a modern method known as the *Theil-Sen estimator*. For Boscovich's data, the Theil-Sen estimate of the slope (the median of the 10 slopes) is

$$b_{1\text{ts}} = 756.6,$$

and the estimate of the intercept is taken to be

$$b_{0\text{ts}} = M_y - b_{1\text{ts}} M_x,$$

where M_x and M_y are the medians of the X and Y values, respectively. Here, $M_x = .4648$, $M_y = 57{,}037$, so

$$b_{0\text{ts}} = 57{,}037 - 756.6(.4648) = 56{,}685.3.$$

Of course, prior to the computer age, the Theil-Sen estimator does not address the problems just described. However, with computers, it is a viable option that has several practical advantages mentioned at the end of this chapter.

2. One complication is that when computing the slopes for all pairs of points, some of the estimates are dependent, making an estimate of the standard error of the average of the slopes difficult at best.

Residuals

There are two major advances that took place when trying to address the two practical problems just described. The first stems from Boscovich who suggested that the fit of a regression line be judged based on what are called residuals. The basic idea is to judge a choice for the slope and intercept based on the overall discrepancy between the observed Y values and those predicted by the regression equation under consideration.

Consider again Boscovich's data and suppose the goal is to judge how well the slope 800.14 and intercept 56,751 perform for all of the available data. For convenience, we write the predicted Y value, given a value for X, as

$$\hat{Y} = b_0 + b_1 X,$$

where the notation $\hat{Y}$ is traditionally used to make a distinction between the observed Y value and the value predicted by the regression equation. So for Boscovich's data, we are considering the regression equation $\hat{Y} = 56{,}751 + 800.14X$, and the goal is to get some overall sense of how well this equation performs. Boscovich's suggestion was to judge the performance of the equation based on what are called *residuals*, which are just the differences between the observed Y values and those predicted by a regression line, namely $\hat{Y}$.

In more formal terms, imagine we have n pairs of points: $(X_1, Y_1), \ldots, (X_n, Y_n)$. Further imagine that consideration is being given to predicting Y with X using the equation $\hat{Y} = b_0 + b_1 X$. Then the residuals corresponding to these n pairs of points are

$$r_1 = Y_1 - \hat{Y}_1, r_2 = Y_2 - \hat{Y}_2, \ldots, r_n = Y_n - \hat{Y}_n.$$

For Boscovich's data, taking the slope and intercept to be 800.14 and 56,751, the resulting values for $\hat{Y}$ and the residuals are:

i	Y_i	$\hat{Y}_i$	r_i (Residuals)
1	56,751	56,737.43	13.57
2	57,037	56,953.52	83.48
3	56,979	57,073.68	−94.68
4	57,074	57,154.27	−80.27
5	57,422	57,344.10	77.90

Boscovich went on to suggest that the overall effectiveness of a regression line be judged by the sum of the absolute values of the residuals. In particular, his suggestion was to choose the slope and intercept to be the values b_1 and b_0 that minimize

$$\sum |r_i|.$$

That is, determine the values for b_1 and b_0 that minimize

$$\sum |Y_i - b_1 X_i - b_0|. \tag{8.2}$$

Today we call this *least absolute value regression*. Boscovich devised a numerical method for determining the slope and intercept, it reflected an important advance over simply averaging the slopes of all pairs of points, but major practical problems remained. It was still relatively difficult to perform the calculations with many pairs of points, and without a computer, computing confidence intervals cannot be done in a practical manner.

Least squares principle

In the year 1809, about 50 years after Boscovich's suggestion to use least absolute value regression to determine the slope and intercept, Legendre came up with a slight variation that would prove to be a major breakthrough. Indeed, Legendre's modification continues to be the most commonly used technique today. His idea was to use the sum of squared residuals when judging how well a regression line performs as opposed to the sum of the absolute values. So Legendre's suggestion is to choose values for the slope (b_1) and intercept (b_0) that minimize

$$\sum r_i^2. \tag{8.3}$$

That is, determine the values for b_1 and b_0 that minimize

$$\sum (Y_i - b_1 X_i - b_0)^2. \tag{8.4}$$

This is an example of what is called the *least squares principle*. Initially it might seem that this modification makes little practical difference, but it greatly simplifies the calculations and it provides a convenient framework for developing a method for computing confidence intervals and testing hypotheses. For convenience, let

$$A = \sum (X_i - \bar{X})(Y_i - \bar{Y})$$

and

$$C = \sum (X_i - \bar{X})^2.$$

Then based on the least squares principle, it can be shown that the slope is given by

$$b_1 = \frac{A}{C} \tag{8.5}$$

and the intercept is given by

$$b_0 = \bar{Y} - b_1 \bar{X}. \tag{8.6}$$

Example 1

Table 8.2 summarizes the computational details using Boscovich's data in table 8.1. As indicated, $A = 283.24$, $C = 0.3915$, so the slope is estimated to be $b_1 = 283.24/0.3915 = 723.5$. Because $\bar{Y} = 57{,}052.6$ and $\bar{X}$=0.43566, the intercept is $b_0 = 57{,}052.6 - 723.5(0.43566) = 56{,}737.4$.

Table 8.2 Computing the least squares slope using Boscovich's data

i	$X_i - \bar{X}$	$Y_i - \bar{Y}$	$(X_i - \bar{X})^2$	$(X_i - \bar{X})(Y_i - \bar{Y})$
1	−0.43566	−301.6	0.18980	131.395
2	−0.13696	−15.6	0.01876	2.137
3	0.02914	−73.6	0.00085	−2.145
4	0.14054	21.4	0.01975	3.008
5	0.40294	369.4	0.16236	148.846
Sums			0.3915	283.24

A conceptual issue

There are several conceptual issues associated with least squares regression that are important to understand. Perhaps the most basic is that least squares regression is attempting to estimate the (conditional) mean of Y given X. To elaborate on what this means, consider a study conducted by E. Sockett and colleagues that deals with children diagnosed with diabetes. One of the goals was to understand how a child's age is related to their C-peptide concentrations. Consider all children age 7 years. For various reasons, C-peptide concentrations among these children will vary. One issue of possible interest is determining the average C-peptide concentration among 7-year-old children. Similarly, one might want to know the average C-peptide concentration among all children age 8, or 7.5, and so on. If we fit a least squares line to the data, it is assumed that the mean C-peptide concentration, given a child's age, is equal to $\beta_0 + \beta_1(\text{Age})$, for some unknown slope (β_1) and intercept (β_0) that are to be estimated based on observations to be made. In the notation of chapter 4, another way to write this is

$$E(Y|X) = \beta_0 + \beta_1 X,$$

which says that the conditional mean of Y (C-peptide concentration in the example), given X (age in the example) is given by $\beta_0 + \beta_1 X$.

Example 2

Imagine you want to buy a house in a particular suburb of Los Angeles. Table 8.3 shows the selling price (in dollars) of homes during the month of May, 1998, plus the size of the home in square feet. Given that you are interested in buying a home with 2,000 square feet, what would you expect to pay? What would you expect to pay if the house has 1,500 square feet instead? It can be seen that the least squares regression line for predicting the selling price, given the size of the house, is $\hat{Y} = 215(X) + 38{,}192$. So an estimate of the average cost of a home having 1,500 square feet is $215(1{,}500) + 38{,}192 = 360{,}692$.

Table 8.3 Sale price of homes (divided by 100,000) versus size in square feet

Home i	Size (X_i)	Sales price (Y_i)	Home i	Size (X_i)	Sales price (Y_i)
1	2,359	510	15	3,883	859
2	3,397	690	16	1,937	435
3	1,232	365	17	2,565	555
4	2,608	592	18	2,722	525
5	4,870	1,125	19	4,231	805
6	4,225	850	20	1,488	369
7	1,390	363	21	4,261	930
8	2,028	559	22	1,613	375
9	3,700	860	23	2,746	670
10	2,949	695	24	1,550	290
11	688	182	25	3,000	715
12	3,147	860	26	1,743	365
13	4,000	1,050	27	2,388	610
14	4,180	675	28	4,522	1,290

Example 3

We saw that for Boscovich's data in table 8.1, the least squares regression estimate of β_1 is $b_1 = 723.44$ and the estimate of β_0 is 56,737.43. This says, for example, that the estimated mean arc length, corresponding to a transformed arc length of 0.4, is $723.44(.4) + 56{,}737.43 = 57{,}026.81$.

In contrast, least absolute value regression, where b_1 and b_0 are chosen to minimize the sum of absolute residuals (as indicated by eq. [7.2]), is not aimed at estimating the mean of Y, given X, but rather the median of Y.

Example 4

For Boscovich's data, least absolute value regression yields $b_1 = 755.62$ and $b_0 = 56{,}686.32$. So the estimated median arc length, corresponding to a transformed arc length of 0.4, is $755.62(.4) + 56{,}686.32 = 56{,}988.57$. In this particular case, there is little difference between the estimated mean of Y (57,026.81) versus the estimated median of Y (56,988.57), but as pointed out in chapter 2, the mean and median can differ substantially because the mean is sensitive to even a single outlier, while many outliers are needed to influence the median. A similar issue occurs here and indeed the two methods can give substantially different results.

Example 5

Figure 8.1 illustrates that least squares regression and least absolute value regression can give substantially different results. Shown is a scatterplot of astronomical data where the goal is to predict the light intensity of stars based

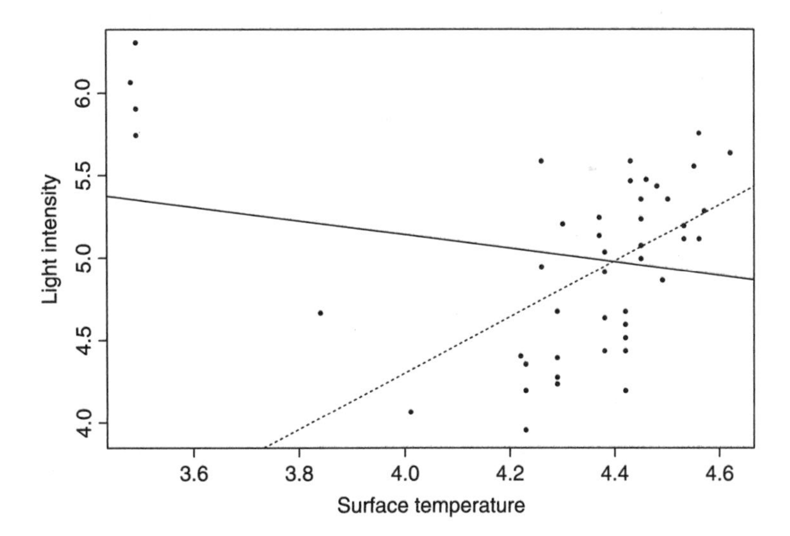

Figure 8.1 A few outliers can drastically alter the least squares regression line. Here, the outliers in the upper left corner result in a regression line having a slightly negative slope. If instead we attempt to estimate the median of Y, rather than the mean, the slope is positive and better reflects the association among the bulk of the points.

on its surface temperature. The solid line with the negative slope is the least squares regression line. The dashed line with the positive slope is the least absolute value regression line.

There are several practical problems that can occur when using least squares regression. One is that a few outliers can render the least squares regression line a poor reflection of the association among the bulk of the points. The last example illustrates this point. It is evident that there is a positive association among the bulk of points in figure 8.1, yet the least squares regression line is negative. The reason is that the four outliers in the upper left corner of figure 8.1 have a tremendous influence on the estimated slope and intercept (just as a single outlier can have a large impact on the mean, as indicated in chapter 2).

A natural reaction is to check for outliers simply by examining a scatterplot, remove any that are found, and then apply least squares regression to the remaining data. In some situations this simple strategy is satisfactory, but in others it is highly unsatisfactory compared to alternative strategies that might be used. (Some of the reasons it is unsatisfactory are given in the final section of this chapter.)

Homoscedasticity

Simply adopting the least squares principle does not address the second general problem that confronts applied researchers: How might confidence intervals for the slope and intercept be computed and how might hypotheses about the slope and intercept be tested? From chapters 5 and 6, a natural guess is that if we assume normality, a solution can be derived. But to get an exact solution, typically an additional assumption is imposed. To describe it, consider again the diabetes study where the goal is to estimate the C-peptide levels based on a child's age. Consider all children who are exactly 7 years of age. Among all these children, there will be some variation among their C-peptide levels. That is, simply knowing that a child is 7 years old does not tells us exactly what their C-peptide level might be; some will have a higher level than others. Similarly, among all 8-year-old children there will be variation among their C-peptide levels, and the same is true for all 9-year-old children as well.

Homoscedasticity, in the context of regression, refers to a situation where the variance of C-peptide levels at age 7 is the same as it is at age 8, or 9, or any age we might pick. If somehow we could measure millions of children at different ages, a plot of the data might look something like figure 8.2 for children who are age 7, 8 or 9. Notice that the variation among the Y values is the same regardless of which age we consider. This is in contrast to *heteroscedasticity*, where the variation differs at two or more age levels. Figure 8.3 illustrates heteroscedasticity.

In a more general and more formal manner, homoscedasticity can be described as follows. Consider any two variables X and Y and let $\text{VAR}(Y|X)$ be the conditional variance of Y, given X. In the diabetes example, Y is C-peptide levels and X is age, so $\text{VAR}(Y|8)$, for example, represents the variance of C-peptide levels among all 8-year-old children. Homoscedasticity means that regardless of what the value for X happens to be, the variance will be the same. Usually, this common (unknown) variance

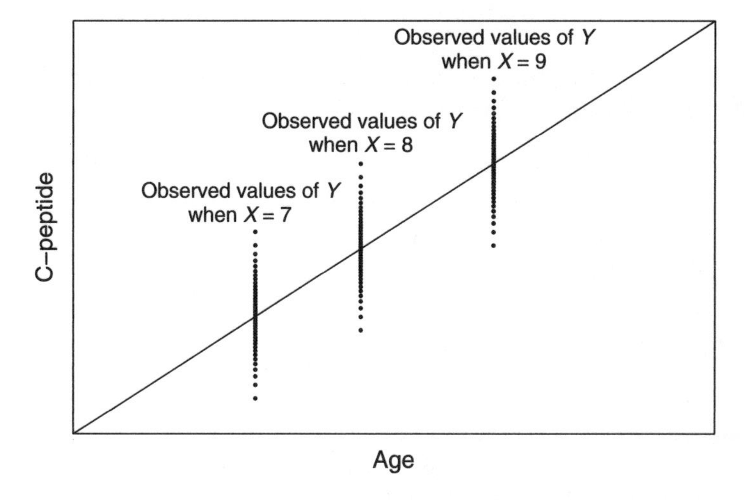

Figure 8.2 An illustration of homoscedasticity: The variation among the Y values does not depends on the value of X.

is represented by σ^2. So homoscedasticity means that regardless of what X we consider, $\text{VAR}(Y|X) = \sigma^2$.

Estimating the assumed common variance

If there is homoscedasticity, how do we estimate the common variance? That is, how do we estimate σ^2? When using the least squares regression the estimate typically used is

$$s_{Y.X}^2 = \frac{1}{n-2} \sum r_i^2. \tag{8.7}$$

That is, estimate the assumed common variance by summing the squared residuals and dividing by $n - 2$ (the number of paired observations minus 2).

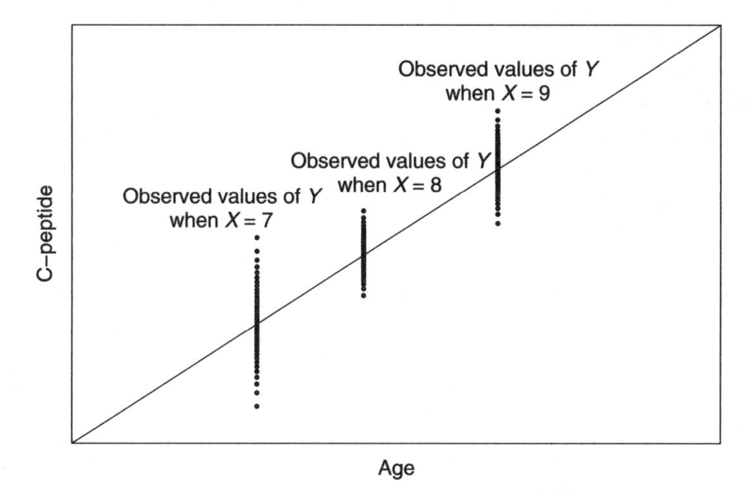

Figure 8.3 An illustration of heteroscedasticity: The variation among the Y values does depends on the value of X.

Example 6

For Boscovich's data in table 8.1, the residuals based on the least squares estimate of the slope and intercept are:

$$13.57393,\ 83.48236,\ -94.68105,\ -80.27228,\ 77.89704.$$

Squaring each of these values and adding the results yields $\sum r_i^2 = 28{,}629.64$. There are five pairs of observations ($n = 5$), so the estimate of the assumed common variance is $s_{Y.X}^2 = 28{,}629.64/3 = 9{,}543.21$.

Extrapolation can be dangerous

Care must be taken when making predictions about Y when using some value for X that lies outside the range of values used to obtain the slope and intercept. If you want to make a prediction about Y using some value for X that is less than any value used to compute the least squares slope and intercept, highly inaccurate results might be obtained. In a similar manner, making predictions for X larger than any value used to compute b_1 and b_0, again highly inaccurate results might result.

Example 7

In example 2 (this section), the sizes of homes range between 688 and 4,870 square feet. What would you estimate the cost of a lot to be with no home on it? From example 2, the least squares regression line is $\hat{Y} = 215(X) + 38{,}192$. An empty lot corresponds to $X = 0$ square feet. The temptation, based on the least squares regression line, might be to estimate the cost of the lot to be $\hat{Y} = 215(0) + 38{,}192 = 38{,}192$. This is an absurd result, however, because for this particular suburb, it is impossible to find any lot this inexpensive.

Comments about assuming the regression line is straight

Caution must be exercised when assuming that a regression line is straight. Close examination of data often reveals that over some intervals of the X values, a linear association between X and Y is a reasonable approximation. That is, a straight line provides a reasonable summary of the data. But over larger ranges, often a straight line becomes unsatisfactory. The next example illustrates this point and provides another example of why extrapolation can be dangerous.

Example 8

Vitamin A is necessary for good health. With too little vitamin A in the diet, serious health problems will occur. Imagine you perform a study relating intake of vitamin A to some measure of good health and find a relationship that appears to be approximately linear. For illustrative purposes, suppose the largest amount of vitamin A used in the study is 4,000 units. Is it reasonable to assume that doubling the amount would have positive health benefits? The answer is, not necessarily. Indeed, if we keep increasing the amount of vitamin A, eventually poor health would result. In fact, a sufficiently high dose of vitamin A can cause death. There are two issues here. One is extrapolation because the study is limited to 4,000 units. The other problem is that over short intervals, a linear association between vitamin A and health is reasonable. But over a sufficiently wide range of vitamin A intake, the association becomes nonlinear.

Problems

1. For the following pairs of points, verify that the least square regression line is $\hat{Y} = 1.8X - 8.5$.

$$X: 5, 8, 9, 7, 14$$

$$Y: 3, 1, 6, 7, 19.$$

2. Compute the residuals using the results from problem 1. Verify that if you square and sum the residuals, you get 47, rounding to the nearest integer.
3. Verify that for the data in problem 1, if you use $\hat{Y} = 2X - 9$, the sum of the squared residuals is larger than 47. Why would you expect a value greater than 47?
4. Suppose that based on $n = 25$ values, $s_x^2 = 12$ and $\sum(X_i - \bar{X})(Y_i - \bar{Y}) = 144$. What is the slope of least squares regression?
5. The following table reports breast cancer rates plus levels of solar radiation (in calories per day) for various cities in the United States. Fit a least squares regression to the data with the goal of predicting cancer rates and comment on what this line suggests.

City	Rate	Daily calories	City	Rate	Daily calories
New York	32.75	300	Chicago	30.75	275
Pittsburgh	28.00	280	Seattle	27.25	270
Boston	30.75	305	Cleveland	31.00	335
Columbus	29.00	340	Indianapolis	26.50	342
New Orleans	27.00	348	Nashville	23.50	354
Washington, DC	31.20	357	Salt Lake City	22.70	394
Omaha	27.00	380	San Diego	25.80	383
Atlanta	27.00	397	Los Angeles	27.80	450
Miami	23.50	453	Fort Worth	21.50	446
Tampa	21.00	456	Albuquerque	22.50	513
Las Vegas	21.50	510	Honolulu	20.60	520
El Paso	22.80	535	Phoenix	21.00	520

6. For the following data, compute the least squares regression line for predicting gpa given SAT.

SAT:	500	530	590	660	610	700	570	640
gpa:	2.3	3.1	2.6	3.0	2.4	3.3	2.6	3.5

7. Compute the residuals for the data used in the previous problem and verify that they sum to zero.

8. For the following data, compute the least squares regression line for predicting Y from X.

X:	40	41	42	43	44	45	46
Y:	1.62	1.63	1.90	2.64	2.05	2.13	1.94

9. In problem 5, what would be the least squares estimate of the cancer rate given a solar radiation of 600? Indicate why this estimate might be unreasonable.

10. Maximal oxygen uptake (mou) is a measure of an individual's physical fitness. You want to know how mou is related to how fast someone can run a mile. Suppose you randomly sample six athletes and get

    mou (milliliters/kilogram): 63.3 60.1 53.6 58.8 67.5 62.5

    time (seconds): 241.5 249.8 246.1 232.4 237.2 238.4

    Compute the least squares regression line and comment on what the results suggest.

11. Verify that for the following pairs of points, the least squares regression line has a slope of zero. Plot the points and comment on the assumption of that the regression line is straight.

    $X : 1\ 2\ 3\ 4\ 5\ 6$

    $Y : 1\ 4\ 7\ 7\ 4\ 1.$

12. Repeat the last problem, only for the points

    $X : 1\ 2\ 3\ 4\ 5\ 6$

    $Y : 4\ 5\ 6\ 7\ 8\ 2.$

13. Vitamin A is required for good health. However, one bite of polar bear liver results in death because it contains a high concentration of vitamin A. Comment on this fact in terms of extrapolation.

14. Sockett et al. (1987) report data related to patterns of residual insulin secretion in children. A portion of the study was concerned with whether age can be used to predict the logarithm of C-peptide concentrations at diagnosis. The observed values are

    Age (X): 5.2 8.8 10.5 10.6 10.4 1.8 12.7 15.6 5.8 1.9 2.2 4.8 7.9 5.2 0.9 11.8 7.9 1.5 10.6 8.5 11.1 12.8 11.3 1.0 14.5 11.9 8.1 13.8 15.5 9.8 11.0 12.4 11.1 5.1 4.8 4.2 6.9 13.2 9.9 12.5 13.2 8.9 10.8

    C-peptide (Y): 4.8 4.1 5.2 5.5 5.0 3.4 3.4 4.9 5.6 3.7 3.9 4.5 4.8 4.9 3.0 4.6 4.8 5.5 4.5 5.3 4.7 6.6 5.1 3.9 5.7 5.1 5.2 3.7 4.9 4.8 4.4 5.2 5.1 4.6 3.9 5.1 5.1 6.0 4.9 4.1 4.6 4.9 5.1

    Create a scatterplot for these data and consider whether a linear rule for predicting Y with X is reasonable.

15. For the data in the last problem, use a computer to verify that a least squares regression line using only X values (age) less than 7 yields $b_1 = 0.247$ and $b_0 = 3.51$. Verify that when using only the X values great than 7 you get $b_1 = .009$

and $b_0 = 4.8$. What does this suggest about using a linear rule for all of the data?

16. For the data in table 8.3, the sizes of the corresponding lots are:

    18,200 12,900 10,060 14,500 76,670 22,800 10,880 10,880 23,090 10,875 3,498 42,689 17,790 38,330 18,460 17,000 15,710 14,180 19,840 9,150 40,511 9,060 15,038 5,807 16,000 3,173 24,000 16,600.

    Use a computer to verify that the least squares regression line for estimating the selling price, based on the size of the lot, is $\hat{Y} = 11X + 436{,}834$.

8.2 Inferences about the slope and intercept

Imagine that the least squares estimate of the slope (b_1) and intercept (b_0) are computed as described in the previous section. In general, these estimates will differ from the true (population) values, roughly meaning the values for the slope and intercept we would get if all individuals of interest could be measured. Again label the true slope and intercept as β_1 and β_0, respectively. In previous chapters we saw how we might make inferences about the population mean μ based on the data available to us. In particular, methods for computing confidence intervals and testing hypotheses were described. The goal in this section is to extend these methods so as to be able to test hypotheses and compute confidence intervals when dealing with the slope and intercept of a regression line. For example, based on the observations available to us, is it reasonable to rule out the possibility that the true (population) slope, β_1, is equal to zero? Is it reasonable to conclude it is at least 1.2?

Classic assumptions

The conventional and most commonly used method for making inferences about the slope and intercept assumes that the Y values corresponding to any X have a normal distribution. In the diabetes study, for example, this means that C-peptide levels among 7-year-old children have a normal distribution, as do the the C-peptide levels among 8-year-old children or any other age we might pick. A second assumption is that there is homoscedasticity. And a third assumption is that if we observe n pairs of points, say $(X_1, Y_1), \ldots, (X_n, Y_n)$, then the Y values form a random sample. So in particular, the $Y_1, \ldots, Y_n$ values are independent.

Estimating the standard errors

With these assumptions in hand, it can be shown that when estimating the slope with the least squares estimator b_1 given by equation (8.5), the squared standard error of b_1 is

$$\frac{\sigma^2}{\sum(X_i - \bar{X})^2} \tag{8.8}$$

That is, b_1 has a sampling distribution just as the sample mean has a sampling distribution as described in chapter 5. So equation (8.8) says that if we were to repeat an experiment millions of time (and in fact infinitely many times), resulting in millions of estimates of the slope, the variance among these slopes would be given by equation (8.8). In practice, we do not know σ^2, but we can estimate σ^2 with $s^2_{Y.X}$, in which case the squared standard error of b_1 is estimated with

$$\frac{s^2_{Y.X}}{\sum(X_i - \bar{X})^2} \tag{8.9}$$

Computing a confidence interval for the slope and intercept

When there is homoscedasticity, random sampling, and when for any X, the corresponding Y values have a normal distribution, exact confidence intervals for the slope (β_1) and intercept (β_0) can be computed using Student's t-distribution introduced in chapter 5. Now the degrees of freedom are $\nu = n - 2$, the confidence interval for the slope is

$$b_1 \pm t\sqrt{\frac{s^2_{Y.X}}{\sum(X_i - \bar{X})^2}}, \tag{8.10}$$

where t is the $1 - \alpha/2$ quantile of Student's t-distribution with $\nu = n - 2$ degrees of freedom. (The value of t is read from table 4 in appendix B.) The confidence interval for the intercept is

$$b_0 \pm t\sqrt{\frac{s^2_{Y.X}\sum X_i^2}{n\sum(X_i - \bar{X})^2}}. \tag{8.11}$$

Testing hypotheses

One way of testing

$$H_0 : \beta_1 = 0, \tag{8.12}$$

the hypothesis that the slope is zero, is to compute a confidence interval for β_1 using equation (8.10) and reject if this interval does not contain the hypothesized value, 0. Alternatively, you can compute

$$T = b_1\sqrt{\frac{\sum(X_i - \bar{X})^2}{s^2_{Y.X}}} \tag{8.13}$$

and reject if

$$|T| \geq t,$$

where again t is the $1-\alpha/2$ quantile of Student's t-distribution with $\nu = n-2$ degrees of freedom. Box 8.1 summarizes the assumptions and computations.

BOX 8.1 Computing confidence intervals and testing hypotheses using least squares regression.

Assumptions

- For any X value, the corresponding Y values have a normal distribution with mean $\beta_0 + \beta_1 X$.
- There is homoscedasticity.
- For the n pairs of observations you observe, $(X_1, Y_1), \ldots, (X_n, Y_n)$, the Y values are independent and represent a random sample.

Confidence Intervals

The $1-\alpha$ confidence interval for the slope is

$$b_1 \pm t\sqrt{\frac{s^2_{Y.X}}{\sum(X_i - \bar{X})^2}}.$$

where t is the $1-\alpha/2$ quantile of Student's t-distribution having $\nu = n-2$ degrees of freedom and read from table 4 in appendix B. The $1-\alpha$ confidence interval for β_0 is

$$b_0 \pm t\sqrt{\frac{s^2_{Y.X}\sum X_i^2}{n\sum(X_i - \bar{X})^2}}.$$

Hypothesis Testing

Reject $H_0 : \beta_1 = 0$ if $|T| > t$ where

$$T = b_1\sqrt{\frac{\sum(X_i - \bar{X})^2}{s^2_{Y.X}}}.$$

Reject $H_0 : \beta_0 = 0$ if $|T| > t$ where now

$$T = b_0\sqrt{\frac{n\sum(X_i - \bar{X})^2}{s^2_{Y.X}\sum X_i^2}}.$$

Example 1

A general goal of a study conducted by G. Margolin and A. Medina was to examine how children's information processing is related to a history of exposure to marital aggression. Results for two of the measures considered are shown in table 8.4. The first, labeled X, is a measure of marital aggression that reflects physical, verbal and emotional aggression during the last year, and Y is a child's score on a recall test. If aggression in the home (X) has a relatively low value, what would we expect a child to score on the recall test (Y)? If the measure of aggression is high, now what would we expect the recall test score to be? As aggression increases, do test scores tend to decrease? Can we be reasonably certain that there is some association between marital aggression and scores on the recall test?

Table 8.4 Measures of marital aggression and recall test scores

Family i	Aggression X_i	Test Score Y_i	Family i	Aggression X_i	Test Score Y_i
1	3	0	25	34	2
2	104	5	26	14	0
3	50	0	27	9	4
4	9	0	28	28	0
5	68	0	29	7	4
6	29	6	30	11	6
7	74	0	31	21	4
8	11	1	32	30	4
9	18	1	33	26	1
10	39	2	34	2	6
11	0	17	35	11	6
12	56	0	36	12	13
13	54	3	37	6	3
14	77	6	38	3	1
15	14	4	39	3	0
16	32	2	40	47	3
17	34	4	41	19	1
18	13	2	42	2	6
19	96	0	43	25	1
20	84	0	44	37	0
21	5	13	57	11	2
22	4	9	46	14	11
23	18	1	47	0	3
24	76	4			

If there is no association between aggression measures and recall test scores, then the slope of the least squares regression line is zero. That is, the hypothesis H_0: $\beta_1 = 0$ is true. Suppose the goal is to test this hypothesis so that the probability of a Type I error is $\alpha = .05$. Then $1-\alpha/2 = .975$. There are $n = 47$ pairs of observations, so the degrees of freedom are $\nu = 47 - 2 = 45$, and the critical value (read from table 4) is $t = 2.01$. The least squares estimate of the slope is $b_1 = -0.0405$, and it can be seen that $\sum(X_1 - \bar{X})^2 = 34{,}659.74$ and that the estimate of the assumed common variance is $s^2_{Y.X} = 14.15$, so the test statistic is

$$T = -0.0405\sqrt{\frac{34{,}659.74}{14.5}} = -1.98.$$

Because $|T| = 1.98 < 2.01$, fail to reject.

Interpreting standard computer output

A practical matter is learning how to read the output of commonly used software. The following examples are intended to help achieve this goal.

Here is a portion of the output you might encounter when using Minitab:

Predictor	Coef	Stdev.Coef	t-ratio
Constant	22.47	10.22	2.20
C1	0.7546	0.1417	5.32

In the first column we see Constant and C1. Constant refers to the intercept, which in our notation is β_0. C1 is the name of the minitab variable that happens to contain the data for the predictor, X. In the next column, headed by Coef, you see 22.47. This is the estimate of the intercept. That is, $b_0 = 22.47$. Under 22.47 you see 0.7546, which is b_1, the estimate of the slope. The next column gives the corresponding estimates of the standard errors. Finally, the last column, headed by t-ratio, gives the values of the test statistic, T, used to test $H_0 : \beta_0 = 0$ and $H_0 : \beta_1 = 0$, respectively. In the example, the test statistic for $H_0 : \beta_1 = 0$ is $T = 5.32$.

SPSS reports the results just described in the following manner:

Variable	B	SE B	T	Sig T
X	0.7546	0.1417	5.32	0.000
Constant	22.47	10.22	2.20	0.033

Again the term constant in the first column refers to the intercept, and here X refers to the predictor. (In reality, you would not see X in this column, but rather the name of the SPSS variable containing the data used to predict Y.) The column headed by B reports the least squares estimates, and the values under SE B are the standard errors. The next column gives the value of the test statistics used to test $H_0 : \beta_1 = 0$ and $H_0 : \beta_0 = 0$. The final column gives the p-values associated with the test statistics. The first p-value is 0.000 indicating that you would reject $H_0 : \beta_1 = 0$ even with α values less than 0.001. If, for example, you want the Type I error to be .01, you would reject $H_0 : \beta_1 = 0$. As for $H_0 : \beta_0 = 0$, the significance level is .033, meaning that the smallest α value for which you would reject is .033. Thus, you would reject if you want the Type I error probability to be .05, but you would not reject if you wanted the Type I error probability to be .01.

Least squares multiple regression

The focus has been on situations where there is a single predictor, but often there is interest in taking into account two or more predictors. For example, how is the the average fuel consumption of a car related to its weight and horsepower? So here we have two predictors (weight and horsepower), what are often called *explanatory variables* or *independent variables*, as opposed to the quantity to be predicted (average fuel consumption), which is often called the *dependent variable* or *outcome variable*.[3] What happens to this association if, in addition to weight and horsepower, we also consider the speed at which the car is driven? How are grades in law school related to undergraduate grade-point average and scores on the LSAT (law school admission test)? What is the association among the heights of college students and their parents?

The classic approach when addressing these questions is based on an extension of the least squares regression method already covered. The usual assumption is that if we are told the values for the predictors, the mean of Y is given by a linear combination of their values. In the law school example, this means that the average grade point in law school is given by

$$\beta_0 + \beta_1(\text{GPA}) + \beta_2(\text{LSAT}),$$

3. It is usually non-statisticians who all predictors independent variables, a convention that some statisticians prefer not to follow.

where GPA is a student's undergraduate grade-point average, and the goal is to use data to determine values for the intercept (β_0) and the two slopes (β_1 and β_2). More generally and more formally, if we have p predictors, $X_1, \ldots, X_p$, then the expected (mean) value of Y, given $X_1, \ldots, X_p$, is typically assumed to be

$$E(Y|X_1, \ldots, X_p) = \beta_0 + \beta_1 X_1 + \beta_2 X_2 + \cdots + \beta_p X_p. \tag{8.14}$$

The least squares principle can be extended to the situation at hand in a straightforward manner. This just means that if we observe the n vectors of observations

$$(X_{11}, \ldots, X_{1p}, Y_1), \ldots, (X_{n1}, \ldots, X_{np}, Y_n),$$

the unknown values of $\beta_0, \beta_1, \cdots, \beta_p$ are estimated with the values $b_0, b_1, \cdots, b_p$ that minimize

$$\sum (Y_i - b_0 - b_1 X_{i1} \cdots - b_p X_{ip})^2, \tag{8.15}$$

respectively. The tedious calculations for determining the values $b_0, b_1, \cdots, b_p$ in this manner can be performed by all of the software packages mentioned in chapter 1.
(For example, when using R or S-PLUS, the function lsfit can be used.) The computational details are not important for present purposes, so they are not described.

Hypothesis testing

A common goal is to test the hypothesis that all of the slope parameters are zero. In symbols, the goal is to test

$$H_0 : \beta_1 = \cdots = \beta_p = 0. \tag{8.16}$$

The classic strategy for accomplishing this goal is based on assumptions similar to the single predictor case ($p = 1$). In particular, regardless of the value of the predictor variables, $X_1, \ldots, X_p$, the corresponding Y values are assumed to have a normal distribution. (In more formal terms, the conditional distribution of Y, given $X_1, \ldots, X_p$, is normal.) An additional assumption is homoscedasticity. That is, the variance of the Y values does not depend on what the value of predictor variables happens to be. Let $\hat{Y} = b_0 + b_1 X_1 + \cdots + b_p X_p$ be the least squares regression line. That is, the values $b_0, \ldots, b_p$ minimize $\sum (Y_i - \hat{Y}_i)^2$, the sum of the squared residuals. The *squared multiple correlation coefficient* is

$$R^2 = 1 - \frac{\sum (Y_i - \hat{Y}_i)^2}{\sum (Y_i - \bar{Y})^2}. \tag{8.17}$$

The classic method for testing the hypothesis given by Equation (8.16) is based on the test statistic

$$F = \left(\frac{n - p - 1}{p} \right) \left(\frac{R^2}{1 - R^2} \right). \tag{8.18}$$

Under normality and homoscedasticity, F has what is called an F distribution with $\nu_1 = p$ and $\nu_2 = n - p - 1$ degrees of freedom. For situation at hand, the null hypothesis is rejected if $F \geq f_{1-\alpha}$, the $1 - \alpha$ quantile of an F distribution with ν_1 and ν_2 degrees of freedom. If the Type I error is to be .1, .05, .025, or .01, the value of $f_{1-\alpha}$ can read from tables 5, 6, 7 and 8, respectively, in appendix B. For example, with $\alpha = .05$, $\nu_1 = 3$, $\nu_2 = 40$, table 6 indicates that the .95 quantile is $f_{.95} = 2.84$. That is, there is a .05

probability of getting an F value greater than 2.84 when in fact the null hypothesis is true and the underlying assumptions are true as well. For $\alpha = .01$, table 8 says that the .99 quantile is 4.31. This means that if you reject when $F \geq 4.31$, the probability of a Type I error will be .01, assuming random sampling, normality and homoscedasticity. All of the commercial software mentioned in chapter 1 contain built-in routines for computing F and determining an appropriate critical value.

Example 2

Imagine a study aimed at predicting depression in young adults. Suppose there are three predictors and based on a sample of 63 participants, $R^2 = .3$, and it is desired to test the hypothesis that all of the slope parameters are zero. Then $n = 63, p = 3$, so

$$F = \left(\frac{63 - 3 - 1}{3}\right)\left(\frac{.3}{1 - .3}\right) = 8.4,$$

the degress of freedom are $\nu_1 = 3$, $\nu_2 = 60$, from table 8 in appendix B, $f_{.95} = 4.13$, and because $8.4 > 4.13$, reject and conclude that not all of the slopes are equal to zero.

Problems

17. Given that $b_1 = -1.5$, $n = 10$, $s^2_{Y.X} = 35$ and $\sum(X_i - \bar{X})^2 = 140$, assume normality and homoscedasticity and find a .95 confidence interval for β_1.
18. Repeat the previous problem, only find a .98 confidence interval.
19. Based on results covered in previous chapters, speculate about why the confidence intervals computed in the the last two problems might be inaccurate. (Comments relevant to this issue will be covered in section 8.4.
20. Assume normality and homoscedasticity and suppose $n = 30$, $\sum X_1 = 15$, $\sum Y_i = 30$, $\sum(X_1 - \bar{X})(Y_i - \bar{Y}) = 30$, and $\sum(X_i - \bar{X})^2 = 10$. Determine the least squares estimates of the slope and intercept.
21. Assume normality and homoscedasticity and suppose $n = 38$, $\bar{Y} = 20$, $\sum X_i^2 = 1922$, $\sum(X_1 - \bar{X})(Y_i - \bar{Y}) = 180$, $\sum(X_i - \bar{X})^2 = 60$ and $s^2_{Y.X} = 121$.

    (a) Determine the least squares estimates of the slope and intercept.
    (b) Test the hypothesis H_0: $\beta_0 = 0$ with $\alpha = .02$
    (c) Compute a .9 confidence interval for β_1.

22. Assume normality and homoscedasticity and suppose $n = 41$, $\bar{Y} = 10$, $\bar{X} = 12$, $\sum(X_1 - \bar{X})(Y_i - \bar{Y}) = 100$, $\sum(X_i - \bar{X})^2 = 400$ and $s^2_{Y.X} = 144$.

    (a) Determine the least squares estimates of the slope and intercept.
    (b) Compute a .9 confidence interval for β_1.

23. Assume normality and homoscedasticity and suppose $n = 18$, $b_1 = 3.1$, $\sum(X_i - \bar{X})^2 = 144$ and $s^2_{Y.X} = 36$. Compute a .95 confidence interval for β_1. Would you conclude that $\beta_1 > 2$?
24. Assume normality and homoscedasticity and suppose $n = 20$, $b_0 = 6$, $\sum X_i^2 = 169$, $s^2_{Y.X} = 25$ and $\sum(X_i - \bar{X})^2 = 90$. Compute a .95 confidence interval for β_0.

8.3 Correlation

The so-called product moment correlation coefficient is

$$r = b_1 \frac{s_x}{s_y}, \tag{8.19}$$

where b_1 is the least squares estimate of the regression slope, and s_x and s_y are the sample standard deviations associated with the X and Y variables, respectively. An alternative way of computing r is

$$r = \frac{A}{\sqrt{CD}}, \tag{8.20}$$

where

$$A = \sum(X_i - \bar{X})(Y_i - \bar{Y}),$$

$$C = \sum(X_i - \bar{X})^2$$

and

$$D = \sum(Y_i - \bar{Y})^2.$$

The value of r is commonly used to summarize the association between two variables and dates back to at least the year 1846 when it was studied by A. Bravais. During the late nineteenth century, Galton made a considerable effort at applying the correlation coefficient. And about this time, Pearson solved some technical problems and provided a mathematical account that helped sway many natural scientists and mathematicians that it has value in the analysis of biological observations. Due to Pearson's efforts, r is often called *Pearson's correlation.*

The coefficient of determination

There is yet one more way to relate Pearson's correlation to least squares regression. Note that if for any individual we do not know X, a natural guess about the value Y for this individual is the average of the Y values, $\bar{Y}$. In the diabetes data, for example, if we were given no information about a child's age or C-peptide level, and we want to guess what the child's C-peptide level happens to be, a reasonable guess would be $\bar{Y}$, the average of all the C-peptide levels available to us. One way of measuring the accuracy of this estimate among the n individuals under study is with $\sum(Y_i - \bar{Y})^2$, the sum of squared differences between the observed Y values and the mean, $\bar{Y}$. Recall that this sum is the numerator of the sample variance of the Y values. In a similar manner, $\sum(Y_i - \hat{Y}_i)^2$ reflects how well we can predict Y using the least squares regression line and X. So the improvement in estimating Y with $\hat{Y}$ over $\bar{Y}$ is the difference between these two sums:

$$\sum(Y_i - \bar{Y})^2 - \sum(Y_i - \hat{Y}_i)^2.$$

So the reduction in error using $\hat{Y}$, relative to using $\bar{Y}$, is

$$r^2 = \frac{\sum(Y_i - \bar{Y})^2 - \sum(Y_i - \hat{Y}_i)^2}{\sum(Y_i - \bar{Y})^2} \tag{8.21}$$

the squared value of Pearson's correlation. The quantity r^2, is called the *coefficient of determination*, which reflects the proportion of variance accounted for using a least squares regression line and X to predict Y. In the diabetes study, for example, $r^2 = .151$ meaning that when using a straight regression line to predict C-peptide levels, given a child's age, 15% of the variation among C-peptide levels is accounted for based on a child's age. Because r has a value between -1 and 1, the coefficient of determination has a value between 0 and 1.

Here is another way of describing the coefficient of determination. The variance of the C-peptide levels is .519. The variance of the residuals is .441. So the fraction of the variance *not* accounted for is $.441/.519 = .850$, or 85%. The fraction that is accounted for is 100%−85%=15%, which is the value of r^2.

Yet another way of describing the coefficient of determination is that it reflects the variation in the predicted Y values (the $\hat{Y}$ values), based on the least squares regression line, relative to the variation in the Y values. In the diabetes example, it can be seen that the variance of the predicted C-peptide levels values is 0.0783. The variance in the observed C-peptide levels is 0.5192, so the proportion of variance accounted for is $0.0783/0.5192 = .15$, which is again equal to r^2.

The population correlation ρ, and its basic properties

Before elaborating on the use and interpretation of r, it is noted that there is a population analog of r, typically written as ρ, where ρ is a lower case Greek rho. Roughly, ρ is the value of r if all individuals of interest could be measured. Usually, not all individuals can be measured, so we estimate ρ with r (in much the same way we estimate the population mean μ with the sample mean $\bar{X}$). It can be shown that the population product moment correlation coefficient, ρ, has a value between -1 and 1. More succinctly,

$$-1 \leq \rho \leq 1.$$

A *fundamental property* of the population correlation coefficient is that if X and Y are independent, then $\rho = 0$. This property is of practical importance because if persuasive empirical evidence, based on r, indicates that $\rho \neq 0$, then it is reasonable to conclude that X and Y are dependent.

It is not necessarily true, however, that if $\rho = 0$, then X and Y are independent. There are, in fact, a variety of ways in which ρ can have a value at or close to zero even when there is, in some sense, a strong association. For example, if $Y = X^2$, there is an exact association between X and Y, yet $\rho = 0$. This illustrates the general result that curvature (simply meaning that the regression line is not straight) can affect ρ. But even when there is a linear association between X and Y, yet $\rho = 0$, it is possible to have dependence. In the context of least squares regression, if there is heteroscedasticity, yet the slope of the regression is zero, then $\rho = 0$, yet knowing X provides information about the possible values for Y.

Testing the hypothesis of a zero correlation

The classic method for testing

$$H_0 : \rho = 0, \tag{8.22}$$

the hypothesis that the population value of Pearson's correlation is zero, is based on the assumption that X and Y are independent and that X or Y has a normal distribution.

The test statistic is

$$T = r\sqrt{\frac{n-2}{1-r^2}}. \tag{8.23}$$

If the null hypothesis is true and the assumptions are met, then T has a Student's t-distribution with $n-2$ degrees of freedom. That is, if the goal is to have the probability of a Type I error equal to α, reject if $|T| \geq t$, where t is the $1-\alpha/2$ quantile of Student's t-distribution, which is read from table 4 in appendix B.

Example 1

For the aggression data in table 8.3, $n = 47$, $r = -0.286$, so $\nu = 45$ and

$$T = -0.286\sqrt{\frac{45}{1-(-0.286)^2}} = -2.$$

With $\alpha = .05$, $1-\alpha/2 = .975$, and from table 4 in appendix B, the critical value is $t = 2.01$, and because $|-2| < 2.01$, we fail to reject. That is, we are unable to conclude that the aggression scores and recall test scores are dependent with $\alpha = .05$.

The intelligent use and interpretation of the hypothesis testing method just described requires a closer look at the underlying assumptions. It can be shown that if two variables are independent, then there is homoscedasticity. Moreover, the assumption of homoscedasticity plays a crucial role in the mathematical derivation of the test statistic T. If there is heteroscedasticity, then the test statistic T is using the wrong standard error, which can have practical implications when interpreting data. For example, it is possible to have $\rho = 0$, yet the probability of rejecting increases as the sample size gets large. That is, regardless how large the sample size might be, you do not control the probability of a Type I error when using T to test the hypothesis that $\rho = 0$. However, if X and Y are independent, the correct standard error is being used and now the probability of a Type I error can be controlled reasonably well. So, in a very real sense, the test statistic T is best described as a test of the hypothesis that X and Y are independent, rather than a test of the hypothesis that $\rho = 0$. Using the wrong standard error also can affect power, the probability of detecting dependence when it exists.

Interpreting r

A good understanding of r requires an understanding not only of what it tells us about an association, but also what it does not tell us. That is, care must be used not to read more into the value of r than is warranted.

Interpreting r is complicated by the fact that various features of the data under study affect its magnitude. Five such features are described here.

Assuming that there is a linear association between and X and Y, the first feature is the magnitude of the residuals. Generally, large residuals tend to result in low values for r. The left panel of figure 8.4 shows a scatterplot of points with $r = .92$. The right panel shows another scatterplot of points, which are centered around the same line as in the left panel, only they are farther from the line. Now $r = .42$.

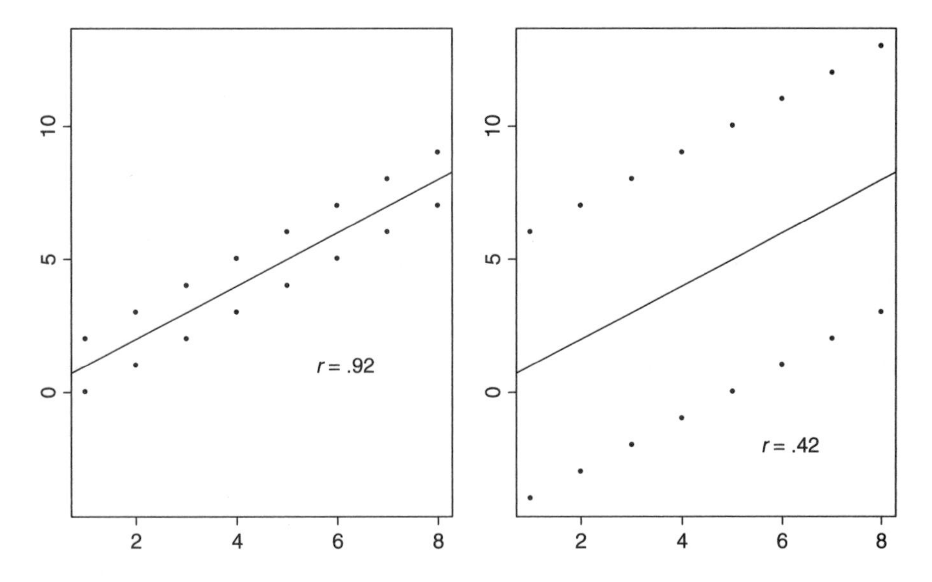

Figure 8.4 An illustration that the magnitude of the Pearson's correlation is influenced by the magnitude of the residuals.

A second feature that affects the magnitude of r is the magnitude of the slope around which the points are centered (e.g., Barrett, 1974; Loh, 1987). The closer the slope happens to be to zero, the lower will be the value of r.

A third feature of data that affects r is outliers. For the star data in figure 8.1, $r = -.21$, which is consistent with the negative slope associated with the least squares regression line. But we have already seen that for the bulk of the points, there is a positive association. Generally, a single unusual value can have a tremendous impact on the value of r resulting in a poor reflection of the association among the majority of points. A result is that regardless of how many observations we might have, slight departures from normality can substantially alter r.

A fourth is restricting the range of the X or Y values. For the star data in figure 8.1, $r = -.21$. If we restrict the range of the X values by considering only X values greater than 3.6, which eliminates the obvious outliers, now $r = .61$. Restricting the range of X or Y can lower r as well.

A fifth feature, already mentioned, is curvature.

In summary, the following features of data influence the magnitude of Pearson's correlation:

- The slope of the line around which points are clustered
- The magnitude of the residuals
- Outliers
- Restricting the range of the X values, which can cause r to go up or down
- Curvature.

Use caution when rejecting the hypothesis that Pearson's correlation is equal to zero

If the hypothesis that $\rho = 0$ is rejected, a common interpretation is that if r is greater than 0, than generally, as X increases, Y increases as well. Similarly, if $r < 0$, a natural speculation is that as X increases, Y decreases. This interpretation is consistent with the least squares regression line because the slope is given by

$$b_1 = r\frac{s_y}{s_x}.$$

So if $r > 0$, the least squares regression line has a positive slope, and if $r < 0$, the reverse is true. *Perhaps this interpretation is usually correct, but this should not be taken for granted.* In figure 8.1, for example, r is negative, but generally, as X increases, Y increases too. Many methods have been derived to get a more exact understanding of how X and Y are related. One simple recommendation is to always plot the data, as well as the least squares regression line, as opposed to relying completely on the value of r to interpret how two variables are related. At a minimum, the least squares regression line should look reasonable when viewed within a plot of the data, but even if it gives a reasonable summary of the data, there can be a considerable practical advantage to using an alternative regression estimator for reasons noted in the next section.

There is a connection between the squared multiple correlation coefficient, R^2, given by equation (8.17), and Pearson's correlation that should be mentioned. R^2 can be seen to be the squared Pearson correlation between the observed Y values (the Y_i values) and the predicted Y values based on the least squares regression line (the $\hat{Y}_i$ values).

A summary of how to use r

To summarize, Pearson's correlation, r, has two useful functions. First, it can be used to establish dependence between two variables by testing and rejecting the hypothesis that ρ is equal to zero. Second, r^2 (the coefficient of determination), reflects the extent to which the least squares regression estimate of Y, namely $\hat{Y}$, improves upon the sample mean, $\bar{Y}$, in terms of predicting Y.

However, even if r^2 is close to one, this does not necessarily mean that the least squares estimate of Y is performing well. It might be, for example, that both $\hat{Y}$ and $\bar{Y}$ perform poorly. And even when Pearson's correlation is very close to one, this does not necessarily mean that the least squares regression line provides a highly accurate estimate of Y, given X. Finally, Pearson's correlation might be relatively ineffective at detecting dependence compared to more modern methods that are now available.

Example 2

S. Mednick conducted a study aimed at understanding the association among some variables related to schizophrenia. With two of the variables he got a Pearson correlation very close to 1. Momentarily he thought he had a major breakthrough. Fortunately, he checked a scatterplot of his data and found an obvious outlier. This one outlier resulted in r being close to one, but with the outlier removed, no association was found.

Establishing independence

A point worth stressing is that rejecting the hypothesis that $\rho = 0$ provides a good empirical argument that there is dependence, but failing to reject is not a compelling reason to conclude that two variables are independent. More modern methods can be sensitive to types of dependence that are difficult to discover when using Pearson's correlation only.

Problems

25. Given the following quantities, find the sample correlation coefficient, r, and test $H_0: \rho = 0$ at the indicated level.

    (a) $n = 27$, $\sum(Y_i - \bar{Y})^2 = 100$, $\sum(X_i - \bar{X})^2 = 625$, $\sum(X_i - \bar{X})(Y_i - \bar{Y}) = 200$, $\alpha = .01$.
    (b) $n = 5$, $\sum(Y_i - \bar{Y})^2 = 16$, $\sum(X_i - \bar{X})^2 = 25$, $\sum(X_i - \bar{X})(Y_i - \bar{Y}) = 10$, $\alpha = .05$.

26. The high school grade-point average (X) and college grade-point (Y) for 29 randomly sampled college freshman yielded the following results: $\sum(Y_i - \bar{Y})^2 = 64$, $\sum(X_i - \bar{X})^2 = 100$, $\sum(X_i - \bar{X})(Y_i - \bar{Y}) = 40$. Test $H_0: \rho = 0$ at the .1 level and interpret the results.

27. For the previous problem, answer the following questions.

    (a) Is it reasonable to conclude that the least squares regression line has a positive slope?
    (b) Is it possible that despite the value for r, as high school grade-point averages increase, college grade-point averages decrease? Explain your answer.
    (c) What might you do, beyond considering r, to decide whether it is reasonable to conclude that as high school grade-point averages increase, college grade-point averages increase as well?

28. Using a computer, determine what happens to the correlation between X and Y if the Y values are multiplied by 3.

29. Repeat the previous problem, only determine what happens to the slope of the least squares regression line.

30. Consider a least squares regression line $Y = .5X + 2 + e$, and where X and e are independent and both have a standard normal distribution. What happens to the correlation between X and Y if instead $Y = .5X + 2 + 2e$? Hint: What happens to the residuals?

31. The numerator of the coefficient of determination is $\sum(Y_i - \bar{Y})^2 - \sum(Y_i - \hat{Y}_i)^2$. Based on the least squares principle, why is this value always greater than or equal to zero?

32. Imagine a study where the correlation between some amount of an experimental drug and liver damage yields a value for r close to zero and the hypothesis H_0: $\rho = 0$ is not rejected. Why might it be unreasonable to conclude that the two variables under study are independent?

33. Suppose $r^2 = .95$.

    (a) Explain why this does not provide convincing evidence that the least squares line provides a good fit to a scatterplot of the points.

(b) If the least squares line provides a poor fit, what does this say about using $\hat{Y}$ versus $\bar{Y}$ to estimate Y.

34. Imagine a situation where points are removed for which the X values are judged to be outliers. Note that this restricts the range of X values. Without looking at the data, can you predict whether Pearson's correlation will increase or decrease after these points are removed?

8.4 Modern advances and insights

Least squares regression and the conventional method for testing hypotheses, outlined in box 8.1, reflect a major advance in our attempts to study associations. In some situations they serve us well, but the reality is that serious practical problems can occur. The goal here is to outline these problems and then comment briefly on how they might be addressed.

Consider a situation where the normality assumption is valid but the assumption of a common (conditional) variance is not. That is, there is heteroscedasticity. For the housing data, for example, imagine that the variation among the selling prices differs depending on how many square feet a house happens to have. For instance, the variation among houses having 1,500 square feet might differ from the variation among homes having 2,000 square feet. Then the standard method for testing hypotheses about the slope, described in box 8.1, might provide poor control over the probability of a Type I error and the corresponding confidence interval can be highly inaccurate. If the distributions are not normal, the situation gets worse. In some cases, the actual probability of a Type I error can exceed .5 when testing at the $\alpha = .05$ level. One fundamental reason for practical problems is that when there is heteroscedasticity, the wrong standard error is being used in box 8.1.

There are methods for testing the hypothesis that all of the conditional variances have a common value, but it is unknown how to tell whether any of these tests have enough power to detect situations where the assumption of homoscedasticity should be discarded. That is, you might be in a situation where heteroscedasticity is a practical concern, yet methods that test the assumption of homoscedasticity fail to alert you to this problem. Even if the conditional variances are equal, nonnormality remains a serious concern. And the combination of nonnormality and heteroscedasticity can be devastating, even with large sample sizes.

Recall that the Theil-Sen estimate of the slope (introduced at the beginning of this chapter) begins by computing the slope for each pair of observations. The median of all these slopes is used to estimate β_1. This approach has practical value in terms of power, our ability to detect a true association, plus it has practical advantages when trying to deal with outliers and heteroscedasticity. Even when the least squares estimate of the slope is nearly equal to the Theil-Sen estimate, testing hypotheses with the least squares estimator can result in relatively low power. That is, using the Theil-Sen estimator, for example, might substantially increase your probability of detecting a true association. The reason is that when there is heteroscedasticity, the least squares estimator can have a relatively large standard error, even under normality. Outliers among the Y values exacerbate this problem.

Currently, when computing a .95 confidence interval or testing hypotheses about the slope based on the least squares estimator, there are two methods that appear to perform well relative to other technique that have been proposed, but no details are given here.[4] The method in box 8.1, can be highly unsatisfactory, but it is important to know.

Outliers and the least squares estimator

As previously pointed out, one or more outliers can greatly influence the sample mean and variance. The same is true when using least squares regression, as already illustrated by figure 8.1. As previously indicated for the data in figure 8.1, the least squares estimator suggests that as surface temperature increases, light intensity decreases, but it is evident that for the majority of the points, the exact opposite is true. One problem is that the four points in the upper left portion of figure 8.1 are outliers that greatly influence the least squares regression line.

A reasonable suggestion is to eliminate any outliers and examine the least squares regression line for the data that remain. If, for example, we simply eliminate the obvious outliers in the upper left corner of figure 8.1, the least squares estimator gives a reasonable summary of the remaining data. There are, however, several practical issues that need to be considered before using this strategy. One is that in many cases, a good outlier detection method is required. Such methods have been derived (Rousseeuw and Leroy, 1987; Rousseeuw and van Zomeren, 1990; Wilcox, 2003), but they require advanced techniques that go well beyond the scope of this book. A natural strategy is to search for outliers by examining a boxplot of the X values and then doing the same for the Y values. Unfortunately, this simple approach can fail to detect outliers that greatly influence the least squares estimator.

Even when a good outlier detection method is used, if the goal is to test hypotheses about the slope and intercept, or to compute confidence intervals, discarding outliers and applying the method in box 8.1 can be highly unsatisfactory. (Similar problems arise when testing hypotheses about Pearson's correlation.) The problem is that you can get an incorrect estimate of the standard error, even when sample sizes are large. (There are exceptions, but expert advice is suggested before considering this strategy.) More advanced books describe how to deal with this problem.

Example 1

L. Doi conducted a study aimed at finding good predictors of reading ability. A portion of her study considered predicting a measure of the ability to identify words (Y) with a measure of speeded naming for digits (X). The values are shown in table 8.5. A boxplot of the Y values indicates no outliers, and a boxplot of the X values reveals six outliers. Figure 8.5 shows a scatterplot of the points. The points that are outliers according to the boxplot are the six right most points marked with a by a 0. The main point here is that more advanced

4. Among the library of R functions mentioned in chapter 1, the function olshc4 uses the so-called HC4 estimate of the standard error and appears to be a relatively good choice for general use. A modified percentile bootstrap methods also performs well and can be applied with the R function Isfitci.

Table 8.5 Reading data

X:	34 49 49 44 66 48 49 39 54 57 39 65 43 43 44 42 71 40 41 38 42 77 40 38 43 42 36 55 57 57 41 66 69 38 49 51 45 141 133 76 44 40 56 50 75 44 181 45 61 15 23 42 61 146 144 89 71 83 49 43 68 57 60 56 63 136 49 57 64 43 71 38 74 84 75 64 48
Y:	129 107 91 110 104 101 105 125 82 92 104 134 105 95 101 104 105 122 98 104 95 93 105 132 98 112 95 102 72 103 102 102 80 125 93 105 79 125 102 91 58 104 58 129 58 90 108 95 85 84 77 85 82 82 111 58 99 77 102 82 95 95 82 72 93 114 108 95 72 95 68 119 84 75 75 122 127

outlier detection methods indicate that there are five additional outliers, which are also marked by an 0 in figure 8.5.[5]

A Glimpse of some modern methods

One general approach when trying to deal with practical problems associated with least squares regression is to use diagnostic tools. That is, check the data to see whether there are indications that using least squares regression might be unsatisfactory. Two concerns with this strategy are that conventional diagnostic tools do not always catch problems that are a practical concern, and if problems are detected, it can be difficult addressing them in an adequate manner when least squares regression is used to the exclusion of all other methods. In fairness, situations arise where diagnostic methods can save least squares regression, but simultaneously, there are general conditions where this is not the case.

Another general approach is to use a method that performs about as well as least squares when the Y values have a normal distribution and there is homoscedasticity,

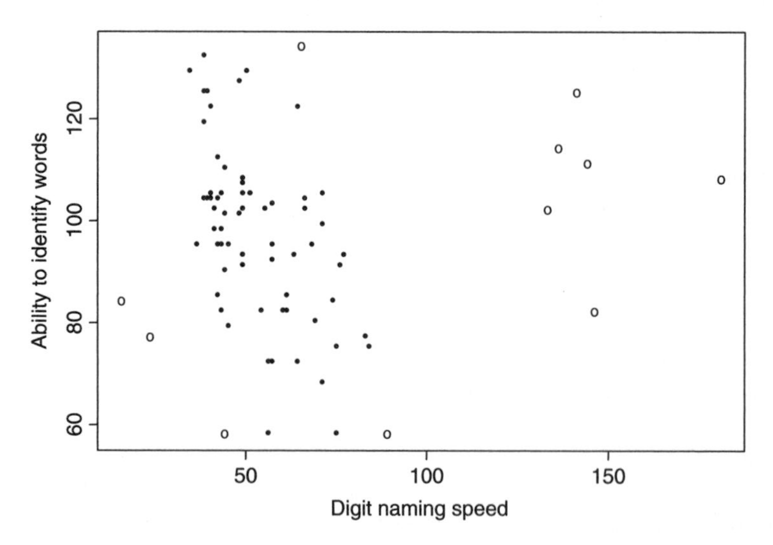

Figure 8.5 The six largest X values are clearly outliers but more advanced methods indicate that there are five additional outliers that might be less obvious.

5. These additional outliers are detected by what is called the minimum volume ellipsoid method.

but which continues to perform well in situations where least squares regression is unsatisfactory. There are a variety of regression estimators designed with this second approach in mind, several of which appear to have considerable practical value. That is, there are situations where the least squares estimator offers a slight advantage, but situations arise where it performs very poorly relative to any one of several alternative methods. One of these is the Theil-Sen estimator mentioned at the beginning of this chapter. It offers protection against the deleterious effects of outliers and it can have a much smaller standard error than the least squares estimator when there is heteroscedasticity or when the Y values have a non-normal distribution. The Theil-Sen estimator does not eliminate all practical problems, and there are methods for improving upon it, but no details are given here. Suffice it to say that in some situations, alternative regression estimators can provide a practical advantage (e.g., Wilcox, 2005).

Currently, the most effective methods for computing confidence intervals are based in part on a variation of what is called a percentile bootstrap method. As mentioned in chapter 6, bootstrap methods offer an alternative to Laplace's strategy for approximating sampling distributions. The most basic version can be used to compute confidence intervals when using the Theil-Sen estimator. Box 8.2 outlines the computations. When using least squares regression, a slight modification of the method in box 8.2 currently seems to be the best technique for general use, but the details are not given here. Unfortunately, commercial software does not come with built-in functions for applying the method, but easy-to-use software is available.[6]

BOX 8.2

Goal

Compute a $1-\alpha$ confidence interval for the slope, β_1, using the Theil-Sen estimator. This is accomplished by approximating the sampling distribution of the Theil-Sen estimator using a particular type of bootstrap method.

You observe n pairs of points, $(X_1, Y_1), \ldots, (X_n, Y_n)$. You obtain a bootstrap sample by randomly sampling with replacement n points from $(X_1, Y_1), \ldots, (X_n, Y_n)$. For this bootstrap sample, compute the Theil-Sen estimate of the slope and label it b^*. Repeat this process B times yielding B bootstrap estimates of the slope: $b_1^*, \ldots, b_B^*$. With $\alpha = .05$, $B = 399$ appears to suffice in terms of achieving an accurate confidence interval. Put these B values in ascending order yielding $b_{(1)}^* \leq \cdots \leq b_{(B)}^*$. Set $L = \alpha B/2$, round L to the nearest integer, and let $U = B - L$. Then a $1-\alpha$ confidence interval for the slope is

$$(b_{(L+1)}^*, b_{(U)}^*).$$

Decision Rule

Reject $H_0 : \beta_1 = 0$ if the confidence interval does not contain zero.

6. See the discussion of the S-PLUS and R functions in chapter 1. The function regci computes confidence intervals when using the Theil-Sen estimator and the function lsfitci is designed for the least squares estimator.

A fundamental issue is whether more modern methods ever make a practical difference versus using the traditional methods described in this chapter. The next example illustrates that the answer is yes.

Example 2

For the reading data in table 8.6, a .95 confidence interval for the slope was computed using the method in box 8.2. It can be seen that the least squares estimates of the slope and intercept are $b_1 = -0.06$ and $b_0 = 65.46$. The .95 confidence interval for the slope, using the method in box 8.1, is $(-0.43, 0.32)$, this interval contains zero, so you would not reject $H_0 : \beta_1 = 0$. (The p-value is 0.76.) So the method most commonly used for making inferences about the slope provides no indication that the true slope differs from zero. Using the method in box 8.2, the .95 confidence interval is $(-0.63, 0.00)$ with a p-value of .035. So now you would reject if you want the probability of a Type I error to be .05. That is, the method in box 8.2 leads to the exact opposite conclusion, the main point being that modern methods can yield substantially different results versus more standard techniques.

Problems

35. If the normality assumption is violated, what effect might this have when computing confidence intervals as described in box 8.1?

36. If the homoscedasticity assumption is violated, what effect might this have when computing confidence intervals as described in box 8.1?

A Summary of Some Key Points

- Least squares regression and Pearson's correlation are the most commonly used methods for studying associations. In some cases the associated hypothesis testing techniques continue to perform well when violating assumptions (normality and homoscedasticity). But at some point, they break down and become highly unsatisfactory.
- Pearson's correlation is useful for establishing dependence, it provides an indication of whether the least squares regression line has a positive or negative slope, but in terms of measuring the strength of an association, it has the potential of being highly misleading.
- Failure to detect an association with Pearson's correlation is not convincing evidence that no association exists, even with large sample sizes.
- There are methods for testing hypotheses about Pearson's correlation and the least squares regression slope that allow heteroscedasticity, but no details are given here.
- The Theil-Sen estimator is just one example of many estimators that have the potential of improving upon the least squares estimator substantially.
- Many new methods have appeared in recent years aimed at describing and detecting non-linear associations. Of particular importance are methods called smoothers. No details are given here, but it is important to at least be aware of them.

8.5 Some concluding remarks

The purpose of this chapter was to introduce basic concepts and to describe standard hypothesis testing methods associated with least squares regression. Another goal was to provide some indication of what might go wrong with standard methods and to briefly outline how some of these problems can be corrected. Generally, it is suggested that the student seek advanced training before attempting any regression analysis. Also, improved techniques continue to emerge. For more about regression, see Li (1985), Montgomery and Peck (1992), Staudte and Sheather et al. (1990), Hampel (1986), Huber (1981), Rousseeuw and Leroy (1987), Belsley et al. (1980), Cook and Weisberg (1992), Carroll and Ruppert (1988), Hettmansperger (1984), Hettmansperger and McKean 1998), Wilcox (2003, 2005).

9

COMPARING TWO GROUPS

Chapters 6 and 7 described how to make inferences about the population mean, and other measures of location, associated with a single population of individuals or things. This chapter extends these methods to situations where the goal is to compare two groups. For example, Table 2.1 reports data from a study on changes in cholesterol levels when participants take an experimental drug. But of fundamental interest is how the changes compare to individuals who receive a placebo instead. Example 4 in section 6.2 described an experiment on the effect of ozone on weight gain among rats. The two groups in this study consisted of rats living in an ozone environment and ones that lived in an ozone-free environment. Do weight gains differ for these groups, and if they do, how might this difference be described? Two training programs are available for learning how to invest in stocks. To what extent, if any, do these training programs differ? How does the reading ability of children who watch thirty hours or more of television per week compare to children who watch ten hours or less? How does the birth weight of newborns among mothers who smoke compare to the birth weight among mothers who do not smoke? In general terms, if we have two independent variables, how might we compare them?

9.1 Comparing the means of two independent groups

When trying to detect and describe differences between groups, by far the most common strategy is to use means. We begin with a classic method designed for two independent groups. By independent groups is meant that the observations in the first group are independent of the observations in the second. In particular, the sample means for the two groups, say $\bar{X}_1$ and $\bar{X}_2$, are independent. So, in the example dealing with weight gain among rats, it is assumed that one group of rats is exposed to an ozone environment, and a separate group of rats, not associated with the first group, is exposed to an ozone-free environment. This is in contrast to using, for example, the same rats under both conditions, or using rats from the same litter, in which case the sample means might be dependent.

The two-sample Student's t-test

The classic and best-known method for comparing the means of two independent groups is called the *two-sample Student's t-test.* Here we let μ_1 and μ_2 represent the two population means, and the corresponding standard deviations are denoted by σ_1 and σ_2. The goal is to test

$$H_0 : \mu_1 = \mu_2, \tag{9.1}$$

the hypothesis that the population means are equal. It turns out that we can get exact control over the probability of a Type I error if the following three assumptions are true:

- Random sampling
- Normality
- Equal variances. That is, $\sigma_1 = \sigma_2$, which is called the *homogeneity of variance* assumption.

Before describing how to test the hypothesis of equal means, first consider how we might estimate the assumed common variance. For convenience, let σ_p^2 represent the common variance and let s_1^2 and s_2^2 be the sample variances corresponding to the two groups. Also let n_1 and n_2 represent the corresponding sample sizes. The typical estimate of σ_p^2 is

$$s_p^2 = \frac{(n_1 - 1)s_1^2 + (n_2 - 1)s_2^2}{n_1 + n_2 - 2}. \tag{9.2}$$

For the special case where the sample sizes are equal, meaning that $n_1 = n_2$, s_p^2 is just the average of the two sample variances. That is,

$$s_p^2 = \frac{s_1^2 + s_2^2}{2}.$$

Now consider the problem of testing the null hypothesis of equal means. Under the assumptions already stated, the probability of a Type I error will be exactly α if we reject the null hypothesis when

$$|T| \geq t, \tag{9.3}$$

where

$$T = \frac{\bar{X}_1 - \bar{X}_2}{\sqrt{s_p^2\left(\frac{1}{n_1} + \frac{1}{n_2}\right)}}, \tag{9.4}$$

and t is the $1 - \alpha/2$ quantile of Student's t-distribution with $\nu = n_1 + n_2 - 2$ degrees of freedom, which is read from table 4 in appendix B. An exact $1 - \alpha$ confidence interval for the difference between the population means, under the same assumptions, is

$$(\bar{X}_1 - \bar{X}_2) \pm t\sqrt{s_p^2\left(\frac{1}{n_1} + \frac{1}{n_2}\right)}. \tag{9.5}$$

Example 1

Salk (1973) conducted a study where the general goal was to examine the soothing effects of a mother's heartbeat on her newborn infant. Infants were

Table 9.1 Weight gain, in grams, for large babies

Group 1 (heartbeat)			
Subject	Gain	Subject	Gain
1	190	11	10
2	80	12	10
3	80	13	0
4	75	14	0
5	50	15	−10
6	40	16	−25
7	30	17	−30
8	20	18	−45
9	20	19	−60
10	10	20	−85

$n_1 = 20, \bar{X}_1 = 18.0, s_1 = 60.1, s_1/\sqrt{n_1} = 13$

Group 2 (heartbeat)							
Subject	Gain	Subject	Gain	Subject	Gain	Subject	Gain
1	140	11	25	−21	−50	31	−130
2	100	12	25	−22	−50	32	−155
3	100	13	25	−23	−60	33	−155
4	70	14	30	−24	−75	34	−180
5	25	15	30	−25	−75	35	−240
6	20	16	30	−26	−85	36	−290
7	10	17	45	−27	−85		
8	0	18	45	−28	−100		
9	−10	19	−45	29	−110		
10	−10	20	−50	30	−130		

$n_2 = 36, \bar{X}_2 = -52.1, s_2 = 88.4, s_2/\sqrt{n_2} = 15$

placed in a nursery immediately after birth and they remained there for four days except when being fed by their mothers. The infants were divided into two groups. The first was continuously exposed to the sound of an adult's heartbeat; the other group was not. Salk measured, among other things, the weight change of the babies from birth to the fourth day. Table 9.1 reports the weight change for the babies weighing at least 3,510 grams at birth. As indicated, the sample standard deviations are $s_1 = 60.1$ and $s_2 = 88.4$. The estimate of the assumed common variance is

$$s_p^2 = \frac{(20-1)(60.1^2) + (36-1)(88.4^2)}{20+36-2} = 6,335.9.$$

So

$$T = \frac{18 - (-52.1)}{\sqrt{6,335.9\left(\frac{1}{20} + \frac{1}{36}\right)}} = \frac{70.1}{22.2} = 3.2.$$

The sample sizes are $n_1 = 20$ and $n_2 = 36$, so the degrees of freedom are $\nu = 20 + 36 - 2 = 54$. If we want the Type I error probability to be $\alpha = .05$, then $1 - \alpha/2 = .975$, and from table 4 in appendix B, $t = 2.01$. Because $|T| = 3.2$, which is greater than 2.01, reject H_0 and conclude that the means differ.

That is, we conclude that among all newborns we might measure, the average weight gain would be higher among babies exposed to the sound of a heartbeat compared to those that are not exposed. By design, the probability that our conclusion is in error is .05, assuming normality and homoscedasticity. The .95 confidence interval for $\mu_1 - \mu_2$, the difference between the population means, is

$$[18 - (-52.1)] \pm 2.01\sqrt{6,335.9\left(\frac{1}{20} + \frac{1}{36}\right)} = (25.5, 114.7).$$

This interval does not contain zero, and it indicates that the difference between the means is likely to be at least 25.5, so again you would reject the hypothesis of equal means.

Violating assumptions: When does Student's *t* perform well?

There are two conditions where the assumption of normality, or equal variances, can be violated and yet Student's *t* appears to continue to perform well in terms of Type I errors and accurate confidence intervals. The homoscedasticity assumption can violated if both distributions are normal and the sample sizes are equal, provided the sample sizes are not overly small, say less than 8 (Ramsey, 1980). As for non-normality, Student's *t* appears to perform well in terms of Type I error probabilities provided the two distributions are identical. That is, not only do they have the same means, they have the same variances, the same amount of skewness, the tails of the distribution are identical, and so on. So if we were to plot the distributions, the plots would be exactly the same. If, for example, you want the probability of a Type I error to be .05, generally, the actual Type I error probability will be less than or equal to .05.[1]

Conditions where Student's *t* performs poorly

If the goal is to test the hypothesis of equal means, without being sensitive to other ways the groups might differ, Student's *t* can be unsatisfactory in terms of Type I errors and accurate confidence intervals when sampling from normal distributions with unequal sample sizes and unequal variances. Problems due to unequal variances are exacerbated when sampling from non-normal distributions instead, and now concerns arise even with equal sample sizes (e.g., Algina, et al., 1994; Wilcox, 1990). When dealing with groups that differ in skewness, again problems with controlling the probability of a Type I error occur, and the combination of unequal variances and different amounts of skewness makes matters worse. Some degree of unequal variances, as well as mild differences in skewness, can be tolerated. But the extent to which this is true, based on the data under study, is difficult to determine in an accurate manner.

1. A key reason is that if we sample an observation from each group, and if the groups have the same skewness, the distribution of the difference between these two observations is symmetric. We saw in chapter 7 that for a symmetric distribution, Type I error probabilities larger than the specified α level can be avoided.

Example 2

Recall from chapter 6 that one of way of determining the distribution of T under normality is to use simulations. That is, to generate data from a normal distribution, compute T, and repeat this process many times. With the aid of a computer, we can extend this method when sampling from two distributions that have equal means but which differ in terms of skewness and have unequal variances. In particular, imagine that we sample 40 observations from a standard normal distribution and 60 observations from the distribution shown in figure 9.1, and then we compute T. Repeating this process 1000 times provides a fairly accurate indication of the distribution of T when the null hypothesis of equal means is true. Figure 9.2 shows a plot of the results plus the distribution of T assuming normality. Under normality, and with a Type I error probability

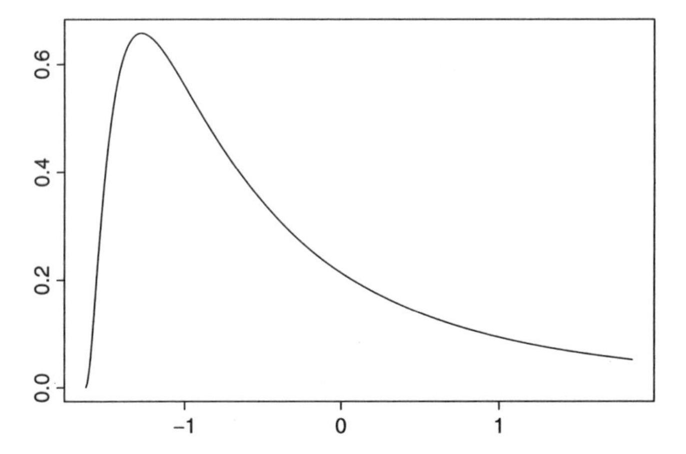

Figure 9.1 A skewed distribution with a mean of zero.

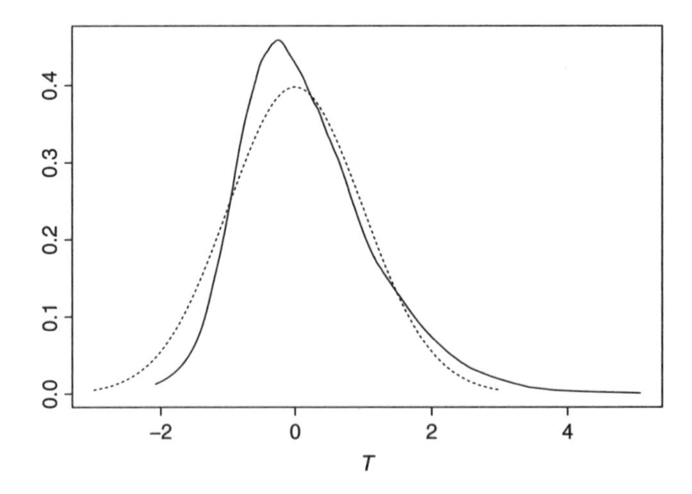

Figure 9.2 The distribution of T when sampling 40 values form a standard normal and 60 values form the distribution in figure 9.1. Also shown is the distribution of T when both groups have normal distribution. This illustrates that differences in skewness can have an important impact on T.

of $\alpha = .05$, Student's T rejects the hypothesis of equal means if $T \leq -2.002$ or if $T \geq 2.002$. But figure 9.2 indicates that we should reject if $T \leq -1.390$ or if $T \geq 2.412$, values that differ substantially from what would be used under normality.

Some authorities might criticize this last example on the grounds that if groups differ in terms of the variances and the amount of skewness they have, surely the means differ as well. That is, they would argue that Type I errors are not an issue in this case. But even if we accept this point of view, this last illustration can be seen to create concerns about power, and it indicates that confidence intervals based on Student's t can be relatively inaccurate.

Usually, a basic requirement of any method is that with sufficiently large sample sizes, good control over the Type I error probability and accurate confidence intervals will be obtained. There are theoretical results indicating that under general conditions, when the goal is to compare the means without being sensitive to other features of the distribution (such as unequal variances), Student's t can be unsatisfactory regardless of how large the sample sizes might be (Cressie and Whitford, 1986). Exceptions are when the sample sizes are equal and when both groups have identical distributions. This means that Student's t provides a valid test of the hypothesis that the distributions are identical, but it can be unsatisfactory when computing confidence intervals for the difference between the means or when testing the hypothesis that groups have equal means.

Finally, in terms of Type II errors and power, Student's t can perform very poorly, compared to alternative techniques, when outliers tend to occur. The presence of outliers does not necessarily mean low power, but the reality is that power might be increased substantially when comparing groups with something other than the means, as will be illustrated. Some additional concerns about Student's t are summarized in Wilcox (2003, 2005). One of these concerns is that unequal variances and differences in skewness can create power problems as well.

Why testing assumptions can be unsatisfactory

Some commercial software now contains a test of the assumption that two groups have equal variances. The idea is that if the hypothesis of equal variances is not rejected, one would then use Student's t. But a *basic principle* is that failing to reject a null hypothesis is not, by itself, compelling evidence that the null hypothesis should be accepted or that the null hypothesis is approximately true. Accepting the null hypothesis is only reasonable if the probability of rejecting (power) is sufficiently high to ensure that differences that have practical importance will be detected. If there is a low probability of detecting a difference that is deemed important, concluding that no difference exists is difficult to defend. In the case of Student's t, would a test of the assumption of equal variances have enough power to detect a situation where unequal variances causes a problem? All indications are that the answer is no (e.g., Markowski and Markowski, 1990; Moser, et al., 1989; Wilcox, et al., 1986; Zimmerman, 2004; Hayes and Cai, 2007). Presumably exceptions occur if the sample sizes are sufficiently large, but it is unclear how we can be reasonably certain when this is the case. Part of the problem is that the extent to which the variances can differ, without having a major impact on the Type I error probability, is a complicated function of the sample sizes, and the extent to which groups differ in terms of skewness, and the likelihood of observing outliers. Testing the hypothesis that data have a normal distribution is another strategy that might be followed. But when do such tests have enough power to detect departures from normality that are a concern? The answer is not remotely clear and so this approach cannot be recommended at this time. A better strategy is to use more modern methods that perform reasonably well under normality, but which continue to perform well under non-normality or when groups have unequal variances.

There are many alternatives to Student's t when comparing groups. Because testing assumptions seems dubious, how can we tell whether some alternative technique might give a substantially different sense about whether and how the groups differ? Currently, the only known strategy that answers this question in an adequate manner is to simply try alternative methods, some of which are outlined later in this chapter. However, a criticism of applying many methods is that control over the probability of at least one Type I error can become an issue. This issue, and methods for dealing with it, are described and illustrated in chapter 11.

Interpreting Student's t when we reject

Despite its many practical problems, Student's t does have a positive feature. If we reject, this is a good indication that the distributions differ in some manner. This is because when the distributions do not differ, it controls the probability of a Type I error fairly well. But even though the method is designed to compare means, in reality it is also sensitive to differences in variances and skewness. As previously noted, some would argue that if the distributions differ, surely the means differ. However, when we reject, it is unclear whether the main reason is due to differences between the means. The main reason could be differences between the variances or skewness. Moreover, rejecting with Student's t raises concerns about whether the confidence interval, given by equation (9.5), is reasonably accurate. In summary, when rejecting with Student's t, it is reasonable to conclude that the groups differ in some manner. But when Student's t indicates that groups differ, there are concerns that the nature of the difference is not being revealed in a reasonably accurate manner. And when Student's t fails to reject, this alone is not compelling evidence that the groups do not differ in any important way.

Dealing with unequal variances: Welch's test

Many methods have been proposed for comparing means when the population variances (σ_1^2 and σ_2^2) differ. None are completely satisfactory. Here we describe one such method that seems to perform reasonably well compared to other techniques that have been derived when attention is restricted to comparing means. Popular commercial software now contains this method, which was derived by Welch (1938).

Recall from chapter 5 that the sampling distribution of the sample mean has variance σ^2/n, which is called the squared standard error of the sample mean. For the situation at hand, the difference between the sample means, $\bar{X}_1 - \bar{X}_2$, also has a sampling distribution, and the corresponding mean of this difference is $\mu_1 - \mu_2$, the difference between the population means. Roughly, this means that if we were repeat a study millions of times, and if we averaged the differences between the sample means resulting from each study, we would get $\mu_1 - \mu_2$, the di.erence between the population means.

Put another way, on average, over many studies, $\bar{X}_1 - \bar{X}_2$ estimates $\mu_1 - \mu_2$. Moreover the variance (or squared standard error) of the difference between the sample means can be shown to be

$$\text{VAR}(\bar{X}_1 - \bar{X}_2) = \frac{\sigma_1^2}{n_1} + \frac{\sigma_2^2}{n_2}.$$

Also recall from chapter 6 that under normality, if we standardize a variable by subtracting its mean, and then dividing by its standard error, we get a standard normal distribution. That is, if a variable has a normal distribution, then in general,

$$\frac{\text{variable} - \text{population mean of the variable}}{\text{standard error of the variable}}, \tag{9.6}$$

will have a standard normal distribution. Here the variable of interest is $\bar{X}_1 - \bar{X}_2$, the difference between the sample means, which has a population mean of $\mu_1 - \mu_2$. Consequently, based on the equation for the squared standard error, $\text{VAR}(\bar{X}_1 - \bar{X}_2)$, it follows that

$$\frac{\bar{X}_1 - \bar{X}_2 - (\mu_1 - \mu_2)}{\sqrt{\frac{\sigma_1^2}{n_1} + \frac{\sigma_2^2}{n_2}}}$$

has a standard normal distribution. *If* the hypothesis of equal means is true, then $\mu_1 - \mu_2 = 0$, in which case this last equation becomes

$$\frac{\bar{X}_1 - \bar{X}_2}{\sqrt{\frac{\sigma_1^2}{n_1} + \frac{\sigma_2^2}{n_2}}},$$

which again has a standard normal distribution. As usual, the population variances are rarely known, but they can be estimated with the sample variances, in which case this last equation becomes

$$W = \frac{(\bar{X}_1 - \bar{X}_2)}{\sqrt{\frac{s_1^2}{n_1} + \frac{s_2^2}{n_2}}}, \tag{9.7}$$

where, as before, s_1^2 and s_2^2 are the sample variances corresponding to the two groups being compared; this is the test statistic used by Welch's test.

When the hypothesis of equal means is true, W will have, approximately, a standard normal distribution if the sample sizes are sufficiently large, thanks to the central limit theorem. That is, we can determine how large W must be to reject the hypothesis of equal means using values in table 1 in appendix B. But in general, W will not have a normal distribution, so some other approximation of an appropriate critical value is required. Welch's approach to this problem is implemented in the following manner. For convenience, let

$$q_1 = \frac{s_1^2}{n_1} \text{ and } q_2 = \frac{s_2^2}{n_2}. \tag{9.8}$$

As was done with Student's t, table 4 in appendix B is used to determine a critical value, t, but now the degrees of freedom are

$$\nu = \frac{(q_1 + q_2)^2}{\frac{q_1^2}{n_1 - 1} + \frac{q_2^2}{n_2 - 1}}. \tag{9.9}$$

Under normality, W has, approximately, a Student's t-distribution with degrees of freedom given by equation (9.9). That is, reject the hypothesis of equal means if $|W| \geq t$. The $1-\alpha$ confidence interval for the difference between the means, $\mu_1 - \mu_2$, is

$$(\bar{X}_1 - \bar{X}_2) \pm t\sqrt{\frac{s_1^2}{n_1} + \frac{s_2^2}{n_2}}. \tag{9.10}$$

Example 3

Tables 2.1 and 2.2 report data on the effectiveness of a drug to lower cholesterol levels. For the data in table 2.1, corresponding to the group that received the experimental drug, the sample size is $n_1 = 171$, the sample variance is $s_1^2 = 133.51$, and the sample mean is $\bar{X}_1 = -9.854$. For the group that received the placebo, $n_2 = 177$, $s_2^2 = 213.97$, and $\bar{X}_2 = 0.124$. To apply Welch's test, compute $q_1 = 133.51/171 = 0.78076$ and $q_2 = 213.97/177 = 1.20887$, in which case the degrees of freedom are

$$\nu = \frac{(0.78076 + 1.20887)^2}{\frac{0.78076^2}{171-1} + \frac{1.2088^2}{177-1}} = 332.99.$$

The test statistic is $W = 7.07$, the $\alpha = .05$ critical value is 1.967, and because $|7.07| \geq 1.967$, reject the null hypothesis.

Student's t versus Welch's test

Some brief comments about the relative merits of Student's t versus Welch's Test should be made. When comparing groups that do not differ in any manner, there is little reason to prefer Student's t over Welch's test. But if the distributions differ in some way, such as having unequal variances. The choice of method can make a practical difference. Welch's test reduces problems with unequal variances, given the goal of comparing means, but it does not eliminate them. Differences in skewness remain a concern, and, as is the case with all methods based on means, outliers can destroy power. So, when rejecting with Welch's test, like Student's t-test, it is reasonable to conclude that the distributions differ in some manner, but there is uncertainty about whether the main reason has to do with differences between the population means; the primary reason could be unequal variances or differences in skewness. And when we fail to reject, this could be because the groups differ by very little, but another possibility is that power is low due to sample sizes that are too small, differences in skewness, or outliers.

In fairness, there are situations where Student's t correctly concludes that groups differ in some manner when Welch's test does not. This can happen because Student's t can be more sensitive to certain types of differences, such as unequal variances.

A positive feature of Welch's method is that with sufficiently large sample sizes, it will control the probability of a Type I error given the goal of comparing means, and it provides accurate confidence intervals as well, assuming random sampling only. This is in contrast to Student's t, which does not satisfy this goal when the sample sizes are unequal and the groups differ in skewness. A rough explanation is that under random sampling, regardless of whether the groups differ, Welch's test uses a correct estimate of the standard error associated with the difference between the means, $\bar{X}_1 - \bar{X}_2$, but there are conditions where this is not the case when using Student's t (Cressie and

Whitford, 1986). As previously noted, an exception is when groups have identical distributions. So again, an argument for considering Student's t is that if it rejects, a good argument can be made that the groups differ in some manner. A very rough rule is that a method that uses the correct standard error is likely to have more power than a method does not. So here, the expectation is that Welch's test will tend to have more power than Student's t, but exceptions are encountered where Student's t rejects and Welch's method does not.

Comments about outliers when comparing means

Any method for comparing groups based on means runs the risk of relatively low power. As noted in previous chapters, outliers can inflate the sample variances which in turn can result in low power, and there is some possibility that the mean will poorly reflect what is typical. Outliers also have other consequences relevant to power that might not be immediately obvious but which are illustrated by the next example.

Example 4

Imagine that an experimental drug is under investigation and that there is concern that it might damage the stomach. For illustrative purposes, suppose the drug is given to a sample of rats, a placebo is given to a control group, and the results are as follows:

Experimental drug: $4, 5, 6, 7, 8, 9, 10, 11, 12, 13$

Placebo: $1, 2, 3, 4, 5, 6, 7, 8, 9, 10$.

The goal is to determine whether the average amount of stomach damage differs for these two groups. The corresponding sample means are $\bar{X}_1 = 8.5$ and $\bar{X}_2 = 5.5$ and $T = 2.22$. With $\alpha = .05$, the critical value is $t = 2.1$, so Student's t would reject the hypothesis of equal means and conclude that the first group has a larger population mean than the second (because the first group has the larger sample mean). Now, if we increase the largest observation in the first group from 13 to 23, the sample mean increases to $\bar{X}_1 = 9.5$. So the difference between $\bar{X}_1$ and $\bar{X}_2$ has increased from 3 to 4 and this would seem to suggest that we have stronger evidence that the population means differ and in fact the first group has the larger population mean. However, increasing the largest observation in the first group also inflates the corresponding sample variance, s_1^2. In particular, s_1^2 increases from 9.17 to 29.17. The result is that T *decreases* to $T = 2.04$ and we no longer reject. That is, increasing the largest observation has more of an effect on the sample variance than the sample mean in the sense that now we are no longer able to conclude that the population means differ. Increasing the largest observation in the first group to 33, the sample mean increases to 10.5, the difference between the two sample means increases to 5 and now $T = 1.79$. So again we do not reject and in fact our test statistic is getting smaller. It is left as an exercise to show that a similar result is obtained when using Welch's test. This illustration provides another perspective on how outliers can mask differences between population means.

Comparing medians

Many methods have been proposed for comparing the medians of two independent groups. Some are based on what are called nonparametric methods, which are discussed in the final chapter of this book. Although nonparametric methods have practical value, as a method for comparing medians, they are unsatisfactory unless rather restrictive assumptions are met (e.g., Fung, 1980). Another possibility is to use what is called a *permutation method*, but as a tool for comparing medians, it is unsatisfactory as well (e.g., Romano, 1990).

Here is a method for comparing medians that is relatively easy to use and which appears to perform fairly well, in terms of controlling the probability of a Type I error, when tied (duplicated) values rarely if ever occur. Let M_1 and M_2 be the sample medians for the two groups and let S_1^2 and S_2^2 be the corresponding McKean–Schrader estimates of the squared standard errors. Then an approximate $1-\alpha$ confidence interval for the difference between the population medians is

$$(M_1 - M_2) \pm c_{1-\alpha/2}\sqrt{S_1^2 + S_2^2},$$

where c is the $1-\alpha/2$ quantile of a standard normal distribution. Alternatively, reject the hypothesis of equal population medians if

$$\frac{|M_1 - M_2|}{\sqrt{S_1^2 + S_2^2}} \geq c.$$

Example 5

Imagine a study aimed at measuring the extent to which men and women are addicted to nicotine. Based on a measure of dependence on nicotine, 30 men are found to have a median value of 4, and 20 women have a median value of 2. If the McKean–Schrader estimates of the corresponding squared standard errors are .8 and .6, then

$$\frac{|M_1 - M_2|}{\sqrt{S_1^2 + S_2^2}} = \frac{|4-2|}{\sqrt{.8+.6}} = 1.69.$$

With $\alpha = .05, 1-\alpha/2 = .975$, and from table 1 in appendix B, the .975 quantile for a standard normal distribution is 1.96. Because 1.69 is less than the critical value, fail to reject.

An important issue: The choice of method can matter

It cannot be emphasized too strongly that, when comparing groups, the choice of method can matter, not only in terms of detecting differences, but in terms of assessing the magnitude of the difference as well. Also, a *basic principle* is that failing to reject when comparing groups does not necessarily mean that any difference between the groups is relatively small or that no difference exists. Different methods provide different perspectives on how groups differ and by how much.

Table 9.2 Self-awareness data

Group 1:	77 87 88 114 151 210 219 246 253
	262 296 299 306 376 428 515 666 1310 2611
Group 2:	59 106 174 207 219 237 313 365 458 497 515
	529 557 615 625 645 973 1065 3215

Example 6

Dana (1990) conducted a study aimed at investigating issues related to self-awareness and self-evaluation. (This study was previously mentioned in connection with example 5 in section 6.5.) In one portion of the study, he recorded the times individuals could keep an apparatus in contact with a specified target. The results, in hundredths of second, are shown in table 9.2. If we compare the groups with Student's t, the p-value is 0.4752, and for Welch's test it is 0.4753. So both methods give very similar results and provide no indication that the groups differ. However, if we compare medians with the method described here, the p-value is 0.0417, so in particular we would reject with the Type I error probability set at $\alpha = .05$. (There are no tied values, suggesting that this method for comparing medians provides reasonably good control over the probability of a Type I error.)

Why is it that we can reject with medians but not even come close to rejecting with means? There are at least two fundamental reasons this can happen. For skewed distributions, means and medians can differ substantially. For the two groups in this last example, the means are 448 and 598, with a difference of $448 - 598 = -150$. The medians are 262 and 497, with a difference of -235, which is bigger than the difference between the means. Also, the estimated standard errors of the means are 136.4 and 157.9 versus 77.8 and 85.02 for the medians. As a result, we conclude that the population medians differ, but we find no evidence that the means differ as well. In fairness, however, we can encounter situations where the difference between the means is larger than the difference between the medians, in which case comparing means might result in more power.

Some practical concerns when comparing medians

It was noted in chapter 6 that when using the McKean–Schrader estimate of the standard error, it can result in highly inaccurate confidence intervals and poor control over the probability of a Type I error due to tied values. This problem extends to the situation here where the goal is to compare the medians of two independent groups. Many alternative methods have been found to suffer from the same problem (Wilcox, 2006). The one method that appears to correct this problem is a percentile bootstrap technique, which is outlined in the final section of this chapter. Another general concern is that comparing medians might result in low power, relative to other methods that might be used, because its standard error can be relatively large when outliers tend to be rare. But, as previously illustrated, its standard error can be relatively low when outliers are common.

Comments on comparing variances

Although the most common approach to comparing two independent groups is to use some measure of location, situations arise where there is interest in comparing variances or some other measure of dispersion. For example, in agriculture, one goal when comparing two crop varieties might be to assess their relative stability. One approach is to declare the variety with the smaller variance as being more stable (e.g., Piepho, 1997). As another example, consider two methods for training raters to assess certain human characteristics. For instance, raters might judge athletic ability or they might be asked to rate aggression among children in a classroom. Then one issue is whether the variance of the ratings differ, depending on how the raters were trained. Also, in some situations, two groups might differ primarily in terms of the variances rather than their means or some other measure of location.

There is a vast literature on comparing variances. The method typically mentioned in an introductory course assumes normality and is based on the ratio of the sample variances, s_1^2/s_2^2. But this approach has long been known to be highly unsatisfactory when distributions are non-normal (e.g., Box, 1953), and so it is not described. (For the most recent results on how to approach this problem, plus methods for comparing other measures of dispersion, see Wilcox, 2003.)

Measuring effect size

As noted in chapter 7, when dealing with hypotheses about a single mean (or any other measure of location), p-values provide an indication of whether some hypothesis should be rejected, but it is unsatisfactory in terms of understanding the extent to which the mean differs from the hypothesized value. A similar result applies when comparing the means of two independent groups. A p-value close to zero provides evidence that the groups differ. For example, if we reject at the .001 level and the first group has a larger sample mean than the second, then we conclude that the first group has the larger population mean. But this tells us nothing about the magnitude of the difference (e.g., Cohen, 1994).

Example 7

An article in *Nutrition Today* (1984, *19*, 22–29) illustrates the importance of this issue. A study was conducted on whether a particular drug lowers the risk of heart attacks. Those in favor of using the drug pointed out that the number of heart attacks in the group receiving the drug was significantly lower than in the group receiving a placebo. As noted in chapter 7, when a researchers reports that a significant difference was found, typically the word significant is being used to signify that a small p-value was obtained. In the study, the hypothesis of equal means was rejected at the $\alpha = .001$ level, and often such a result is described as '*highly significant*'. However, critics of the drug argued that the difference between the number of heart attacks was trivially small. They concluded that because of the expense and side effects of using the drug, there is no compelling evidence that patients with high cholesterol levels should be put on this medication. A closer examination of the data revealed that the standard errors corresponding to the two groups were very small, so it was possible to get a statistically significant result that was clinically unimportant.

Acquiring a good understanding of how groups differ can be a nontrivial problem that might require several perspectives. One possibility is to simply use the difference between some measure of location, with the usual choice being the means, and in some situations this simple approach will suffice. But if we limit our choice to the means, there is a practical concern: Different measures of location can provide a different sense about how much the groups differ. As already illustrated, if we switch to medians, this can alter our sense about whether there is a large difference between the groups. Although we have already seen serious negative features associated with the mean, this is not to suggest that this approach should be abandoned. Rather, the issue is, if we limit ourselves to means, are there important details that are being missed? Often the answer is yes.

Currently, there is a method for measuring effect size, based in part on the means, that is commonly used by many researchers. The method assumes that the two groups have a common variance, which we again label σ_p^2. That is, $\sigma_1^2 = \sigma_2^2 = \sigma_p^2$ is assumed. Then the so-called standardized difference between the groups is

$$\Delta = \frac{\mu_1 - \mu_2}{\sigma_p}, \tag{9.11}$$

where Δ is an upper case Greek delta. Assuming normality, Δ can be interpreted in a simple manner. For example, if $\Delta = 2$, then the difference between the means is two standard deviations, and for normal distributions we have some probabilistic sense of what this means. A common practice is to interpret Δ values of .2, .5 and .8 as small, medium, and large effect sizes, respectively. It is known, however, that under general conditions, this oversimplifies the issue of measuring effect size. For instance, we encounter situations where $\Delta = .2$, which supposedly would be interpreted as a small effect size, when in fact there is a large difference based on plots of the data. The usual estimate of Δ is

$$d = \frac{\bar{X}_1 - \bar{X}_2}{s_p}$$

and is often called *Cohen's d*.

Many new methods have been derived with the goal of better understanding how groups differ (e.g., Agina, et al., 2005; Wilcox, 2003), but these methods are not covered here. A simple yet useful approach is to plot the data for both groups. Boxplots are often used, and kernel density estimators, mentioned in chapter 3, are often a good choice. Some illustrations are given in the final section of this chapter.

The main point, which cannot be stressed too strongly, is that a single numerical quantity, aimed at assessing how groups differ and by how much, can be unsatisfactory and too simplistic. This is not always the case, but assuming that a single measure of effect size is adequate is a strategy that cannot be recommended. The general issue of assessing effect size in a satisfactory manner is a complex problem that might require advanced techniques.

Comparing two binomial distributions

In many situations, comparing groups corresponds to comparing two binomial distributions. For example, if the probability of surviving an operation using method 1 is p_1, and if the probability of surviving using method 2 is p_2, do p_1 and p_2 differ, and if they do differ, by how much? As another example, how does the proportion of women who believe the President of the United States is an effective leader compare to the corresponding proportion for men?

An appropriate test statistic can be derived using the same strategy used by Welch's method, as represented by equation (9.6). From chapters 4 and 5, if $\hat{p}$ indicates the proportion of successes, the corresponding squared standard error is $p(1-p)/n$. For two independent proportions, $\hat{p}_1$ and $\hat{p}_2$, the variance or squared standard error of their difference can be shown to be

$$\mathrm{VAR}(\hat{p}_1-\hat{p}_2)=\frac{p_1(1-p_1)}{n_1}+\frac{p_2(1-p_2)}{n_2},$$

and an estimate of this squared standard error is simply

$$\frac{\hat{p}_1(1-\hat{p}_1)}{n_1}+\frac{\hat{p}_2(1-\hat{p}_2)}{n_2}.$$

So an appropriate test of

$$H_0: p_1=p_2 \tag{9.12}$$

is

$$Z=\frac{\hat{p}_1-\hat{p}_2}{\sqrt{\frac{\hat{p}_1(1-\hat{p}_1)}{n_1}+\frac{\hat{p}_2(1-\hat{p}_2)}{n_2}}}, \tag{9.13}$$

and the null hypothesis is rejected if

$$|Z|\geq c,$$

where c is the $1-\alpha/2$ quantile of a standard normal distribution, which is read from table 1 in appendix B. A $1-\alpha$ confidence interval for the difference between the two probabilities is

$$(\hat{p}_1-\hat{p}_2)\pm c\sqrt{\frac{\hat{p}_1(1-\hat{p}_1)}{n_1}+\frac{\hat{p}_2(1-\hat{p}_2)}{n_2}} \tag{9.14}$$

The method for comparing binomial distributions, just described, illustrates basic principles and so is particularly appropriate for an introductory course. A positive feature is that it performs reasonably well when the probability of success for the two groups is not too close to 0 or 1. Many improved methods have been proposed for comparing binomial distributions, comparisons of which are reported by Storer and Kim (1990) and Beal (1987). For more recent suggestions, see Berger (1996) and Coe and Tamhane (1993).[2]

Example 8

Imagine that a new treatment for alcoholism is under investigation and that a method has been agreed upon for assessing whether the treatment is deemed effective after 12 months. An issue is whether the effectiveness of the treatment differs for men versus women. For illustrative purposes, suppose for 7 out 12 males the treatment is a success, so $\hat{p}_1=7/12$, and that for 25 women there

2. Easy-to-use R and S-PLUS functions are available for applying these methods; see chapter 1.

are 22 successes, in which case $\hat{p}_2 = 22/25$. Then simple calculations, based on equation (9.13), show that $Z = -1.896$. If we want the probability of a Type I error to be .05, the critical value is $c = 1.96$, and because $|-1.896| < 1.96$, we fail to reject.

Example 9

Table 2.3 reported data on how many sexual partners undergraduate males want during the next 30 years. Of the 105 males, 49 said that they want one partner. So the estimated probability of answering 1 is $\hat{p}_1 = 49/105$. In the same study, 101 of 156 females also responded that they want one sexual partner during the next 30 years, so $\hat{p}_2 = 101/156$. For the entire population of undergraduates, and assuming random sampling, do the corresponding population probabilities, p_1 and p_2, differ? And if the answer is yes, what can be said about the magnitude of the difference? It can be seen that $Z = -2.92$, and the null hypothesis would be rejected with a Type I error of $\alpha = .01$. So the conclusion would be that males differ from females. The estimated difference between the two probabilities is -0.181 and the .99 confidence interval is $(-0.340, -0.021)$. That is, we can be reasonably certain that the difference is at least -0.021, and it could be as large as -0.34.

Problems

1. Suppose that the sample means and variances are $\bar{X}_1 = 15$, $\bar{X}_2 = 12$, $s_1^2 = 8$, $s_2^2 = 24$ with sample sizes $n_1 = 20$ and $n_2 = 10$. Verify that $s_p^2 = 13.14$, $T = 2.14$ and that Student's t-test rejects the hypothesis of equal means with $\alpha = .05$.
2. For two independent groups of subjects, you get $\bar{X}_1 = 45$, $\bar{X}_2 = 36$, $s_1^2 = 4$, $s_2^2 = 16$ with sample sizes $n_1 = 20$ and $n_2 = 30$. Assume the population variances of the two groups are equal and verify that the estimate of this common variance is 11.25.
3. Still assuming equal variances, test the hypothesis of equal means using Student's t-test and the data in the last problem. Use $\alpha = .05$.
4. Repeat the last problem, only use Welch's test for comparing means.
5. Comparing the results for the last two problems, what do they suggest regarding the power of Welch's test versus Student's t-test when the sample variances differ sufficiently.
6. For two independent groups of subjects, you get $\bar{X}_1 = 86$, $\bar{X}_2 = 80$, $s_1^2 = s_2^2 = 25$, with sample sizes $n_1 = n_2 = 20$. Assume the population variances of the two groups are equal and verify that Student's t rejects with $\alpha = .01$.
7. Repeat the last problem using Welch's method.
8. Comparing the results of the last two problems, what do they suggest about using Student's t versus Welch's method when the sample variances are approximately equal?
9. For $\bar{X}_1 = 10$, $\bar{X}_2 = 5$, $s_1^2 = 21$, $s_2^2 = 29$, $n_1 = n_2 = 16$, compute a .95 confidence interval for the difference between the means using Welch's method and state whether you would reject the hypothesis of equal means.
10. Repeat the last problem, only use Student's t instead.

11. Two methods for training accountants are to be compared. Students are randomly assigned to one of the two methods. At the end of the course, each student is asked to prepare a tax return for the same individual. The amounts of the refunds reported by the students are

$$\text{Method 1}: 132, 204, 603, 50, 125, 90, 185, 134$$

$$\text{Method 2}: 92, -42, 121, 63, 182, 101, 294, 36.$$

Using Welch's test, would you conclude that the methods differ in terms of the average return? Use $\alpha = .05$.

12. Responses to stress are governed by the hypothalamus. Imagine you have two groups of subjects. The first shows signs of heart disease and the other does not. You want to determine whether the groups differ in terms of the weight of the hypothalamus. For the first group of subjects with no heart disease, the weights are

$$11.1, 12.2, 15.5, 17.6, 13.0, 7.5, 9.1, 6.6, 9.5, 18.0, 12.6.$$

For the other group with heart disease, the weights are

$$18.2, 14.1, 13.8, 12.1, 34.1, 12.0, 14.1, 14.5, 12.6, 12.5, 19.8, 13.4, 16.8, 14.1, 12.9.$$

Determine whether the groups differ based on Welch's test. Use $\alpha = .05$.

13. The .95 confidence interval for the difference between the means, using Student's t, is (2.2, 20.5). What are the practical concerns with this confidence interval?

14. For the first of two binomial distributions, there are 15 successes among 24 observations. For the second, there are 23 successes among 42 observations. Test H_0: $p_1 = p_2$ with a Type I error probability of .05.

15. A film producer wants to know which of two versions of a particular scene is more likely to be viewed as disturbing. One group of 98 individuals views the first version and 40 say that it is disturbing. The other group sees the second version and 30 of 70 people say that it is disturbing. Test the hypothesis that the two probabilities are equal, using $\alpha = .05$ and compute a .95 confidence interval.

16. It is found that of 121 individuals who take a training program on investing in commodities, 20 make money during the next year and the rest do not. With another training program, 15 of 80 make money. Test the hypothesis that the probability of making money is the same for both training programs, using a Type I error probability of .05.

17. In a study dealing with violence between factions in the Middle East, one goal was to compare measures of depression for two groups of young males. In the first group, no family member was wounded or killed by someone belonging to the opposite faction, and the measures were

$$22, 23, 12, 11, 30, 22, 7, 42, 24, 33, 28, 19, 4, 34, 15, 26, 50, 27, 20, 30, 14, 42.$$

The second group consisted of young males who had a family member killed or wounded. The observed measures were

$$17, 22, 16, 16, 14, 29, 20, 20, 19, 14, 10, 8, 26, 9, 14, 17, 21, 16, 14, 11,$$

$$14, 11, 29, 13, 4, 16, 16, 7, 21.$$

Test the hypothesis of equal means with Student's t test with $\alpha = .05$ and compute a .95 confidence interval.

18. For the data in the last problem, the difference between the medians is -7.5 and the corresponding McKean–Schrader estimate of the standard error is $\sqrt{S_1^2 + S_2^2} = 3.93$. Verify that you do not reject the hypothesis of equal medians with $\alpha = .05$.

19. Referring to the previous two problems, the hypothesis of equal means is rejected, but the hypothesis of equal medians is not. Comment on why this is not surprising.

20. Does the consumption of alcohol limit our attention span? An article in the July 10, 2006 *Los Angeles Times* described a study conducted at the University of Washington where 23 people were given enough alcohol to reach a blood alcohol level of 0.04% (half the legal limit in many states). A second group of 23 people drank no alcohol. The researchers then showed members of both groups a 25 second video clip in which two teams passed a ball back and forth and asked them to count the number of times one team passed the ball. During the clip, a person in a gorilla suit walks through the crowd, thumps its chest, and walks off. Researchers found that 11 of the 23 participants in the control group saw the gorilla, versus 10 in the alcohol group. Verify that you reject the hypothesis that the two groups have the same probability of seeing the gorilla using $\alpha = .05$.

9.2 Comparing two dependent groups

Welch's test and the two-sample Student's t-test, described in the previous section, both assume that the groups being compared are independent. But often dependent groups are compared instead. Imagine, for example, that a training program for increasing endurance is under investigation, the endurance of participants is measured before training starts, they undergo the training program for four weeks, and then their endurance is measured again. An issue is whether there has been a change in the average endurance, but because the same participants are measured both before and after training, it is unreasonable to assume that these two measures are independent. And if they are dependent, Welch's test and the two-sample Student's t-test for comparing means, described in section 9.1, are no longer valid.

As another example, consider again the study where the goal is to assess the effects of ozone on weight gain among rats. Now, however, rather than randomly sampling rats, pairs of rats from the same litter are sampled, one is assigned to the ozone-free group and the other is exposed to ozone. Because these pairs of rats are related, it is unreasonable to assume that their reactions to their environment are independent.

As a final example, consider the issue of whether men differ from women in terms of their optimism about the future of the economy. If married couples are sampled and measured, it is unreasonable to assume that a woman's response is independent of her husband's views.

Assuming normality, there is a simple method for comparing dependent groups. Roughly, randomly sample n pairs of observations, for each pair compute their difference, and then apply the one-sample version of Student's t covered in chapter 7. This can be described in a more formal manner as follows. Denote the n pairs of observations by

$$(X_{11}, X_{12})$$
$$\vdots$$
$$(X_{n1}, X_{n2}).$$

So $X_{11}, \ldots, X_{n1}$ represent the n observations in the first group and $X_{12}, \ldots, X_{n2}$ represent the n observations in the second. Compute all pairwise differences, which is denoted by

$$D_1 = X_{11} - X_{12}$$
$$\vdots$$
$$D_n = X_{n1} - X_{n2}.$$

It can be seen that the population mean of the D values, say μ_D, is equal to difference between the means of the two dependent groups, $\mu_1 - \mu_2$. That is, $\mu_D = \mu_1 - \mu_2$. Consequently, testing the hypothesis of equal means is accomplished by testing the hypothesis that the D values have a population mean of zero. To test $H_0 : \mu_D = 0$, compute the mean and variance of the D values:

$$\bar{D} = \frac{1}{n} \sum_{i=1}^{n} D_i$$

and

$$s_D^2 = \frac{1}{n-1} \sum_{i=1}^{n} (D_i - \bar{D})^2.$$

Next, compute

$$T_D = \frac{\bar{D}}{s_D/\sqrt{n}}.$$

The critical value is t, the $1-\alpha/2$ quantile of Student's t-distribution with $\nu = n-1$ degrees of freedom. The hypothesis of equal means is rejected if $|T_D| \geq t$. This is called the *paired t-test*. A $1-\alpha$ confidence interval for $\mu_D = \mu_1 - \mu_2$, the difference between the means, is

$$\bar{D} \pm t \frac{s_D}{\sqrt{n}}.$$

Example 1

A company wants to know whether a particular treatment reduces the amount of bacteria in milk. To find out, counts of bacteria were made before and after the treatment is applied resulting in the outcomes shown in table 9.3. For instance, based on the first sample, the bacteria count before treatment is $X_{11} = 6.98$, after the treatment it is $X_{12} = 6.95$, and so the reduction in the bacteria count is $D_1 = .03$. The goal is to determine whether, on average, the reduction differs from zero. The sample mean of the difference scores (the D_i values) is $\bar{D} = .258$. The sample variance of the difference scores is $s_D^2 = .12711$. And because the sample size is $n = 12$, $T = 2.5$ and the degrees of freedom are $\nu = 12 - 1 = 11$. With $\alpha = .05$, the critical value is $t = 2.2$, so reject. That is, there is evidence that the treatment results in a reduction in the bacteria count, on average.

Table 9.3 Bacteria counts before and after treatment

Sample (i)	Before treatment (X_{i1})	After treatment (X_{i2})	Difference $D_i = X_{i1} - X_{i2}$
1	6.98	6.95	0.03
2	7.08	6.94	0.14
3	8.34	7.17	1.17
4	5.30	5.15	0.15
5	6.26	6.28	−0.02
6	6.77	6.81	−0.04
7	7.03	6.59	0.44
8	5.56	5.34	0.22
9	5.97	5.98	−0.01
10	6.64	6.51	0.13
11	7.03	6.84	0.19
12	7.69	6.99	0.70

Violating the normality assumption

As was the case when comparing two independent groups, if the two groups have identical distributions, the paired Student's t-test performs well in terms of avoiding Type I error probabilities well above the specified level, given the goal of comparing the means. Roughly, the reason is that for this special case, the distribution of D is symmetric about the value zero, so we have a situation similar to the one described in section 7.3. But if the groups differ in terms of skewness, practical problems can occur in terms of Type I errors when comparing the means, and the corresponding confidence intervals might be inaccurate. And when there are outliers, again, power might be relatively poor. With a sufficiently large sample size, practical problems become negligible, but it can be difficult knowing when this is the case.

Using medians

A simple alternative to using means, when comparing dependent groups, is to use medians. For example, you can simply test the hypothesis that the population median of the difference scores has a value equal to zero using the methods covered in chapter 7. There is, however, an issue that should be made clear. When working with means, a little algebra shows that the difference between the sample means is equal to the mean of the difference scores. That is, $\bar{D} = \bar{X}_1 - \bar{X}_2$. Moreover, the same is true when working with the population means: $\mu_D = \mu_1 - \mu_2$. Consequently, testing H_0: $\mu_D = 0$ is the same as testing H_0: $\mu_1 - \mu_2 = 0$. However, the same is not true when working with medians. There are exceptions, but under general conditions, the median of the difference scores is not equal to the difference between the medians. In symbols, if we let M_D be the sample median of the difference scores, and we let M_1 and M_2 be the sample medians corresponding to the two dependent groups, then usually, $M_D \neq M_1 - M_2$.

Example 2

Consider the data in table 9.3. The median bacteria count before the treatment is $M_1 = 6.875$. The median bacteria count after the treatment is $M_2 = 6.55$. And the median of the difference scores, given in the final column of table 9.3, is $M_D = .145$. As is evident, $M_D \neq M_1 - M_2$.

There are methods for testing the hypothesis that the medians of two dependent groups are equal, rather than testing the hypothesis that the median of the difference scores is zero. But the details are too involved to cover in an introductory course.[3]

Problems

21. For 49 pairs of sisters, a researcher wanted to know whether the older sisters differ, on average, from the younger sisters, based on a test of social anxiety. It was found that $\bar{D} = 3$ and $s_D = 4$. Test the hypothesis of equal means with a Type I error probability of .05. Discuss the interpretation of the result.

22. For the previous problem, compute a .95 confidence interval for the difference between the means. What concerns, if any, might there be about this confidence interval?

23. A course aimed at improving an understanding of good nutrition is attended by 28 students. Before the course began, students took a test on nutrition and got the following scores:

$$72, 60, 56, 41, 32, 30, 39, 42, 37, 33, 32, 63, 54, 47, 91,$$
$$56, 79, 81, 78, 46, 39, 32, 60, 35, 39, 50, 43, 48.$$

After the course was completed, they were tested again yielding the scores

$$66, 53, 57, 29, 32, 35, 39, 43, 40, 29, 30, 45, 46, 51, 79,$$
$$68, 65, 80, 55, 38, 35, 30, 50, 37, 36, 34, 37, 54.$$

Test the hypothesis that the means of the before scores are equal to the means after training using $\alpha = .05$. Also compute a .95 confidence interval.

24. The median of 20 difference scores is found to be $M_D = 5$, there are no tied values, and the McKean–Schrader estimate of the standard error is $S_D = 2$. Verify that you would reject the hypothesis that the difference scores have a population median of zero with $\alpha = .05$.

25. Referring to the results of the previous problem, suppose the medians corresponding to these two groups are 7 and 4, respectively. Is it reasonable to conclude that the hypothesis of equal medians would also be rejected with $\alpha = .05$?

9.3 Some modern advances and insights

Welch's test and Student's t-test were designed to be sensitive to differences between the population means, but in reality they can be sensitive to other features, such as unequal variances or a difference in skewness, as already explained. Consequently, when these tests indicate that groups differ, there can be some doubt about how they differ and by how much. Welch's test avoids this problem with sufficiently large sample sizes, but there is uncertainty about how large the sample sizes must be if the goal is to be sensitive

3. For information about how this is done, see Wilcox (2005).

to the means only. So a basic issue is whether we can find methods based on measures of location that correct this problem.

If the goal is to compare the population means, currently, some type of bootstrap t-method, combined with Welch's test, seems to be among the best methods available. As previously explained, the basic idea is to perform simulations on the data with the goal of determining the distribution of Welch's test statistic, W, when the null hypothesis is true. In essence, you use the data to estimate appropriate critical values rather than assume normality. But it should be noted that even this technique can be sensitive to differences between the groups beyond differences between the population means. This is not to suggest all results based on means are suspect; at a minimum they can help establish that groups differ, but care must be taken when interpreting the results.

Of course, one could compare medians rather than means, in various situations this strategy performs well, but there are exceptions as already described and illustrated. A common recommendation for dealing with non-normality is to use some type of nonparametric method, (see chapter 13). We will see that they have practical value and provide a useful perspective on how groups compare, but awareness of recent advances, relevant to these methods, is strongly recommended.

Yet another approach is to use a compromise amount of trimming, and as previously noted, 20% trimming is often a good choice. One method for comparing 20% trimmed means was derived by Yuen (1974) and reduces to Welch's test when there is no trimming. The computational details, assuming 29% trimming, are as follows.

Let h_1 and h_2 be the number of observations left after trimming when computing the 20% trimmed mean for the first and second groups, respectively. Let $\bar{X}_{t1}$ and $\bar{X}_{t1}$ be the 20% trimmed means and let s_{w1}^2 and s_{w2}^2 be the corresponding 20% Winsorized variances. Let

$$d_1 = \frac{(n_1 - 1)s_{w1}^2}{h_1(h_1 - 1)}$$

and

$$d_2 = \frac{(n_2 - 1)s_{w2}^2}{h_2(h_2 - 1)}.$$

Yuen's test statistic is

$$T_y = \frac{\bar{X}_{t1} - \bar{X}_{t2}}{\sqrt{d_1 + d_2}}. \tag{9.15}$$

and the degrees of freedom are

$$\nu_y = \frac{(d_1 + d_2)^2}{\frac{d_1^2}{h_1 - 1} + \frac{d_2^2}{h_2 - 1}}.$$

The $1 - \alpha$ confidence interval for $\mu_{t1} - \mu_{t2}$, the difference between the population trimmed means, is

$$(\bar{X}_{t1} - \bar{X}_{t2}) \pm t\sqrt{d_1 + d_2}, \tag{9.16}$$

where t is the $1 - \alpha/2$ quantile of Student's t-distribution with ν_y degrees of freedom. The hypothesis of equal trimmed means ($H_0 : \mu_{t1} = \mu_{t2}$) is rejected if

$$|T_y| \geq t.$$

In terms of Type I errors and power, Yuen's method offers a substantial advantage over Welch's test. In principle, differences between means can be larger than differences between trimmed means or medians, which might mean more power when comparing means. Also, methods based on means can be sensitive to differences between groups, such as differences in skewness, which are missed when comparing 20% trimmed means. Generally, however, means usually offer little advantage in terms of power, and trimmed means often provide a substantial advantage. Wu (2002) compared the power of a variety of methods for comparing two independent groups using data from 24 dissertations. No single method was alway best and in some cases methods based on means performed well. But in general, methods based on means had the poorest power and typically the best power was achieved with methods based on 20% trimmed means.

Example 1

Consider again the self-awareness data in example 6 of section 9.1. We saw that the hypothesis of equal medians was rejected with a Type I error probability of $\alpha = .05$; the p-value was .047. In contrast, comparisons based on the means did not come close to rejecting. Applying Yuen's test, it can be seen that $\bar{X}_{t1} = 282.7$, $\bar{X}_{t2} = 444.8$, $\nu = 23$, $T_y = 2.044$, and with $\alpha = .05$, $t = 2.069$, so Yuen's method does not quite reject. (The p-value is .052.) So in this particular case, methods for comparing 20% trimmed means and medians give similar results (their p-values differ by very little), but both of these methods differ considerably, in terms of p-values, compared to using Student's t and Welch's method.

A basic principle underlying this last example is that when groups do not differ in any manner, meaning that they have identical distributions, the choice of method for comparing the groups makes little difference. But if the groups differ, situations are encountered where the the choice of method can make a considerable difference.

The percentile bootstrap method

Chapters 6 and 7 outlined the basics of the percentile bootstrap method, which is readily extended to comparing two groups. For illustrative purposes, imagine that the goal is to compare the medians of two independent groups. Using a computer, we generate a bootstrap sample from the first group. That is, we randomly sample, with replacement, n_1 observations from the first group. The median, based on this bootstrap sample, is labeled M_1^*. We proceed in the same fashion for the second group yielding M_2^*. Then one of three outcomes will occur: $M_1^* < M_2^*$, $M_1^* = M_2^*$, or $M_1^* > M_2^*$. Next, we repeat this process B times, and for the sake of illustration, we use $B = 1000$. Let A be the number of times $M_1^* < M_2^*$ and let C be the number of times $M_1^* = M_2^*$. Proceeding in a manner similar to what was done in section 7.3, let $Q = (A + .5C)/B$ and set P equal to Q or $1 - Q$, whichever is smaller. Then a p-value for testing H_0: $\theta_1 = \theta_2$, the hypothesis that the population medians are equal, is $p = 2P$. Currently, this is the only method for comparing medians that has been found to provide reasonably accurate control over the probability of a Type I error when tied (duplicated) values are likely to occur.

A confidence interval for the difference between the population medians can be computed as well. Let $U^* = M_1^* - M_2^*$. That is, based on a bootstrap sample from each group, U^* is the difference between the resulting medians. Repeating this process

1000 times yields 1000 U^* values, which we label as $U_1^*, \ldots, U_{1000}^*$. If we put these 1000 values in ascending order, the middle 95% provide a .95 confidence interval for the difference between the population medians.

Example 2

Using a computer, we find that among 1000 bootstrap samples from each of two independent groups, there were 900 instances where $M_1^* < M_2^*$, and there were 26 instances where $M_1^* = M_2^*$. That is, $B = 1000$, $A = 900$ and $C = 26$. Consequently, $Q = (A + .5C)/B = (900 + .5(26))/1000 = .913$, $P = 1 - .913 = .087$, and the p-value is $2 \times .087 = .174$.

The percentile bootstrap method does not perform well when the goal is to compare means. But it does perform very well when comparing 20% trimmed means.

Comments on plotting data

Various graphical methods can help provide perspective on how groups differ and by how much. That is, they can help assess effect size. Among some researchers, one popular graphical tool is called an *error bar*, which is just a plot of the means for the two groups under study plus an interval around each mean based on the standard error. Examples of such a graph are shown in figure 9.3 using the data from example 6 in section 9.1. The circles indicate the means. In the left panel, the ends of the lines extending above the circles indicate the value of the sample mean plus one standard error. Similarly, the lines extending below the circles indicate the value of the sample mean minus one standard error. In symbols, the top and bottom ends of the lines indicate the value of $\bar{X} + s/\sqrt{n}$ and $\bar{X} - s/\sqrt{n}$, respectively. For the first group, the sample mean is $\bar{X} = 448.1$, the sample standard deviation is $s = 353624$, the sample size is 19, so the estimated standard error is $s/\sqrt{n} = 34.33$. Consequently, the bottom end of the first vertical line marks the value

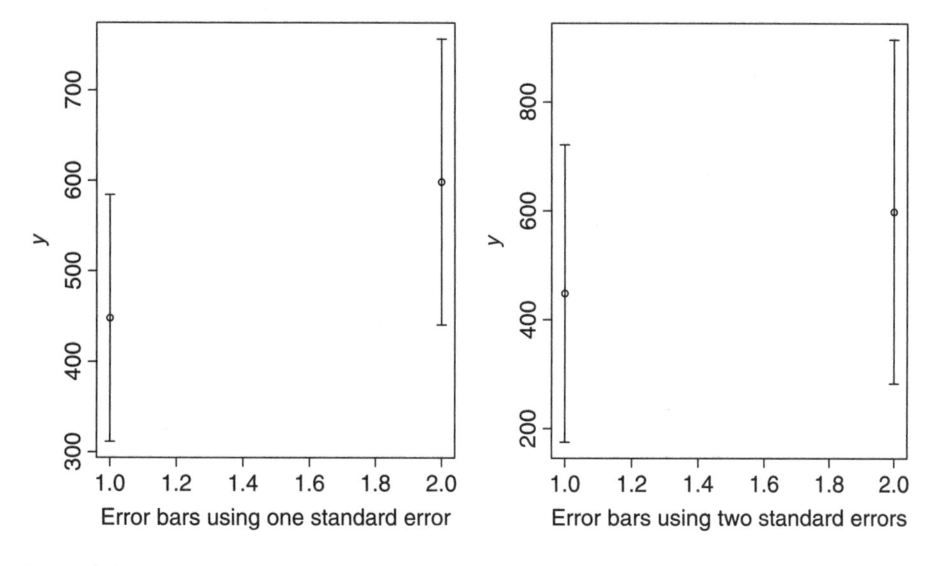

Figure 9.3 Example of error bars using the self awareness data in example 6 of section 9.1.

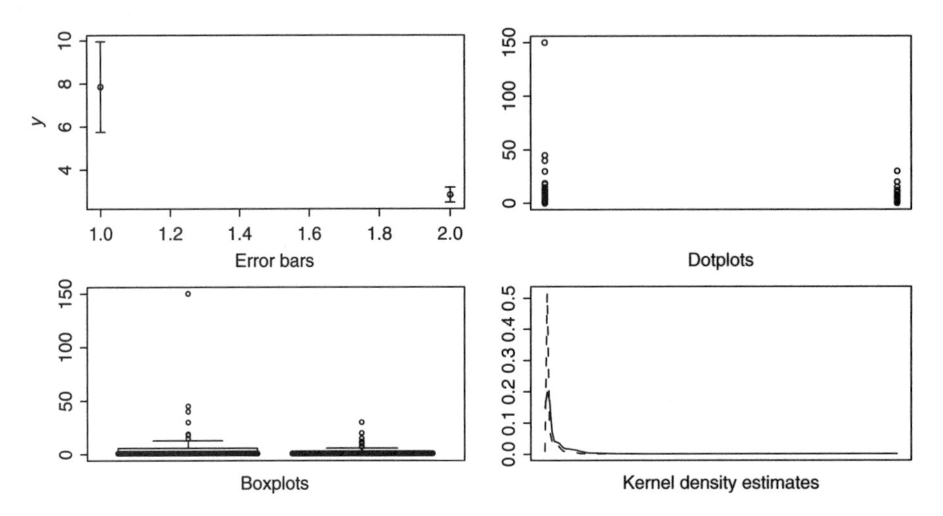

Figure 9.4 Four graphical summaries of the sexual attitude data (with the extreme outlier among the male response removed).

of $\bar{X}-s/\sqrt{n}=413.77$ and the top end indicates the value $\bar{X}+s/\sqrt{n}=482.43$. That is, the ends of the lines indicate a confidence interval for the population mean. In the left panel, the error bars were computed using two standard errors about the mean rather than one.

Consider again the sexual attitude data used in example 9 of section 9.1. The upper left corner of figure 9.4 shows the resulting error bars (with the extreme outlier removed). Another graphical approach is to simply create a dot plot for each group. That is, simply plot points corresponding to each of the values under study. The upper right corner shows such a plot for the sexual attitude data. (The dots on the left reflect the responses given by males.) Boxplots and kernel density estimates are other possibilities. The lower left corner shows a boxplot of the sexual attitude data and the lower right panel shows a kernel density estimate for each group. (The solid line is the plot for males.)[4]

Although error bars provide useful information, note that compared to the other graphs, error bars can be relatively uninformative about the overall nature of the data. Also, in Figure 9.4, plots of the means suggest a large difference between the two groups, in contrast to the other three plots. Yet one more concern about error bars should be noted. Imagine that the error bars are constructed so as to indicate a .95 confidence interval for each mean. If the intervals do not overlap, as is the case in figure 9.4, this might suggest that the hypothesis of equal means should be rejected. It can be seen, however, that even if both intervals have exactly a .95 probability of containing the mean, this approach does not control the probability of a Type I error when testing the hypothesis of equal means (e.g., Schenker and Gentleman, 2001; Wilcox, 2003, section 8.4). Even with large sample sizes, poor control over the probability of a Type I error results, and so in terms of testing hypotheses, error bars should not be used.

4. The plots based on the kernel density estimators were created with the R function g2plot, which belongs to a library of functions mentioned in chapter 1.

A Summary of Some Key Points

- Currently, the most commonly used method for comparing groups is Student's *t*-test for means.
- Different methods can provide different conclusions about whether groups differ and by how much. There are exceptions, but often reliance on a single method for understanding how, and if, groups differ can be unsatisfactory. (This suggests using multiple methods for comparing groups, but there are some technical issues that should be taken into account when performing multiple tests. Details will be covered in chapter 11.)
- Methods based on medians can reduce power problems associated with means; medians can provide more accurate confidence intervals and better power if outliers are common. But if distributions are normal or outliers are relatively rare, some other method might offer more power. Tied values can destroy the ability of some methods for comparing medians, in terms of controlling the probability of a Type I error or providing accurate confidence intervals, but all indications are that a percentile bootstrap method effectively deals with this problem.
- One strategy for reducing problems that can occur, when using means or medians is to compare groups, is to use a compromise amount of trimming, and as usual, a 20% trimmed mean appears to be a good choice in many instances. (Chapter 13 will describe yet another approach that deserves serious consideration.)
- There are other graphical methods for comparing groups that can be useful but which are not covered here. One of these is called a *Q-Q plot* and another is called a *shift function*. Both deal with comparing all of the quantiles. So included as a special case, they display the difference between the medians and the quartiles. An advantage of the shift function is that confidence intervals for the difference between all of the quantiles are provided, assuming random sampling only. The result is a detailed indication of where and how groups differ.[5]

Problems

26. Despite any problems it might have, summarize how you would justify using Student's t-test to compare two independent groups.

27. Summarize any practical concerns about Student's t-test and comment on how they might be addressed.

28. Summarize the relative merits of comparing groups with medians.

29. For two independent groups, 1000 bootstrap samples are generated and it is found that there are 10 instances where the bootstrap sample trimmed mean for the first group is less than the bootstrap sample trimmed mean for the second. And there 2 instances where the bootstrap sample trimmed means are equal. Determine a p-value when testing the hypothesis of equal population trimmed means.

30. Verify that for the self-awareness data in table 9.1, when applying Yuen's method, the test statistic is $T_y = 2.044$.

5. For appropriate software, see Wilcox, 2003.

10

COMPARING MORE THAN TWO GROUPS

Chapter 9 described methods for comparing two groups, but quite often more than two groups are of interest. For example, a researcher might have four methods for treating schizophrenia, in which case there is the issue of whether the choice of method makes a difference. As another example, several drugs might be used to control high blood pressure. Do the drugs differ in terms of side effects?

Imagine that the goal is to compare J groups having population means $\mu_1, \ldots, \mu_J$. The most common strategy is to begin by testing

$$H_0 : \mu_1 = \cdots = \mu_J, \tag{10.1}$$

the hypothesis that all J groups have equal means. The immediate goal is to describe methods aimed at testing this hypothesis when dealing with independent groups. Then, at various points, the relative merits of these methods are discussed.

10.1 The ANOVA *F* test for independent groups

We begin with a classic, routinely used method for testing the hypothesis given by equation (10.1) when dealing with independent groups. As in chapter 9, independent groups means that the observations in any two groups are independent. In the schizophrenia example, this requirement would be met if we randomly sample participants who will undergo one of the treatments, and then these participants are randomly assigned to one and only one treatment. If, for instance, all four treatment methods were applied to the same participants, independence would no longer be a reasonable assumption and the methods covered here would be inappropriate.

The method assumes

- Random sampling
- Normality
- All J groups have the same variance.

Letting $\sigma_1^2, \ldots, \sigma_J^2$ denote the population variances, this last assumption means that

$$\sigma_1^2 = \sigma_2^2 = \cdots = \sigma_J^2, \tag{10.2}$$

which is called the *homogeneity of variance assumption*. For convenience, this assumed common variance is denoted by σ_p^2. *Heteroscedasticity* refers to a situation where not all the variances are equal.

Outline of the classic method

The basic strategy for testing equation (10.1), which was derived by Sir Ronald Fisher, arises as follows. First imagine that the null hypothesis of equal means is true. Then the sample means corresponding to the J groups under study are all attempting to estimate the same quantity, yet their individual values will vary. A rough characterization of the classic method is that it attempts to determine whether the variation among the sample means is sufficiently large to reject the hypothesis of equal population means. The more variation among the sample means, the stronger the evidence that the null hypothesis should be rejected.

Let $\bar{X}_1, \ldots, \bar{X}_J$ be the sample means and momentarily consider the case where all groups have the same sample size, which is denoted by n. We begin by computing the average of the sample means,

$$\bar{X}_G = \frac{1}{J} \sum \bar{X}_j,$$

which is called the *grand mean*. Next, we measure the variation among these means in much the same way the sample variance, s^2, measures the variation among n observations. The variation among the sample means is given by

$$V = \frac{1}{J-1} \sum (\bar{X}_j - \bar{X}_G)^2.$$

In words, V is computed by subtracting the grand mean from each of the individual means, squaring the results, adding these squared differences, and then dividing by $J - 1$, the number of groups minus one. It can be shown that when the null hypothesis of equal means is true, V estimates σ_p^2/n, the assumed common variance divided by the sample size. Multiplying V by the sample size, n, yields what is called the *mean squares between groups*, which estimates the assumed common variance σ_p^2 *when the null hypothesis is true.*[1] Put another way, the mean squares between groups is given by

$$\text{MSBG} = \frac{n}{J-1} \sum (\bar{X}_j - \bar{X}_G)^2.$$

However, when the null hypothesis is false, MSBG does not estimate σ_p^2, it estimates σ_p^2 plus a quantity that reflects how much the population means differ. That is,

1. A commonly used term is sums of squares between groups, which is $\text{SSBG}=n\sum(\bar{X}_j - \bar{X}_G)^2$.

the more unequal the population means happen to be, the larger will be MSBG, on average.

To say this in a more precise way, let

$$\bar{\mu} = \frac{1}{J} \sum \mu_j$$

be the average of the population means and let

$$\sigma_\mu^2 = \frac{\sum(\mu_j - \bar{\mu})^2}{J-1},$$

which represents the variation among the population means. In general, MSBG is estimating

$$\sigma_p^2 + n\sigma_\mu^2.$$

This says that the value of MSBG is affected by two quantities: the variation within each group, represented by σ_p^2, plus the variation among the population means, which is reflected by σ_μ^2. When the means are all equal, there is no variation among the population means, meaning that $\sigma_\mu^2 = 0$. That is, MSBG is estimating σ_p^2.

Next, let $s_1^2, \ldots, s_J^2$ represent the sample variances corresponding to the J groups. By assumption, all of these sample variances estimate the common (population) variance σ_p^2. As was done in chapter 9, we simply average the sample variances to get a single estimate of σ_p^2, still assuming equal sample sizes. This average is called the *mean squares within groups* and is given by

$$\text{MSWG} = \frac{1}{J} \sum s_j^2.$$

A key result is that when the hypothesis of equal means is true, both MSBG and MSWG are attempting to estimate the same quantity, namely, the assumed common variance. When the null hypothesis is false, MSWG continues to estimate the assumed common variance, but now MSBG is estimating something larger, meaning that on average, MSBG will tend to be larger than MSWG.

Based on the results just summarized, we reject the hypothesis of equal means if MSBG is sufficiently larger than MSWG. A convenient way of measuring the extent to which they differ is with

$$F = \frac{\text{MSBG}}{\text{MSWG}}. \tag{10.3}$$

Note that if each group has n observations, the total number of observations in all J groups is $N = nJ$. The distribution of F, when the null hypothesis is true, is called an *F distribution* with degrees of freedom

$$\nu_1 = J - 1,$$

and

$$\nu_2 = N - J.$$

That is, the F distribution depends on two quantities: the number of groups being compared, J, and the total number of observations in all of the groups, N.

Moreover, the hypothesis of equal means is rejected if $F \geq f$, where f is the $1-\alpha$ quantile of an F distribution with $\nu_1 = J-1$ and $\nu_2 = N-J$ degrees of freedom. (This is the same F distribution introduced in chapter 8.) Tables 5, 6, 7 and 8 in appendix B report critical values, f, for $\alpha = .1$, .05, .025 and .01 and various degrees of freedom. For example, with $\alpha = .05$, $\nu_1 = 6$, $\nu_2 = 8$, table 6 indicates that the .95 quantile is $f = 3.58$. That is, there is a .05 probability of getting a value for F that exceeds 3.58 when in fact the population means are equal. For $\alpha = .01$, table 8 says that the .99 quantile is 6.32. This means that if you reject when $F \geq 6.32$, the probability of a Type I error will be .01, assuming normality and that the groups have equal variances.

The method just described for comparing means is called an *analysis of variance* or *ANOVA F* test. Although the goal is to compare means, it goes by the name analysis of variance because the method is based on the variation of the sample means relative to the variation within each group. Table 10.1 outlines what is called an analysis of variance summary table; it is a classic way of summarizing the computations. Table 10.2 summarizes the analysis based on the data in table 10.1.

Example 1

Imagine a study aimed at comparing four groups with ten observations in each group. That is, $J = 4$ and $n = 10$, so the total sample size is $N = 40$. Consequently, the degrees of freedom are $\nu_1 = 3$ and $\nu_2 = 36$. Figure 10.1 shows the distribution of F when the hypothesis of equal means is true. As indicated, the .95 quantile is $f = 2.87$, meaning that the hypothesis of equal means is rejected if $F \geq 2.87$.

The description of the ANOVA F test was done with equal sample sizes primarily for convenience. The method is readily applied when the sample sizes are unequal using the steps in box 10.1. Many software packages perform the calculations. What is perhaps more important is understanding the relative merits of the method, which is a topic discussed momentarily. But in case it helps, the following example illustrates the computational steps.

Table 10.1 Analysis of variance (ANOVA) summary table

Source of variation	Degrees of freedom	Sum of squares	Mean square	F
Between groups	J-1	SSBG	MSBG	$F = \frac{\text{MSBG}}{\text{MSWG}}$
Within groups	N-J	SSWG	MSWG	
Totals	N-1	SSBG + SSWG		

Table 10.2 ANOVA summary table for the data in table 10.1

Source of variation	Degrees of freedom	Sum of squares	Mean square	F
Between groups	3	0.36	0.12	1.51
Within groups	36	2.8542	0.0793	
Totals	39	3.2142		

BOX 10.1 **Summary of the ANOVA F test with or without equal sample sizes.**

Notation

X_{ij} refers to the ith observation from the jth group, $i = 1, \ldots, n_j$; $j = 1, \ldots, J$. (There are n_j observations randomly sampled from the jth group.)

Computations

$$A = \sum \sum X_{ij}^2$$

(In words, square each value, add the results, and call it A.)

$$B = \sum \sum X_{ij}$$

(In words, sum all the observations and call it B.)

$$C = \sum_{j=1}^{J} \frac{1}{n_j} \left(\sum_{i=1}^{n_j} X_{ij} \right)^2$$

(Sum the observations for each group, square the result, divide by the sample size, add the results corresponding to each group.)

$$N = \sum n_j$$

$$\text{SST} = A - \frac{B^2}{N}$$

$$\text{SSBG} = C - \frac{B^2}{N}$$

$$\text{SSWG} = \text{SST} - \text{SSBG} = A - C$$

$$\nu_1 = J - 1$$

$$\nu_2 = N - J$$

$$\text{MSBG} = \text{SSBG}/\nu_1$$

$$\text{MSWG} = \text{SSWG}/\nu_2$$

Test Statistic

$$F = \frac{\text{MSBG}}{\text{MSWG}}.$$

Decision Rule

Reject H_0 if $F \geq f$, the $1 - \alpha$ quantile of an F distribution with $\nu_1 = J - 1$ and $\nu_2 = N - J$ degrees of freedom.

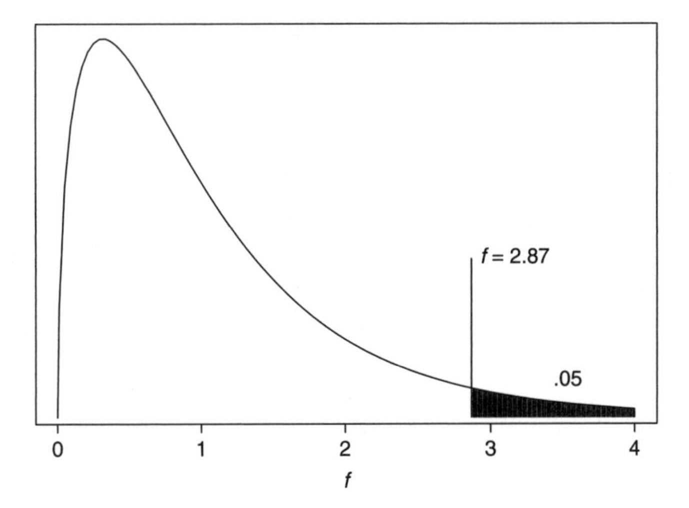

Figure 10.1 The distribution of F when comparing four groups with 10 observations in each group and the hypothesis of equal means is true. That is, $J = 4$, $n = 10$, so $\nu_1 = 4 - 1 = 3$ and $\nu_2 = 4(10) - 4 = 36$. If the Type I error is to be $\alpha = .05$, reject if $F \geq 2.87$.

Example 2

The computations in box 10.1 are illustrated with the following data:

Group 1: 7, 9, 8, 12, 8, 7, 4, 10, 9, 6
Group 2: 10, 13, 9, 11, 5, 9, 8, 10, 8, 7
Group 3: 12, 11, 15, 7, 14, 10, 12, 12, 13, 14.

We see that

$A = 7^2 + 9^2 + \cdots + 14^2 = 3{,}026,$

$B = 7 + 9 + \cdots + 14 = 290,$

$C = \dfrac{(7 + 9 + \cdots + 6)^2}{10} + \dfrac{(10 + 13 + \cdots + 7)^2}{10} + \dfrac{(13 + 11 + \cdots + 14)^2}{10} = 2{,}890,$

$N = 10 + 10 + 10 = 30,$

$\text{SST} = 3{,}026 - 290^2/30 = 222.67,$

$\text{SSBG} = 2{,}890 - 290^2/30 = 86.67,$

$\text{SSWG} = 3{,}026 - 2890 = 136,$

$\text{MSBG} = 86.67/(3 - 1) = 43.335,$

$\text{MSWG} = 136/(30 - 3) = 5.03,$

so

$$F = \frac{43.335}{5.03} = 8.615$$

The degrees of freedom are $\nu_1 = 3 - 1 = 2$ and $\nu_2 = 30 - 3 = 27$. With $\alpha = .01$ we see from table 8 in appendix B, the critical value is $f = 5.49$. Because $8.165 \geq 5.49$, reject the hypothesis of equal means.

When does the ANOVA *F* test perform well?

Similar to the two-sample Student's t-test in chapter 9, if the J groups under study do not differ in any manner, meaning that all J groups have the same distribution, the ANOVA F test performs well in the sense that the actual Type I error probability will not exceed the specified level by very much. So if the goal is to have the Type I error probability equal to .05, the actual Type I error probability will, in general, be less than or equal to .05 when sampling from distributions that are non-normal but otherwise identical. In particular, not only do the groups have equal means, they do not differ in terms of their variances or the amount of skewness. In practical terms, given the goal of controlling the probability of a Type I error, the ANOVA F test provides a satisfactory test of the hypothesis that J groups have identical distributions, meaning that if we reject, it is reasonable to conclude that two or more groups differ in some manner.

But if the goal is to develop a test that is sensitive to means, without being sensitive to other ways the groups might differ, the ANOVA F test can be highly unsatisfactory. Indeed, the ANOVA F test suffers from problems similar to those described in chapter 9 when using Student's t-test. In fact, as the number of groups increases, practical problems are exacerbated. For example, in chapter 9, it was noted that under normality, if the sample sizes are equal but the population variances differ, Student's t controls the probability of a Type I error fairly well except possibly for very small sample sizes. But when comparing four groups with the ANOVA F test, this is no longer true.

Example 3

For instance, with equal sample sizes of 50 for each group, if all four groups have a normal distribution with a mean of 0, the first group has a standard deviation of 4, and the rest have a standard deviation of 1, the actual Type I error probability is approximately .088 when testing at the $\alpha = .05$ level. Non-normality only makes matters worse. Although the seriousness of a Type I error will vary from one study to the next, it has been suggested that if the goal is to have the Type I error probability equal .05, at a minimum, the actual Type I error probability should not exceed .075 (e.g., Bradley, 1978).

Perhaps a more serious issue has to do with the ability of the ANOVA F test to detect true differences. There are situations where it performs well, such as when its underlying assumptions are true. But for a variety of reasons, it can have poor power relative to other methods that might be used. In fairness, we encounter situations where it makes little difference which method is used, and but we also encounter situations where the ANOVA F test does not come close to rejecting, yet more modern methods detect a difference, a result that was already illustrated in chapter 9 when using Student's t. Roughly, the more groups being compared, the more likely it is that relatively poor power will result when using the ANOVA F test due to differences in skewness, unequal variances, and outliers.

Example 4

Consider the following values for four groups.

$$\text{Group } 1: -0.89, 1.22, 1.03, 2.02, 1.41, 1.69, -1.48, 1.96, 0.44, 0.41, 1.71, 3.04, -0.63, 1.77, 2.50.$$

$$\text{Group } 2: 2.75, 0.26, 0.75, -1.52, 1.15, 1.00, 0.78, -0.01, 0.92, -0.68, 0.91, -0.41, 0.44, 0.43, 1.34.$$

$$\text{Group } 3: 0.12, -0.45, 0.37, -1.75, 1.31, 0.23, -0.53, 0.73, 0.54, -1.08, -0.65, -1.89, -0.53, 0.20, 0.09.$$

$$\text{Group } 4: 1.49, 0.10, 2.21, 0.90, -0.17, -0.84, 1.18, -0.64, -1.32, -0.89, 0.27, -0.93, 0.56, -0.27, -0.27.$$

These values were generated from normal distributions where the first group has a population mean of 1, and the other three groups have population means equal to 0. It can be seen that the ANOVA F test rejects with a Type I error probability set equal to $\alpha = .01$. (The p-value can be seen to be .008.) That is, the F test makes a correct decision about whether the groups differ. But suppose one of the values in the fourth group happens to be an outlier. In particular, what if by chance the first observation in this group were 6 rather than 1.49. Now the ANOVA F test no longer rejects with a Type I error probability set to $\alpha = .05$. Yet, had we compared the first three groups, ignoring the fourth, we would again reject with $\alpha = .01$. This illustrates that a single outlier in any group has the potential of masking true differences among the other groups under study, even if the other groups have normal distributions.

Example 5

Consider a situation where four normal distributions are to be compared, each has variance 1, the first group has mean 1 and the rest have means equal to 0. If all four groups have a sample size of 15 and the Type I error probability is set at .05, then the power of the F test is, approximately, .71. But now consider the same situation, only the fourth group has the contaminated normal distribution described in section 4.6. So the population mean is again 0, but now the power of the F test is only .38. This illustrates that if the normality assumption is violated in even one group, true differences among the remaining groups can be missed as a result, even if they have normal distributions with a common variance.

Currently, the practical concerns about power, when using the ANOVA F test, are rarely discussed or illustrated in an introductory course. But it is easily the most commonly used method when comparing multiple groups, and so it is important to be aware of the method and have some sense of its relative merits. An argument for using the ANOVA F might be that there are conditions where it has relatively high power, possibly because it can be sensitive to differences among the groups that other methods tend to miss. Nevertheless, it is unrealistic to assume that the ANOVA F will have high power—often situations arise where some alternative method is much more likely to detect true differences.

One suggestion for trying to salvage the F test is to first test the hypothesis of equal variances, and if this hypothesis is not rejected, assume equal variances and use F. Chapter 9 pointed out that this strategy is known to fail when comparing two groups and it continues to fail when using the ANOVA F, even when dealing with normal distributions (Markowski and Markowski, 1990; Moser et al. 1989; Wilcox et al. 1986; Zimmerman, 2004). The basic problem is that tests for equal variances do not have enough power to detect situations where violating the assumption of equal variances causes practical problems. As in chapter 9, presumably this is no longer true if the sample sizes are sufficiently large, but it is unclear just how large the sample sizes must be, particularly when dealing with non-normal distributions.

Dealing with unequal variances: Welch's test

Many methods have been proposed for testing the equality of J means without assuming equal variances (e.g., Chen and Chen, 1998; Mehrotra, 1997; James, 1951; Krutchkoff, 1988; Alexander and McGovern, 1994; Fisher, 1935, 1941; Cochran and Cox, 1950; Wald, 1955; Asiribo and Gurland, 1989; Lee and Ahn, 2003; Scariano and Davenport, 1986; Matuszewski and Sotres, 1986; Pagurova, 1986; Weerahandi, 1995.) Unfortunately, all of the methods just cited, plus many others, have been found to have serious practical problems (e.g., Keselman, Wilcox, Taylor and Kowalchuk, 2000; Keselman and Wilcox, 1999). If the goal is to use a method that is sensitive to means, without being sensitive to other ways the groups might differ, the method described here is not completely satisfactory, but it performs reasonably well under normality and heteroscedasticity and it forms the basis of a technique that deals with non-normality. The method is due to Welch (1951) and the computational details are described in box 10.2.

BOX 10.2 Computations for Welch's method

Goal

Without assuming equal variances, test $H_0 : \mu_1 = \mu_2 = \cdots = \mu_J$, the hypothesis that J independent groups have equal means.

Computations

Let

$$w_1 = \frac{n_1}{s_1^2}, w_2 = \frac{n_2}{s_2^2}, \ldots, w_J = \frac{n_J}{s_J^2}.$$

Next, compute

$$U = \sum w_j$$

$$\tilde{X} = \frac{1}{U} \sum w_j \bar{X}_j$$

$$A = \frac{1}{J-1} \sum w_j (\bar{X}_j - \tilde{X})^2$$

$$B = \frac{2(J-2)}{J^2-1} \sum \frac{(1-\frac{w_j}{U})^2}{n_j - 1}$$

$$F_w = \frac{A}{1+B}.$$

When the null hypothesis is true, F_w has, approximately, an F distribution with

$$\nu_1 = J - 1$$

and

$$\nu_2 = \left[\frac{3}{J^2-1} \sum \frac{(1-w_j/U)^2}{n_j - 1} \right]^{-1}$$

degrees of freedom.

Decision Rule

Reject H_0 if $F_w \geq f$, where f is the $1-\alpha$ quantile of the F distribution with ν_1 and ν_2 degrees of freedom.

Example 6

Consider again, the data used in Example 3, where all four groups have normal distributions with a common variance. That is, the assumptions underlying the ANOVA F test are true. Computing Welch's test statistic, F_w, as described in box 10.2, it can be seen that $F_w = 3.87$, the degrees of freedom are $\nu_1 = 3$ and $\nu_2 = 30.9$ and with $\alpha = .01$, the critical value is approximately 4.486. So unlike the ANOVA F test, the hypothesis of equal means is not rejected. However, with $\alpha = .02$, now Welch's test rejects. The p-value for Welch's test is .018 versus .008 when using the ANOVA F. If the assumptions of the ANOVA F test are true, it will have better power than Welch's test, but generally the improvement is not that striking. However, when there is more variability in some groups versus the others, Welch's test can detect a true difference among the means in situations where the ANOVA F does not, even when sampling from normal distributions.

Example 7

The following data illustrate that the ANOVA F and Welch's test can yield substantially different p-values, which can alter your decision about whether to reject the hypothesis of equal means:

Group 1: 53, 2, 34, 6, 7, 89, 9, 12
Group 2: 7, 34, 5, 12, 32, 36, 21, 22
Group 3: 5, 3, 7, 6, 5, 8, 4, 3.

The ANOVA F test yields $F = 2.7$ with a critical value of 3.24 when the Type I error is taken to be $\alpha = .05$, so you do not reject. (The p-value is .09.) In contrast, Welch's test yields $F_w = 8$ with a critical value of 4.2, so now you reject. (The p-value is .009.)

Comparing groups based on medians

Many methods for comparing the medians of multiple groups have been derived.[2] One general approach is to begin by testing

$$H_0 : \theta_1 = \cdots = \theta_J, \tag{10.4}$$

where $\theta_1, \ldots, \theta_J$ are the population medians. Relatively simple techniques have been proposed, but they can be rather unsatisfactory. There are two methods that seem to perform tolerably well except possibly when tied values occur, but the details are too involved to give here. Only one method has been found to be generally satisfactory, even when tied values occur, and it is based on a simple extension of the percentile bootstrap method outlined in chapter 9; the details are outlined in chapter 11.

Dealing with dependent groups

Often, the groups to be compared are dependent. Consider, for example, a study aimed at investigating the relationship between environment and aggression. To shed light on this issue, a psychologist randomly samples litters of mice, and then four mice from each litter are selected, with each mouse placed in one of four different environments. After some time, the aggressiveness of each mouse is measured and the goal is to determine whether the groups differ based on this measure. Because mice from the same litter are used, it is unreasonable to assume that the observations in any two groups are independent, and special methods that allow dependence are required.

As another example, an investigator might want to know how a particular drug affects cholesterol levels over time. If cholesterol levels are measured before the study begins, and then measured again every month for four months, there are five groups to be compared corresponding to the five times measures were taken. Because the same individuals are measured at all five time points, independence cannot be assumed. Such studies are called *within-subjects designs* or *repeated-measures designs*.

It is briefly noted that there is a classic variation of the ANOVA F test aimed at testing the hypothesis of equal means among dependent groups. This classic approach requires, in addition to normality, that the variances and covariance satisfy a property called *sphericity* (e.g. Kirk, 1995), but complete details are not provided. (If, for example, all the groups have the same variance, and all pairs of groups have the same Pearson correlation, sphericity is achieved.) It is merely remarked that violating this assumption creates practical problems when trying to control the probability of a Type I error. Under normality, this problem can be reduced by using what is called the Hyunh-Feldt correction, which is reported by some of the more popular software packages (such as SPSS). But like all methods based on means, non-normality can wreak havoc on power, and if the goal is to be sensitive to differences among the means, without being sensitive

2. For a recent comparison of various methods, see Wilcox (2006).

to other ways the groups might differ, again there are general situations where this ANOVA F test can be unsatisfactory.

Problems

1. For the following data,

Group 1	Group 2	Group 3
3	4	6
5	4	7
2	3	8
4	8	6
8	7	7
4	4	9
3	2	10
9	5	9

$\bar{X}_1 = 4.75 \quad \bar{X}_2 = 4.62 \quad \bar{X}_3 = 7.75$
$s_1^2 = 6.214 \quad s_2^2 = 3.982 \quad s_3^2 = 2.214,$

assume that the three groups have a common population variance, σ_p^2. Estimate σ_p^2.

2. For the data in the previous problem, test the hypothesis of equal means using the ANOVA F. Use $\alpha = .05$.

3. For the data in problem 1, verify that Welch's test statistic is $F_w = 7.7$ with degrees of freedom $\nu_1 = 2$ and $\nu_2 = 13.4$. Then verify that you would reject the hypothesis of equal means with $\alpha = .01$.

4. Construct an ANOVA summary table using the following data, as described in section 10.1, then test the hypothesis of equal means with $\alpha = .05$.

Group 1	Group 2	Group 3	Group 4
15	9	17	13
17	12	20	12
22	15	23	17

5. In the previous problem, what is your estimate of the assumed common variance?

6. For the data used in the last two problems, verify that for Welch's test, $F_w = 3.38$ with $\nu_1 = 3$ and $\nu_2 = 4.42$.

7. Based on the results of the previous problem, would you reject the hypothesis of equal means with $\alpha = .1$?

8. Why would you not recommend the strategy of testing for equal variances, and if not significant, using the ANOVA F test rather than Welch's method?

9. Five independent groups are compared with $n = 15$ observations for each group. Fill in the missing values in the following summary table.

Source of variation	Degrees of freedom	Sum of squares	Mean square	F
Between groups	–	50	–	–
Within groups	–	150	–	

10. Referring to box 10.2, verify that for the following data, MSBG = 14.4 and MSWG = 12.59.

G1	G2	G3
9	16	7
10	8	6
15	13	9
	6	

11. Consider five groups ($J = 5$) with population means 3, 4, 5, 6, and 7, and a common variance $\sigma_p^2 = 2$. If the number of observations is 10 ($n = 10$), indicate what is being estimated by MSBG, and based on the information given, determine its value. That is, if the population means and common variance were known, what would the value of MSBG be if it were giving perfectly accurate information? How does this differ from the value estimated by MSWG?

12. For the following data, verify that you do not reject with the ANOVA F testing with $\alpha = .05$, but you do reject with Welch's test.

    Group 1: 10 11 12 9 8 7
    Group 2: 10 66 15 32 22 51
    Group 3: 1 12 42 31 55 19

    What might explain the discrepancy between the two methods?

13. Consider the following ANOVA summary table:

Source of variation	Degrees of freedom	Sum of squares	Mean square	F
Between Groups	3	300	100	10
Within Groups	8	80	10	
Total	11	428		

    Verify that the number of groups is $J = 4$, the total number of observations is $N = 12$ and that with $\alpha = .025$ the critical value is 5.42.

14. A researcher reports a p-value of .001 with the ANOVA F test. Describe what conclusions are reasonable based on this result.

15. Summarize the reasons you might fail to reject with the ANOVA F test.

16. Someone tests for equal variances and fails to reject. Does this justify the use of the ANOVA F test?

17. A researcher reports that a test for normality was performed, and that based on this test, no evidence of non-normality was found. Why might it be unreasonable to assume normality despite this result?

18. Outline how you might construct an example where sampling is from normal distributions, Welch's test rejects, but the ANOVA F test does not.

10.2 Two-way ANOVA

An important and commonly used generalization of the ANOVA F test has to with what are called two-way designs. *One-way designs* refer to studies where multiple levels of a single factor are of interest. For example, a farmer might want to determine how different amounts of fertilizer affect the growth of her crops. If five amounts of fertilizer are under consideration, this reflects a one-way design because a single variable, amount of fertilizer, is being manipulated. There are five *levels* for this factor, which correspond to the five amounts of fertilizer that are under consideration. One of the examples given at the beginning of this chapter had to do with four methods for treating schizophrenia. The factor here is method of treatment, and there are four levels because four types of treatment are to be compared.

A two-way design is like a one-way design, only two factors are of interest rather than just one. Consider, for example, a study aimed at understanding behavior when individuals diet in order to lose weight. In a study described by Atkinson et al. (1985, pp. 324–325), two groups of individuals were identified; those who were on a diet to lose weight and those who were not. So far, this is a one-way design with two levels. But the researchers wanted to understand the effects of a second factor—forced eating. Some participants were forced to consume two milk shakes, others had one milk shake, and a third group consumed no milk shakes at all. So now we have a two-way design, where the second factor has three levels and reflects how many milk shakes were consumed. After consuming the milk shakes, the participants sampled several flavors of ice cream and were encouraged to eat as much as they wanted. The outcome measure of interest is the amount of ice cream consumed. It was found that the more milk shakes consumed by individuals not on a diet, the less ice cream they would consume later. In contrast, the dieters who had two milk shakes ate more ice cream than those who had drank one milk shake or none.

Here, the basic concepts of a two-way design are illustrated with a study where the goal is to understand the effect of diet on weight gains in rats. Specifically, four diets are considered which differ in: (1) amounts of protein (high and low) and (2) the source of the protein (beef versus cereal). So this is a two-way design with two levels for each factor. The results for these four groups are reported in table 10.3 and are taken from

Table 10.3 Weight gains (in grams) of rats on one of four diets

Beef		Cereal	
Low	High	Low	High
90	73	107	98
76	102	95	75
90	118	97	56
64	104	80	111
86	81	98	95
51	107	74	88
72	100	74	82
90	87	67	77
95	117	89	86
78	111	58	92
$\bar{X}_1 = 79.2$	$\bar{X}_2 = 100$	$\bar{X}_3 = 83.9$	$\bar{X}_4 = 85.9$

Table 10.4 Depiction of the population means for four diets

	Source	
Amount	Beef	Cereal
High	μ_1	μ_2
Low	μ_3	μ_4

Snedecor and Cochran (1967). Different rats were used in the four groups, so the groups are independent. The first column gives the weight gains of rats fed a low protein diet with beef the source of protein. The next column gives the weight gains for rats on a high protein diet again with beef the source of protein, and the next two columns report results when cereal is substituted for beef.

It is convenient to depict the population means as shown in table 10.4. Table 10.4 indicates, for example, that μ_1 is the population mean associated with rats receiving a low protein diet from beef. That is, μ_1 is the average weight gain if all of the millions of rats we might study are fed this diet. Similarly, μ_4 is the population mean for rats receiving a low protein diet from cereal.

Rather than just compare the means of the four groups, often it is desired to compare the levels of each factor, ignoring the other. For example, you might want to compare the rats receiving a high versus low protein diet, *ignoring* the source of the protein. To illustrate how this might be done, imagine that the values of the population means are as follows:

Amount	Source	
	Beef	Cereal
High	$\mu_1 = 45$	$\mu_2 = 60$
Low	$\mu_3 = 80$	$\mu_4 = 90$

For rats on a high protein diet, the mean is 45 when consuming beef versus 60 when consuming cereal instead. If you want to characterize the typical weight gain for a high-protein diet ignoring source, a natural strategy is to average the two population means yielding $(45 + 60)/2 = 52.5$. That is, the typical rat on a protein diet gains 52.5 grams. For the more general situation depicted by table 10.4, the typical weight gain on a high protein diet would be $(\mu_1 + \mu_2)/2$, the average of the means over source of protein. Similarly, the typical weight gain for a rat on a low-protein diet would be $(\mu_3 + \mu_4)/2$, which in the example is $(80 + 90)/2 = 85$ grams. Of course, you can do the same when dealing with source of protein, ignoring amount. The typical weight gain for a rat eating beef, ignoring amount of protein, is $(45 + 80)/2 = 62.5$, and for cereal it is $(60 + 90)/2 = 75$.

What is needed is some way of testing the hypothesis that weight gain is different for a high protein diet versus a low protein diet, ignoring source of protein. One way of doing this is to test

$$H_0: \frac{\mu_1 + \mu_2}{2} = \frac{\mu_3 + \mu_4}{2},$$

the hypothesis that the average of the populations means when the source of protein is beef is equal to the average for cereal. If this hypothesis is rejected, then there is said to be a *main effect* for the amount of protein. More generally, a main effect for the first factor (amount) is said to exist if

$$\frac{\mu_1+\mu_2}{2} \neq \frac{\mu_3+\mu_4}{2}.$$

Similarly, you might want to compare source of protein ignoring amount. One approach is to test

$$H_0: \frac{\mu_1+\mu_3}{2} = \frac{\mu_2+\mu_4}{2},$$

the hypothesis that the average of the means in the column for beef is equal to the average for the column headed by cereal. If this hypothesis is rejected, then there is said to be a **main effect** for the source of protein. More generally, a main effect for the second factor is said to exist if

$$\frac{\mu_1+\mu_3}{2} \neq \frac{\mu_2+\mu_4}{2}.$$

Interactions

There is one other important feature of a two-way design. Consider again a 2 by 2 design where the goal is to compare high and low protein diets in conjunction with two protein sources. Suppose the *population* means associated with the four groups have the values previously indicated. Now look at the first row (high amount of protein) and notice that the weight gain for a beef diet is 45 grams versus a weight gain of 60 for cereal. As is evident, there is an increase of 15 grams. In contrast, with a low protein diet, switching from beef to cereal results in an increase of 10 grams on average. That is, in general, switching from beef to cereal results in an increase for the average amount of weight gained, but the increase differs depending on whether we look at high or low protein. This is an example of what is called an *interaction.*

More formally, an *interaction* is said to exist if

$$\mu_1-\mu_2 \neq \mu_3-\mu_4.$$

In words, an interaction exists if for the first level of factor A the difference between the means is not equal to the difference between the means associated with the second level of factor A. *No interaction* means that these differences are equal. In symbols, no interaction means that

$$\mu_1-\mu_2 = \mu_3-\mu_4.$$

The basic ideas just described have been extended to more complex situations where the first factor has J levels and the second has K. Consider, for example, a study of survival times of animals given one of three poisons. This is the first factor, which has $J=3$ levels. Further imagine that four methods aimed at treating the animals are of interest. This is the second factor, which has $K=4$ levels. This type of design is often called a J-by-K design. In the example, we have a 3-by-4 design.

To describe a common approach to analyzing J by K designs, imagine that the population means are labeled as shown in table 10.5. So, for example, μ_{11} is the mean of the group associated with the first level of first and second factors. The mean

Table 10.5 Commonly used notation for the means in a J-by-K ANOVA

	Factor B			
	μ_{11}	μ_{12}	$\cdots$	μ_{1K}
	μ_{21}	μ_{22}	$\cdots$	μ_{2K}
Factor A	$\vdots$	$\vdots$	$\vdots$	$\vdots$
	μ_{J1}	μ_{J2}	$\cdots$	μ_{JK}

corresponding to the third level of factor A and the fourth level of factor B is denoted by μ_{34}. The following example provides some indication of how the groups are often compared.

Example 1

Consider again the 3 by 4 design design where the first factor corresponds to three types of poison and the second factor reflects four types of treatments. Using the notation in table 10.5, μ_{11} is the mean survival time when an animal is exposed to the first poison and is given the first treatment. Similarly, μ_{12} is mean survival time when an animal is exposed to the first poison and is given the second treatment. The *marginal mean* survival time when exposed to the first poison is just the average of the means among the four treatments. That is, the marginal mean is

$$\bar{\mu}_{1.} = \frac{\mu_{11} + \mu_{12} + \mu_{13} + \mu_{14}}{4}.$$

In a similar manner, the marginal means for the second and third poisons are

$$\bar{\mu}_{2.} = \frac{\mu_{21} + \mu_{22} + \mu_{23} + \mu_{24}}{4}$$

and

$$\bar{\mu}_{3.} = \frac{\mu_{31} + \mu_{32} + \mu_{33} + \mu_{34}}{4}.$$

And a common way of comparing the levels of Factor A, ignoring factor B, is to test the hypothesis that the marginal means are equal. In symbols, test

$$H_0 : \bar{\mu}_{1.} = \bar{\mu}_{2.} = \bar{\mu}_{3.}.$$

When this hypothesis is rejected, it is said that there is a main effect for factor A. In the example, this means that typical survival times, ignoring treatment, differ.

As is probably evident, comparing levels of factor B, ignoring factor A, can be handled in a similar manner. Now, for each level of factor B, the strategy is to consider the average of the population means over the levels of factor A. Then the goal is to test the hypothesis that the resulting marginal means are equal.

Example 2

Consider again the previous example, only now the goal is to compare treatments, ignoring the type of poison. For each type of treatment, the average survival times over the type of poison are

$$\bar{\mu}_{.1} = \frac{\mu_{11} + \mu_{21} + \mu_{31}}{3},$$
$$\bar{\mu}_{.2} = \frac{\mu_{12} + \mu_{22} + \mu_{32}}{3},$$
$$\bar{\mu}_{.3} = \frac{\mu_{13} + \mu_{23} + \mu_{33}}{3},$$

and

$$\bar{\mu}_{.4} = \frac{\mu_{14} + \mu_{24} + \mu_{34}}{3}.$$

Then a natural way of comparing treatments, ignoring the type of poison, is to test

$$H_0 : \bar{\mu}_{.1} = \bar{\mu}_{.2} = \bar{\mu}_{.3} = \bar{\mu}_{.4}.$$

Finally, there is the issue of interactions for the more general case of a J-by-K ANOVA design. An interaction is said to exist if there is an interaction for any two levels of factor A and any two levels of factor B.

Example 3

Continuing the previous two examples, consider the first and third levels of factor A. That is, we focus on survival times when dealing with the first and third types of poison. Simultaneously, consider the second and fourth types of treatments. So we are considering four means:

$$\begin{matrix} \mu_{12} & \mu_{14} \\ \mu_{32} & \mu_{34}. \end{matrix}$$

Extending the earlier description of an interaction in an obvious way, if $\mu_{12} - \mu_{14} = \mu_{32} - \mu_{34}$, there is no interaction among these four means. If instead $\mu_{12} - \mu_{14} \neq \mu_{32} - \mu_{34}$, there is an interaction. More generally, no interaction is said to exist if for any two levels of factor A, say i and i', and any two levels of factor B, say say k and k',

$$\mu_{ik} - \mu_{ik'} = \mu_{i'k} - \mu_{i'k'}.$$

Example 4

Imagine that unknown to us, the population means are

		Factor B			
		Level 1	Level 2	Level 3	Level 4
	Level 1	40	40	50	60
Factor A	Level 2	20	20	50	80
	Level 3	20	30	10	40

Looking at level 1 of factor A, we see that the means increase by 0 as we move from level 1 of factor B to level 2. The increase for level 2 of factor A is again 0. In particular, the difference between the first two means is the same as the difference between the other two, so there is no interaction for these four means. However, looking at level 1 of factor A, we see that the means increase by 10 as we move from level 1 to level 3 of factor B. In contrast, there is an increase of 30 for level 2 of factor A, which means that there is an interaction. In formal terms, the difference between the first two means is not equal to the difference between the other two.

The focus has been on understanding what is meant by 'no main effects' and 'no interactions' in a two-way ANOVA design. But nothing has been said about how these hypotheses might be tested. Suffice it so say that again, main effects and interactions can be tested, assuming normality and equal variances but, for brevity, the tedious computational details are not provided. However, a brief outline of the quantities routinely computed might be of some use. Recall that in a one-way design, a quantity is computed that reflects the variation among the means; it is called the sum of squares between groups, and multiplying this quantity by the sample size yields an estimate of the assumed common variance called the mean squares between groups (MSBG). Similar quantities are computed in a two-way design. When the means associated with the first factor are equal, we get a quantity called the mean squares for factor A, MSA, which estimates the assumed common variance. Similarly, the second factor is typically called factor B, and when the means are equal, a quantity called the mean squares for factor B, MSB, estimates the assumed common variance. Finally, when the hypothesis of no interaction is true, a quantity called the *mean squares interaction*, MSINTER, also estimates the assumed common variance. When any of these three hypotheses are false, the corresponding mean squares term tends to be larger than the mean squares within groups (MSWG). The test statistic for factor A is

$$F = \frac{\text{MSA}}{\text{MSWG}},$$

and the degrees of freedom are $\nu_1 = J - 1$ and $\nu_2 = N - JK$. For factor B the test statistic is

$$F = \frac{\text{MSB}}{\text{MSWG}},$$

with degrees of freedom $\nu_1 = K - 1$ and $\nu_2 = N - JK$. And for the hypothesis of no interactions, the test statistic is

$$F = \frac{\text{MSINTER}}{\text{MSWG}},$$

with degrees of freedom $\nu_1 = (J - 1)(K - 1)$ and $\nu_2 = N - JK$.

Example 5

Consider again the study of survival times where animals are given one of three poisons and of interest are four methods aimed at treating the animals. So $J = 3$, $K = 4$, and for illustrative purposes, assume 10 animals are assigned to each of these 12 groups. So the total sample size is $N = 3 \times 4 \times 10 = 120$. Imagine that a computer program reports that for the first factor, MSA=200 and MSWG=103. Then F=1.94, and the degrees of freedom are $\nu_1 = 2$ and $\nu_2 = N - JK = 108$. If we want the Type I error probability to be $\alpha = .05$, from table 6 in appendix B, the critical value is $f = 3.08$, and because F is less than 3.08, fail to reject. If MSINTER=250, then $F = 250/103 = 2.42$, the degrees of freedom are $\nu_1 = 6$ and $\nu_2 = N - JK = 108$, the critical value is 2.18, and because F exceeds this critical value, reject the hypothesis of no interaction.

Violating assumptions

The two-way ANOVA F test assumes normality and equal variances. As was the case for the one-way design, violating these assumptions can result in the test being sensitive to features other than differences among the means. Problems do not always arise, but it is unrealistic to assume that this issue can be ignored. So as before, the ANOVA F can provide an indication that groups differ, but it might not adequately isolate the nature of the difference. There are methods aimed at allowing unequal variances, which include extensions of Welch's test, but no details are given here. And as usual, when comparing groups with any method based on means, there is some risk of relatively low power.

A Summary of Some Key Points

- The methods in this chapter are aimed at comparing groups based on means. In terms of controlling the probability of a Type I error, they perform well when all groups have identical distributions. That is, not only are the means equal, the variances are equal, groups have the same amount of skewness, and so on.
- If the methods in this chapter reject, it is reasonable to conclude that two or more of the groups differ in some manner. But because the methods are sensitive to more than just differences among the means, there is uncertainty about how groups differ when we reject. Also, the methods in this chapter do not indicate which groups differ. (This issue is addressed in chapter 11.)
- Generally, the more groups that are compared, the more sensitive the ANOVA F becomes to differences among the groups beyond the means. That is, if the goal is control Type I errors when all groups have equal means, the more groups we have, the more difficult it is to achieve this goal.

Continued

A Summary of Some Key Points (*cont'd*)

- If the goal is to compare groups in a manner that is sensitive to a measures of location (such as the median), without being sensitive to other differences that might exist, more advanced techniques should be used. These advanced methods help isolate how groups differ and by how much.
- As is the case with all methods based on means, when using the methods in this chapter, power might be low relative to other methods that have been developed in recent years. But the methods described here are routinely used, so they are important to know.

Problems

19. Consider a 2-by-2 design with population means

		Factor B	
		Level 1	Level 2
Factor A	Level 1	$\mu_1 = 110$	$\mu_2 = 70$
	Level 2	$\mu_3 = 80$	$\mu_4 = 40$

State whether there is a main effect for factor A, for factor B, and whether there is an interaction.

20. Consider a 2-by-2 design with population means

		Factor B	
		Level 1	Level 2
Factor A	Level 1	$\mu_1 = 10$	$\mu_2 = 20$
	Level 2	$\mu_3 = 40$	$\mu_4 = 10$

State whether there is a main effect for factor A, for factor B, and whether there is an interaction.

21. A computer program reports that for a 2-by-3 ANOVA, with 15 observations in each group, MSA = 400, MSB = 200, MSINTER = 200, MSWG = 50. Perform the tests of no main effects and no interaction with $\alpha = .01$.

22. A computer program reports that for a 4-by-5 ANOVA, with 10 observations in each group, MSA = 600, MSB = 400, MSINTER = 300, MSWG = 100. Perform the tests of no main effects and no interaction with $\alpha = .05$.

10.3 Modern advances and insights

There is a variety of methods aimed at improving upon the ANOVA F test and Welch's method for comparing multiple groups. Many of these modern methods help isolate the nature of any differences that might exist. For example, some methods are designed to be sensitive to differences in measures of location, without being sensitive to other ways the groups might differ. That is, when you reject, there is more certainty that this is due to differences in measures of location. When the ANOVA F test rejects, the primary reason might be differences among the variances or different amounts of skewness. Perhaps a more serious concern with both the ANOVA F test and Welch's method is that they can have poor power, relative to more modern techniques. As in previous chapters, some of the best methods are based on 20% trimmed means. Methods for dependent groups as well as a two-way ANOVA design are available as well. Bootstrap methods also have practical value. However, the details of all of these methods are much too involved to give here.

11

MULTIPLE COMPARISONS

Chapter 10 described how to test the hypothesis that two or more groups have a common mean. In symbols, the goal was to test

$$H_0 : \mu_1 = \cdots = \mu_J. \tag{11.1}$$

But typically, one wants to know more about how the groups compare: which groups differ, how do they differ, and by how much?

Note that rather than test the hypothesis given by equation (11.1), another approach is to test the hypothesis of equal means for all pairs of groups. For example, if there are four groups ($J = 4$), methods in chapter 9 could be used to test the hypothesis that the mean of the first group is equal to the mean of the second, the mean of first group is equal to the mean of third, and so on. In symbols, the goal is to test

$$H_0 : \mu_1 = \mu_2,$$

$$H_0 : \mu_1 = \mu_3,$$

$$H_0 : \mu_1 = \mu_4,$$

$$H_0 : \mu_2 = \mu_3,$$

$$H_0 : \mu_2 = \mu_4,$$

$$H_0 : \mu_3 = \mu_4.$$

There is, however, a technical issue that needs to be taken into account. Suppose there are no differences among the groups, in which case none of the six null hypotheses just listed should be rejected. To keep things simple for the moment, assume all four groups have normal distributions with equal variances, in which case Student's t-test in chapter 9 provides exact control over the probability of a Type I error when testing any single hypothesis. Further assume that each of the six hypotheses just listed is tested with $\alpha = .05$. So for *each* hypothesis, the probability of a Type I error is .05, but the probability of *at least one* Type I error, when performing all six tests, is approximately .2. That is, there is about a 20% chance of erroneously concluding that two or more groups differ when in fact none of them differ at all. With six groups, there are 15 pairs of means to be compared, and now the probability of at least one Type I error is about .36. In general, as the number of groups increases, the more likely we are to find a difference when none exists if we simply compare each pair of groups using $\alpha = .05$. To deal with this problem, many methods have been proposed with the goal of controlling what is called the familywise error rate (FWE). The *familywise error rate* (sometimes called the *experimentwise error rate*) is the probability of making at least one Type I error when performing multiple tests.

11.1 Classic methods for independent groups

We begin by describing classic methods for comparing the means of independent groups that assume normality and that the groups have equal population variances. So for J groups, homoscedasticity is assumed meaning that

$$\sigma_1^2 = \cdots = \sigma_J^2.$$

These methods are generally called multiple comparison procedures, which simply means multiple hypotheses are to be tested. Concerns with these classic methods have been known for some time, but they are commonly used, so it is important to be aware of them and their relative merits.

Fisher's least significant difference method

One of the earliest strategies for comparing multiple groups is the so-called *least significant difference* (LSD) *method* due to Sir Ronald Fisher. Assuming normality and homoscedasticity, first perform the ANOVA F test in chapter 10. If the hypothesis of equal means is rejected, apply Student's t to all pairs of means, but unlike the approach in chapter 9, typically the assumption of equal variances is taken advantage of by using the data from all J groups to estimate the assumed common variance when any two groups are compared with Student's t. Under normality and homoscedasticity, this has the advantage of increasing the degrees of freedom, which in turn can mean more power.

To elaborate, suppose the ANOVA F test rejects with a Type I error probability of $\alpha = .05$ and let MSWG (the mean squares within groups described in chapter 10) be the estimate of the assumed common variance. (So for equal sample sizes, MSWG is just the average of the sample variances among all J groups.) Consider any two groups, say groups j and k. The goal is to test

$$H_0 : \mu_j = \mu_k, \tag{11.2}$$

the hypothesis that the mean of the jth group is equal to the mean of the kth group. In the present context, the test statistic is

$$T = \frac{\bar{X}_j - \bar{X}_k}{\sqrt{MSWG\left(\frac{1}{n_j} + \frac{1}{n_k}\right)}}. \tag{11.3}$$

When the assumptions of normality and homoscedasticity are met, T has a Student's t-distribution with $\nu = N - J$ degrees of freedom, where J is the number of groups being compared and $N = \sum n_j$ is the total number of observations in all J groups. So when comparing the jth group to the kth group, you reject the hypothesis of equal means if

$$|T| \geq t_{1-\alpha/2},$$

Table 11.1 Hypothetical data for three groups

G1	G2	G3
3	4	6
5	4	7
2	3	8
4	8	6
8	7	7
4	4	9
3	2	10
9	5	9

where $t_{1-\alpha/2}$ is the $1-\frac{\alpha}{2}$ quantile of Student's t-distribution with $N-J$ degrees of freedom.

Example 1

The method is illustrated using the data in table 11.1. Assume that the probability of at least one Type I error is to be .05. It can be seen that MSWG $=4.14$, the sample means are $\bar{X}_1=4.75$, $\bar{X}_2=4.62$, and $\bar{X}_3=7.75$, and the F test rejects when the Type I error probability is $\alpha=.05$. So according to Fisher's LSD method, you would proceed by comparing each pair of groups with Student's t-test performed with $\alpha=.05$. For the first and second groups ($j=1$ and $k=2$),

$$T=\frac{|4.75-4.62|}{\sqrt{4.14(\frac{1}{8}+\frac{1}{8})}}=.128.$$

The degrees of freedom are $\nu=21$, and with $\alpha=.05$, table 4 in appendix B says that the critical value is $t=2.08$. Therefore, you fail to reject. That is, the F test indicates that there is a difference among the three groups, but Student's t suggests that the difference does not correspond to groups 1 and 2. For groups 1 and 3,

$$T=\frac{|4.75-7.75|}{\sqrt{4.14(\frac{1}{8}+\frac{1}{8})}}=2.94,$$

and because 2.94 is greater than the critical value, 2.08, reject. That is, conclude that groups 1 and 3 differ. In a similar manner, you conclude that groups 2 and 3 differ as well because $T=3.08$.

When the assumptions of normality and homoscedasticity are true, Fisher's method controls FWE when $J=3$. That is, the probability of at least one Type I error will be less than or equal to α. However, when there are more than three groups ($J>3$), this is not necessarily true (Hayter, 1986). To gain some intuition as to why, suppose four groups are to be compared, the first three have equal means, but the mean of the fourth group is so much larger than the other three that power is close to one. That is, with near certainty, you will reject with the ANOVA F test and proceed to compare all pairs of means with Student's t at the α level. So in particular you will test

$$H_0: \mu_1 = \mu_2,$$

$$H_0: \mu_1 = \mu_3,$$

$$H_0: \mu_2 = \mu_3,$$

each at the α level, all three of these hypotheses are true, and the probability of at least one Type I error among these three tests will be greater than α.

The Tukey-Kramer method

Tukey was the first to propose a method that controls FWE, the probability of at least one Type I error. He assumed normality and homoscedasticity (equal variances) and obtained an exact solution when all J groups have equal sample sizes. Kramer (1956) proposed a generalization that provides an approximate solution when the sample sizes are unequal and Hayter (1984) showed that when the groups have equal population variances and sampling is from normal distributions, Kramer's method is conservative. That is, it guarantees that FWE will be less than or equal to α.

When comparing the jth group to the kth group, the Tukey-Kramer $1-\alpha$ confidence interval for the difference between the means, $\mu_j - \mu_k$, is

$$(\bar{X}_j - \bar{X}_k) \pm q\sqrt{\frac{\text{MSWG}}{2}\left(\frac{1}{n_j} + \frac{1}{n_k}\right)}, \tag{11.4}$$

where n_j is the sample size of the jth group, MSWG is again the mean square within groups, which estimates the assumed common variance, and q is a constant read from table 9 in appendix B, which depends on the values of α, J (the number of groups being compared), and the degrees of freedom,

$$\nu = N - J,$$

where again N is the total number of observations in all J groups. When testing $H_0: \mu_j = \mu_k$, the test statistic is

$$T = \frac{\bar{X}_j - \bar{X}_k}{\sqrt{\frac{\text{MSWG}}{2}\left(\frac{1}{n_j} + \frac{1}{n_k}\right)}}$$

and the hypothesis is rejected in $|T| \geq q$. Alternatively, reject if the confidence interval, given by equation (11.4), does not contain zero. Under normality, equal variances and equal sample sizes, the probability of at least one Type I error, when no two groups differ, is exactly α.

Example 2

Table 11.2 shows some hypothetical data on the ratings of three methods for treating migraine headaches. Each method is rated by a different sample of individuals. The total number of participants is $N = 23$, so the degrees

Table 11.2 Ratings of methods for treating migraine headaches

Method 1	Method 2	Method 3
5	6	8
4	6	7
3	7	6
3	8	8
4	4	7
5	5	
3	8	
4	5	
8		
2		

of freedom are $\nu = N - J = 23 - 3 = 20$, the sample means are $\bar{X}_1 = 4.1$, $\bar{X}_2 = 6.125$, and $\bar{X}_3 = 7.2$, and the estimate of the common variance is MSWG $= 2.13$. If the probability of at least one Type I error is to be $\alpha = .05$, table 9 in appendix B indicates that $q = 3.58$. When comparing groups 1 and 3 ($j = 1$ and $k = 3$), the test statistic is

$$T = \frac{4.1 - 7.2}{\sqrt{\frac{2.13}{2}\left(\frac{1}{10} + \frac{1}{5}\right)}} = -5.48.$$

Because $|-5.48| \geq 3.58$, the hypothesis of equal means is rejected. The confidence interval for $\mu_1 - \mu_3$, the difference between the means corresponding to groups 1 and 3, is

$$(4.1 - 7.2) \pm 3.58\sqrt{\frac{2.13}{2}\left(\frac{1}{10} + \frac{1}{5}\right)} = (-5.12, -1.1).$$

This interval does not contain zero, so again you reject the hypothesis that the typical ratings of methods 1 and 3 are the same. You can compare methods 1 to 2 and methods 2 to 3 in a similar manner, but the details are left as an exercise.

Some important properties of the Tukey-Kramer method

There are some properties associated with Tukey-Kramer method that are important to keep in mind. First, in terms of controlling the probability of at least one Type I error, the method does not assume, nor does it require, that you must first reject with the ANOVA F test. This is in contrast to Fisher's method, which requires that you first reject with the ANOVA F test. Second, if the Tukey-Kramer method is applied only if the ANOVA F test rejects, its properties are changed. For example, if we ignore the F test and simply apply the Tukey-Kramer method with $\alpha = .05$, then under normality and when the population variances and sample sizes are equal, the probability of at least one Type I error, when none of the groups differ, is exactly .05. But if the Tukey-Kramer method is applied only after the F test rejects, this is no longer true, the probability of at least one Type I error will be less than .05 (Bernhardson, 1975). In practical terms, when it comes to controlling the probability of at least one Type I error, there is no need to first reject with the ANOVA F test to justify using the Tukey-Kramer method. And if the Tukey-Kramer method is used only after the F test rejects, power can be reduced. Currently, however, the usual practice is to use the Tukey-Kramer method only if the F test rejects. That is, the insight reported by Bernhardson (1975) is not yet well known.

It was previously noted that if, for example, three groups differ and have normal distributions, but there is an outlier in a fourth group, the F test can fail to reject and consequently miss the true differences among the first three groups. Notice that even if the Tukey-Kramer method is used without first requiring that the F test rejects, this problem is not necessarily corrected. The reason is that it is based on the sample variances of all the groups. For instance, when comparing groups 1 and 2, the sample variances from the other groups are used to compute MSWG. The point is that if there are outliers in any of these other groups, they can inflate the corresponding variance, which in turn can inflate MSWG, which can result in low power when comparing groups 1 and 2. More generally, outliers in any group can can result in low power among groups where no outliers occur.

Example 3

Consider the following data for three groups

$$G1: 268, 114, -21, 313, 128, 239, 227, 59, 379, 100$$

$$G2: -209, -37, -10, 151, 126, -151, 41, 158, 59, 22$$

$$G3: 32, 187, -21, -54, 14, 169, -17, 304, 134, -103.$$

It can be seen that MSWG is 15,629 and that when comparing groups 1 and 2 with the Tukey-Kramer method, $T = 4.2$. If the probability of at least one Type I error is to be .05, table 9 in appendix B indicates that the critical value is $q = 3.5$, approximately. Because the absolute value of T exceeds 3.5, reject the hypothesis that groups 1 and 2 have equal means. But suppose the first two observations in the third group are increased to 400 and 500, respectively. Then MSWG increases to 24,510, and now, when comparing groups 1 and 2, $T = 3.34$, and the hypothesis of equal means is no longer rejected, even though none of the values in groups 1 and 2 were altered. A way of avoiding the problem just illustrated is to switch to one of the methods described in the next section of this chapter.

Scheffé's method

There is one more classic method that should be mentioned that can be used to compare all pairs of groups and which is designed to control FWE (the probability of at least one Type I error). It is called *Scheffé's method*, which assumes normality and that groups have equal variances. The computational details are not given here, but a property of the method should be described. If the goal is to compare all pairs of groups, Scheffé (1959) shows that the Tukey-Kramer method is preferable because if the underlying assumptions are true, Scheffé's method will have less power versus Tukey-Kramer. The reason is that Scheffé's method is too conservative in terms of Type I errors. That is, FWE will not be exactly .05, but less than .05 to the point that the Tukey-Kramer method will be more likely to detect true differences. Both the Tukey-Kramer and Scheffé's method can be used for more complex comparisons that are not covered here. It turns out that for some situations, Scheffé's method is preferable to the Tukey-Kramer method, but the details go beyond the scope of this book. But, like the Tukey-Kramer method, when the assumptions of equal variances and normality are not true, this method can perform poorly relative to more modern techniques.

Problems

1. Assuming normality and homoscedasticity, what problem occurs when comparing multiple groups with Student's t-test?

2. For five independent groups, assume that you plan to do all pairwise comparisons of the means and you want FWE to be .05. Further assume that $n_1 = n_2 = n_3 = n_4 = n_5 = 20$, $\bar{X}_1 = 15$, $\bar{X}_2 = 10$, $s_1^2 = 4$ and $s_2^2 = 9$, $s_3^2 = s_4^2 = s_5^2 = 15$, test $H_0 : \mu_1 = \mu_2$ using Fisher's method, assuming the ANOVA F test rejects.

3. Repeat the previous problem, only use the Tukey-Kramer method

4. For four independent groups, assume that you plan to do all pairwise comparisons of the means and you want FWE to be .05. Assume $n_1 = n_2 = n_3 = n_4 = n_5 = 10$ $\bar{X}_1 = 20$, $\bar{X}_2 = 12$, $s_1^2 = 5$, $s_2^2 = 6$, $s_3^2 = 4$, $s_4^2 = 10$, and $s_5^2 = 15$. Test $H_0 : \mu_1 = \mu_2$ using Fisher's method.

5. Repeat the previous problem, only use the Tukey-Kramer method

6. Imagine you compare four groups with Fisher's method and you reject the hypothesis of equal means for the first two groups. If the largest observation in the fourth group is increased, what happens to MSWG? What does this suggest about power when comparing groups 1 and 2 with Fisher's method?

7. Repeat the previous problem but with the Tukey-Kramer method.

11.2 Methods that allow unequal population variances

All of the methods in the previous section are known to be unsatisfactory, in terms of Type I errors, when groups have unequal population variances, even when the normality assumption is true (e.g., Dunnett, 1980a; Jeyaratnam and Othman, 1985). As in previous chapters, when groups do not differ in any manner, meaning that they have identical distributions, the Tukey-Kramer and Scheffé methods are satisfactory. But in terms of computing accurate confidence intervals and achieving relatively high power, they can be highly unsatisfactory under general conditions. Important advances toward more satisfactory solutions are methods that allow unequal variances. Many such methods have been proposed that are designed to control the probability of at least one Type I error (FWE), comparisons of which were made by Dunnett (1980b), assuming normality. Still assuming that independent groups are to be compared, two of the methods that performed well in Dunnett's study are described here.

Dunnett's T3 method

Dunnett's so-called T3 procedure is just Welch's method described in chapter 9, but with the critical value adjusted so that FWE is approximately equal to α when sampling from normal distributions. Let s_j^2 be the sample variance for the jth group, let n_j be the sample size, and set

$$q_j = \frac{s_j^2}{n_j}.$$

When comparing group j to group k, the degrees of freedom are

$$\hat{\nu}_{jk} = \frac{(q_j + q_k)^2}{\frac{q_j^2}{n_j - 1} + \frac{q_k^2}{n_k - 1}}.$$

The test statistic is

$$W = \frac{\bar{X}_j - \bar{X}_k}{\sqrt{q_j + q_k}},$$

and $H_0 : \mu_j = \mu_k$, the hypothesis that groups j and k have equal means, is rejected if $|W| \geq c$, where the critical value, c, is read from table 10 in appendix B.[1] A confidence interval for $\mu_j - \mu_k$, the difference between the means of groups j and k, is given by

$$(\bar{X}_j - \bar{X}_k) \pm c\sqrt{\frac{s_j^2}{n_j} + \frac{s_k^2}{n_k}}.$$

When using table 10, you need to know the total number of comparisons you plan to perform. When performing all pairwise comparisons, the total number of comparisons is

$$C = \frac{J^2 - J}{2}.$$

In the illustration, there are three groups ($J = 3$), so the total number of comparisons to be performed is

$$C = \frac{3^2 - 3}{2} = 3.$$

If you have four groups *and* you plan to perform all pairwise comparisons, $C = (4^2 - 4)/2 = 6$.

1. Table 10, appendix B provides the .05 and .01 quantiles of what is called the Studentized maximum modulus distribution.

Example 1

Four methods for back pain are being investigated. For each method, 10 randomly sampled individuals are treated for two weeks and then they report the severity of their back pain. The results are as follows:

Method 1	Method 2	Method 3	Method 4
5	2	3	4
2	0	0	4
3	0	0	7
3	4	0	3
0	3	1	2
0	0	8	3
0	2	2	2
1	4	0	4
0	0	2	1
13	3	0	9

Tedious calculations yield the following results:

Method	Method	$\|W\|$	c	ν
1	2	0.655	3.09	12.10
1	3	0.737	2.99	15.11
1	4	0.811	3.00	14.82
2	3	0.210	2.98	15.77
2	4	2.249	2.97	16.06
3	4	2.087	2.93	17.98

For example, when comparing methods 1 and 2, the absolute value of the test statistic is $|W| = 0.655$ and the critical value is $c = 3.09$, based on degrees of freedom $\nu = 12.1$, so the hypothesis of equal means is not rejected. For all other pairs of groups, again the hypothesis of equal means is not rejected.

Games-Howell method

An alternative to Dunnett's T3 is the Games and Howell (1976) method. When comparing the jth group to the kth group, you compute the degrees of freedom, $\hat{\nu}_{jk}$, exactly as in Dunnett's T3 procedure, and then you read the critical value, q, from table 9 in appendix B. (table 9 reports some quantiles of what is called the Studentized range distribution.) The $1-\alpha$ confidence interval for $\mu_j - \mu_k$ is

$$(\bar{X}_j - \bar{X}_k) \pm q\sqrt{\frac{1}{2}\left(\frac{s_j^2}{n_j} + \frac{s_k^2}{n_k}\right)}.$$

You reject $H_0 : \mu_j = \mu_k$ if this interval does not contain zero, which is the same as rejecting if

$$\frac{|\bar{X}_j - \bar{X}_k|}{\sqrt{\frac{1}{2}\left(\frac{s_j^2}{n_j} + \frac{s_k^2}{n_k}\right)}} \geq q.$$

Under normality, the Games-Howell method appears to perform better than Dunnett's T3, in terms of Type I errors, when all groups have a sample size of at least fifty. A close competitor under normality is Dunnett's (1980b) C method, but no details are given here.

Example 2

Imagine you have three groups with $\bar{X}_1 = 10.4$, $\bar{X}_2 = 10.75$,

$$\frac{s_1^2}{n_1} = .11556,$$

$$\frac{s_2^2}{n_2} = .156.$$

Then $\hat{\nu} = 19$ and with $\alpha = .05$, $q = 3.59$, so the confidence interval for the difference between the population means, $\mu_1 - \mu_2$, is

$$(10.4 - 10.75) \pm 3.59\sqrt{\frac{1}{2}(.11556 + .156)} = (-.167,\ 0.97).$$

This interval contains 0, so you do not reject the hypothesis of equal means.

ANOVA versus multiple comparison procedures

Dunnett's T3 and the Games-Howell method do not require that you first test the hypothesis of equal means with a method in chapter 10 in order to control the probability of at least one Type I error (FWE). Indeed, they are designed to control FWE when applied as just described. If they are used contingent upon rejecting the hypothesis of equal means with the methods in chapter 10, their properties, in terms of Type I errors, are altered. This raises the practical issue of when and why the methods in chapter 10 should be used. If we assume normality and that the groups have equal variances, there are methods for performing all pairwise comparisons, called step-down and step-up techniques, that make use of the methods in chapter 10 and which can increase power. Variations that allow unequal variances could be used as well. However, under non-normality, these methods can have exceptionally poor power when comparing means and therefore are not described. (Switching to measures of location that perform well under non-normality might correct practical concerns about power.)

Comparing medians

There is a simple extension of the T3 method that can be used to compare medians rather than means. Simply put, for any two groups, test the hypothesis of equal medians by applying the method in section 9.1, which is based on the sample median and the McKean-Schrader estimate of the standard error. In symbols, let M_j and M_k be the sample medians corresponding to groups j and k and let S_j^2 and S_k^2 be the McKean-Schrader estimates of the squared standard errors. The hypothesis of equal medians is rejected when

$$\frac{|M_j - M_k|}{\sqrt{S_j^2 + S_k^2}} \geq c,$$

where c is read from table 10 with degrees of freedom $\nu = \infty$ in order to control the probability of at least one Type I error. As noted in chapter 9, when comparing two groups, this approach appears to perform reasonably well, in terms of controlling the probability of a Type I error, provided there are no tied (duplicated) values within either group. But with tied values in any group, the actual Type I error probability might be considerably larger than intended. (A method for handling tied values is described in the final section of this chapter.)

Example 3

Imagine that for five groups, all pairs of groups are to be compared using medians. Then the total number of tests to be performed is $C = (5^2 - 5)/2 = 10$. If the familywise error rate (FWE) is to be .05, table 10 says that the critical value is $c = 2.79$. So if for groups 1 and 3, $M_1 = 12$, $M_3 = 4$, $S_1^2 = 16$ and $S_2^2 = 22$, the test statistic is

$$\frac{12 - 4}{\sqrt{16 + 22}} = 1.3,$$

which is less than 2.79, so we fail to detect a difference between the medians corresponding to groups 1 and 3.

Dealing with two-way ANOVA designs

The multiple comparison procedures described in this section can be generalized to a two-way ANOVA design. (Independent groups are being assumed.) But before continuing, it helps to quickly review Welch's test covered in chapter 9, which is aimed at comparing the means of two independent groups. The goal was to test

$$H_0 : \mu_1 - \mu_2 = 0,$$

the hypothesis that the two means are equal. The strategy was to estimate the difference between the means with the difference between the sample means: $\bar{X}_1 - \bar{X}_2$. Then this difference was divided by an estimate of the standard error of $\bar{X}_1 - \bar{X}_2$, which is given

$$\sqrt{\frac{s_1^2}{n_1} + \frac{s_2^2}{n_2}},$$

and which yields the test statistic

$$W = \frac{\bar{X}_1 - \bar{X}_2}{\sqrt{\frac{s_1^2}{n_1} + \frac{s_2^2}{n_2}}},$$

where s_1^2 and s_2^2 are the sample variances corresponding to the two groups being compared, and n_1 and n_2 are the sample sizes. In words, the estimate of the squared standard error of the first sample mean is given by the corresponding sample variance divided by the sample size. Similarly, the estimated squared standard error of the second sample mean is equal to the corresponding sample variance divided by the sample size. Finally, the estimated squared standard error of the difference between the means is given by the sum of the estimated squared standard errors.

To explain how Welch's test can be extended to a two-way design, it helps to begin with the simplest case: a 2-by-2 design. As already explained in chapter 10, we are dealing with a situation that can be depicted as follows:

	Factor B	
Factor A	μ_1	μ_2
	μ_3	μ_4

Recall from chapter 10 that when dealing with factor A, the goal is to test

$$H_0: \frac{\mu_1+\mu_2}{2} = \frac{\mu_3+\mu_4}{2}.$$

Chapter 10 described how this hypothesis can be tested, but another important approach is described here. Normality is assumed, but unlike the F test in chapter 10, the method is designed to handle unequal variances. For convenience, note that this last hypothesis can be stated as

$$H_0: \mu_1+\mu_2 = \mu_3+\mu_4,$$

which in turn can be written as

$$H_0: \mu_1+\mu_2-\mu_3-\mu_4 = 0.$$

Notice the similarity between this hypothesis and the hypothesis tested by Welch's method as described in chapter 9. Both deal with a certain linear combination of the means. Welch's test is concerned with H_0: $\mu_1 - \mu_2 = 0$, the hypothesis that the difference between the means is zero. In general, when dealing the jth group, an estimate of the squared standard error of the sample mean, $\bar{X}_j$, is given by

$$q_j = \frac{s_j^2}{n_j},$$

the sample variance divided by the sample size. With Welch's test, we get an estimate of the squared standard error of the difference between the sample means by adding the estimates of the corresponding squared standard errors. That is, we use $q_1 + q_2$. It can be shown that for independent groups, if we add some means together and subtract out others, again the squared standard error is estimated simply by summing the corresponding estimates of the squared standard errors of the sample means. So here, when dealing with $H_0: \mu_1+\mu_2-\mu_3-\mu_4 = 0$, we estimate $\mu_1+\mu_2-\mu_3+\mu_4$, which is an example of what is called a *linear contrast*, with

$$\bar{X}_1+\bar{X}_2-\bar{X}_3-\bar{X}_4,$$

and an estimate of squared standard error is given by

$$q_1+q_2+q_3+q_4.$$

The resulting test statistic is

$$W = \frac{\bar{X}_1+\bar{X}_2-\bar{X}_3-\bar{X}_4}{\sqrt{q_1+q_2+q_3+q_4}}.$$

When using an extension of Welch's method, the degrees of freedom are based on two quantities. The first is

$$V_1 = (q_1 + q_2 + q_3 + q_4)^2,$$

which is the square of the sum of the estimated squared standard errors. The second is

$$V_2 = \frac{q_1^2}{n_1 - 1} + \frac{q_2^2}{n_2 - 1} + \frac{q_3^2}{n_3 - 1} + \frac{q_4^2}{n_4 - 1}.$$

The degrees of freedom are $\nu = V_1/V_2$ and when dealing with a *single* hypothesis, the null hypothesis is rejected if $|W| \geq t$, where t is the $1 - \alpha/2$ quantile read from table 4 in appendix B. (When dealing with multiple hypotheses, use table 10, instead, as illustrated momentarily.)

Example 4

Imagine that we have four independent groups with means $\bar{X}_1 = 4$, $\bar{X}_2 = 8$ $\bar{X}_3 = 2$ $\bar{X}_4 = 6$. Then an estimate of $\mu_1 + \mu_2 - \mu_3 - \mu_4$ is simply $\bar{X}_1 + \bar{X}_2 - \bar{X}_3 - \bar{X}_4 = 4$. If the sample variances are $s_1^2 = 24$, $s_2^2 = 32$, $s_3^2 = 48$, and $s_4^2 = 36$, and if the sample sizes are $n_1 = 12$, $n_2 = 8$, $n_3 = 12$, and $n_4 = 6$, then $q_1 = 24/12 = 2$, $q_2 = 32/8 = 4$, $q_3 = 4$ and $q_4 = 6$. Consequently, the estimated squared standard error of $\bar{X}_1 + \bar{X}_2 - \bar{X}_3 - \bar{X}_4$ is $q_1 + q_2 + q_3 + q_4 = 2 + 4 + 4 + 6 = 16$, so the estimated standard error is $\sqrt{16} = 4$, and an appropriate test statistic for $H_0: \mu_1 + \mu_2 - \mu_3 - \mu_4 = 0$ is

$$W = \frac{4 + 8 - 2 - 6}{4} = 1.$$

We see that $V_1 = 16^2 = 256$,

$$V_2 = \frac{2^2}{11} + \frac{4^2}{7} + \frac{4^2}{11} + \frac{6^2}{5} = 11.3,$$

so the degrees of freedom are $\nu = 256/11.3 = 22.65$, in which case, if the probability of a Type I error is to be $\alpha = .05$, then table 4 in appendix B indicates that the critical value is approximately $t = 2.07$. Because $|W| < 2.07$, you fail to reject the hypothesis of no main effect for factor A.

The method for dealing with factor A extends immediately to factor B. Now the goal is to test

$$H_0: \frac{\mu_1 + \mu_3}{2} = \frac{\mu_2 + \mu_4}{2},$$

which is the same as testing

$$H_0: \mu_1 + \mu_3 - \mu_2 - \mu_4 = 0.$$

So now our test statistic is

$$W = \frac{\bar{X}_1 + \bar{X}_3 - \bar{X}_2 - \bar{X}_4}{\sqrt{q_1 + q_2 + q_3 + q_4}}$$

and the degrees of freedom are computed as before.

Example 5

We continue the previous example, only now we focus on factor B. We have that

$$W = \frac{4+2-8-8}{4} = -2.5.$$

The degrees of freedom are again $\nu = 22.65$, the critical is again $t = 2.07$ (still assuming that the probability of a Type I error is to be $\alpha = .05$), and because $|W| \geq 2.07$, reject and conclude there is a main effect of factor B,

Interactions are handled in the same manner. The main difference is that now the goal is to test

$$H_0 : \frac{\mu_1 - \mu_3}{2} = \frac{\mu_2 - \mu_4}{2},$$

which is the same as testing

$$H_0 : \mu_1 - \mu_3 - \mu_2 + \mu_4 = 0.$$

The test statistic is

$$W = \frac{\bar{X}_1 - \bar{X}_3 - \bar{X}_2 + \bar{X}_4}{\sqrt{q_1 + q_2 + q_3 + q_4}}$$

and the the degrees of freedom remain the same.

Example 6

Continuing the last two examples, now we have

$$W = \frac{4-2-8+8}{4} = 0.5,$$

again the critical value (for $\alpha = .05$) is $t = 2.07$, and because $|W| < 2.07$, we fail to detect an interaction.

Controlling the probability of at least one Type I error

In the previous three examples, each test was performed with the probability of a Type I error set at .05. How might we control the probability of at least one Type I error? A simple strategy, assuming normality, is to note that a total of $C = 3$ hypotheses are to be performed, and then simply read a critical value from table 10 in appendix B rather than table 4.

Example 7

The last three examples tested three hypotheses: (1) no main effects for factor A, (2) no main effects for factor B, and (3) no interaction. So when referring to table 10 in appendix B, the total number of hypotheses to be tested in $C = 3$. And for each of these tests, the degrees of freedom are $\nu = 22.65$. If we want the probability of at least one Type I error to be .05, then table 10 indicates that the critical value is approximately $t = 2.58$. In each case, $|W|$ is less than 2.58, so now none of the three hypotheses would be rejected.

Extension to a J-by-K design

It is noted that the method for testing hypotheses in a 2-by-2 design, just described and illustrated, can be extended to the general case of any J-by-K design. Here the goal is to briefly outline how this is done.

Consider a 3-by-4 design and for convenience, label the populations means among these 12 groups as follows:

	Factor B			
	μ_1	μ_2	μ_3	μ_4
Factor A	μ_5	μ_6	μ_7	μ_8
	μ_9	μ_{10}	μ_{11}	μ_{12}

Following the description of the two-way ANOVA in chapter 10, we write the marginal means for factor A as

$$\bar{\mu}_{1.} = \frac{\mu_1 + \mu_2 + \mu_3 + \mu_4}{4},$$

$$\bar{\mu}_{2.} = \frac{\mu_5 + \mu_6 + \mu_7 + \mu_8}{4},$$

and

$$\bar{\mu}_{3.} = \frac{\mu_9 + \mu_{10} + \mu_{11} + \mu_{12}}{4}.$$

As was explained, no differences among the levels of factor A is taken to mean that

$$H_0 : \bar{\mu}_{1.} = \bar{\mu}_{2.} = \bar{\mu}_{3.}$$

is true. But when this hypothesis is rejected, there is no indication of which levels of factor A differ.

We begin by rewriting the null hypotheses in a more convenient form in much the same manner as was done in the 2-by-2 design. To illustrate the process, focus on testing

$$H_0 : \bar{\mu}_{1.} = \bar{\mu}_{3.}.$$

A little algebra shows that this is the same as testing

$$H_0 : \mu_1 + \mu_2 + \mu_3 + \mu_4 - \mu_9 - \mu_{10} - \mu_{11} - \mu_{12} = 0.$$

As usual, we estimate this linear contrast by replacing the population means with the sample means yielding

$$\bar{X}_1 + \bar{X}_2 + \bar{X}_3 + \bar{X}_4 - \bar{X}_9 - \bar{X}_{10} - \bar{X}_{11} - \bar{X}_{12}.$$

An estimate of the squared standard error is given by

$$q_1 + q_2 + q_3 + q_4 + q_9 + q_{10} + q_{11} + q_{12}.$$

The test statistic is

$$W = \frac{\bar{X}_1 + \bar{X}_2 + \bar{X}_3 + \bar{X}_4 - \bar{X}_9 - \bar{X}_{10} - \bar{X}_{11} - \bar{X}_{12}}{\sqrt{q_1 + q_2 + q_3 + q_4 + q_9 + q_{10} + q_{11} + q_{12}}}.$$

The degrees of freedom are

$$\nu = \frac{(q_1+q_2+q_3+q_4+q_9+q_{10}+q_{11}+q_{12})^2}{\frac{q_1^2}{n_1-1}+\frac{q_2^2}{n_2-1}+\frac{q_3^2}{n_3-1}+\frac{q_4^2}{n_4-1}+\frac{q_9^2}{n_9-1}+\frac{q_{10}^2}{n_{10}-1}+\frac{q_{11}^2}{n_{11}-1}+\frac{q_{12}^2}{n_{12}-1}},$$

and the null hypothesis is rejected if $|W| \geq t$, the $1-\alpha/2$ quantile of Student's t-distribution with ν degrees of freedom. As before, if more that one test is to be performed and it is desired to control FWE, read the critical value from table 10 in appendix B rather than table 4.

Example 8

Imagine that two medications for treating hypertension are under investigation and that researchers want to take into account ethnic background. Here, three ethnic backgrounds are considered and labeled A, B, and C. Further imagine that the results are as follows:

		Ethnicity		
		A	B	C
Medication	1	$\bar{X}_1=24$ $s_1^2=48$ $n_1=8$	$\bar{X}_2=36$ $s_2^2=56$ $n_2=7$	$\bar{X}_3=28$ $s_3^2=60$ $n_3=12$
	2	$\bar{X}_4=14$ $s_4^2=64$ $n_4=8$	$\bar{X}_5=24$ $s_5^2=25$ $n_5=5$	$\bar{X}_6=20$ $s_6^2=40$ $n_6=10$

The goal is to compare medications, ignoring ethnicity. Then the hypothesis of interest is H_0: $\mu_1+\mu_2+\mu_3-\mu_4-\mu_5-\mu_6=0$. We see that $q_1=s_1^2/n_1=6$, $g_2=8$, $q_3=5$, $q_4=8$, $q_5=5$, and $q_6=4$. The test statistic is

$$W=\frac{24+36+28-14-24-20}{\sqrt{6+8+5+8+5+4}}=5.$$

The degrees of freedom are

$$\nu=\frac{(6+8+5+8+5+4)^2}{\frac{6^2}{7}+\frac{8^2}{6}+\frac{5^2}{11}+\frac{8^2}{7}+\frac{5^2}{4}+\frac{4^2}{9}}=36.8.$$

With $\alpha=.05$, the critical value (from table 4 in appendix B) is approximately 2.03, and because $|W|>2.06$, reject and conclude that the medications differ, ignoring ethnicity.

Problems

8. For five independent groups, assume that you plan to do all pairwise comparisons of the means and you want FWE to be .05. Further assume that $n_1=n_2=n_3=n_4=n_5=20$, $\bar{X}_1=15$, $\bar{X}_2=10$, $s_1^2=4$ and $s_2^2=9$, $s_3^2=s_4^2=s_5^2=15$, test $H_0: \mu_1=\mu_2$ using Dunnett's T3.
9. Repeat the previous problem, only use Games-Howell.

10. For four independent groups, assume that you plan to do all pairwise comparisons of the means and you want FWE to be .05. Further assume that $n_1 = n_2 = n_3 = n_4 = n_5 = 10$ $\bar{X}_1 = 20$, $\bar{X}_2 = 12$, $s_1^2 = 5$, $s_2^2 = 6$, $s_3^2 = 4$, $s_4^2 = 10$, and $s_5^2 = 15$. Test $H_0 : \mu_1 = \mu_2$ using Dunnett's T3.
11. Repeat the previous problem, only use Games-Howell.
12. For four groups, you get sample medians $M_1 = 34$, $M_2 = 16$, $M_3 = 42$, $M_4 = 22$, $S_1^2 = 33$, $S_2^2 = 64$, $S_3^2 = 8$, $S_4^2 = 5$. Assuming that the goal is to test the hypothesis of equal medians for all pairs of groups such that FWE is .05, determine whether you would reject when comparing groups 2 and 4.
13. In the previous problem, comment on the results if there are tied values in the first group but not the other three.
14. Referring to example 8 in this section, compare ethnic groups A and B, ignoring type of medication. Assume this is the only hypothesis to be tested and that the goal is to have a Type I error probability of .05.
15. In the previous problem, imagine that the goal is to compare all pairs of groups for factor B and that the goal is to have the probability of at least one Type I error equal to .05. What is the critical value you would use when comparing ethnicity groups B and C?
16. Referring to example 8 in this section, imagine the goal is to check for interactions when dealing with ethnicity groups A and B. Test the hypothesis of no interaction for this special case assuming all other interactions are to be ignored and that the Type I error probability is to be .05.

11.3 Methods for dependent groups

The methods described in the previous section take advantage of the independence among the groups when trying to control the probability of at least one Type I error. When comparing dependent groups instead, alternative methods must be used.

Bonferroni method

One of the simplest and earliest methods for comparing dependent groups is based on what is known as the Bonferroni inequality. The strategy is simple: If you plan to test C hypotheses and want FWE to be α, test each of the individual hypotheses at the α/C level. For the special case where all pairs of groups are to be compared,

$$C = \frac{J^2 - J}{2}.$$

In terms of ensuring that FWE does not exceed α, the only requirement is that the individual tests ensure that the Type I error probability does not exceed α/C. So for the case where means are compared with the paired t-test in section 9.2, if five tests are to be performed, and the goal is to have FWE equal to .05, then each paired t-test would be performed with $\alpha = .05/5 = .01$.

Example 1

As a simple illustration, imagine that the means for all pairs of four groups are to be compared with FWE equal to .03. So a total of $C = 6$ tests are to be performed, meaning that each test should be performed with $\alpha = .03/6 = .005$. If, for example, each group has a sample size of 11, then from chapter 9, the critical value would be the .995 quantile of a Student's t-distribution with degrees of freedom being $\nu = 11 - 1 = 10$, read from table 4 in appendix B. In particular, the critical value is $t_{.995} = 3.169$. So, for example, if the paired t-test is applied to groups 1 and 2 yielding $T = -2.9$, because $|-2.9|$ is less than 3.169, fail to reject, and the other five tests would be performed in a similar manner.

Statistical software routinely reports p-values when comparing any two groups. An alternative description of the Bonferroni method is to reject the null hypothesis if the p-value is less than or equal to α/C, where again C is the total number of tests to be performed.

Example 2

Imagine that all pairs of four dependent groups are to be compared. Then $C = 6$ hypotheses are to tested and, for illustrative purposes, imagine that the p-values are as follows:

Number	Test	p-value
1	$H_0: \mu_1 = \mu_2$	$p_1 = .006$
2	$H_0: \mu_1 = \mu_3$	$p_2 = .025$
3	$H_0: \mu_1 = \mu_4$	$p_3 = .003$
4	$H_0: \mu_2 = \mu_3$	$p_4 = .540$
5	$H_0: \mu_2 = \mu_4$	$p_5 = .049$
6	$H_0: \mu_3 = \mu_4$	$p_6 = .014$

So, for example, the p-value for the first hypothesis is $p_1 = .006$, and the p-value for the second hypothesis is $p_2 = .025$ If the probability of at least one Type I error is to be .1, the Bonferroni method says that each individual test would be rejected if its p-value is less than or equal to $0.1/6 = 0.017$. So here, hypotheses 1, 3, and 6 would be rejected.

Rom's method

Several improvements on the Bonferroni method have been published and one that stands out is a so-called sequentially rejective method derived by Rom (1990), which has been found to have good power relative to several competing techniques (e.g., Olejnik et al., 1997). To apply it, compute a p-value for each of the C tests to be performed and label them $p_1, \ldots, p_C$. (Standard software reports p-values.) Next, put the p-values in descending order, which are now labeled $p_{[1]} \geq p_{[2]} \geq \cdots \geq p_{[C]}$. So $p_{[1]}$ is the largest p-value, $p_{[2]}$ is the next largest, and $p_{[C]}$ is the smallest. Proceed as follows:

1. Set k =1.
2. If $p_{[k]} \leq d_k$, where d_k is read from table 11.3, stop and reject all C hypotheses; otherwise, go to step 2.

Table 11.3 Critical values, d_k, for Rom's method

k	$\alpha = .05$	$\alpha = .01$
1	.05000	.01000
2	.02500	.00500
3	.01690	.00334
4	.01270	.00251
5	.01020	.00201
6	.00851	.00167
7	.00730	.00143
8	.00639	.00126
9	.00568	.00112
10	.00511	.00101

3. Increment k by 1. If $p_{[k]} \leq d_k$, stop and reject all hypotheses having a p-values less than or equal d_k
4. If $p_{[k]} > d_k$, repeat step 3.
5. Continue until you reject or all C hypotheses have been tested.

An advantage of Rom's method is that its power is greater than or equal to the Bonferroni approach. In fact, Rom's method always rejects as many, or more, hypotheses. A negative feature is that confidence intervals are not readily computed.

Example 3

A large company is considering four variations of a baked good for mass distribution. An issue is whether potential customers rate the baked goods differently depending on which variation is used. To find out, randomly sampled individuals are asked to try a baked good produced by each method. That is, each individual consumes a baked good produced by each of the four variations. Imagine that all pairwise comparisons among the resulting four dependent groups are performed yielding the p-values shown in table 11.4. Further assume that you want FWE to be .05. The largest p-value is .62, this is greater than .05, so you fail to reject the corresponding hypothesis, $H_0 : \mu_2 = \mu_3$. The next largest p-value is .130, this is greater than $d_2 = .025$ (read from table 11.3), so fail to reject $H_0 : \mu_2 = \mu_4$. The next largest is .015, this is less than $d_3 = .0167$, so you stop and reject the corresponding hypothesis as well as those having smaller p-values.

Table 11.4 An illustration of Rom's method

Number	Test	p-value	
1	$H_0 : \mu_1 = \mu_2$	$p_1 = .010$	$p_{[5]}$
2	$H_0 : \mu_1 = \mu_3$	$p_2 = .015$	$p_{[3]}$
3	$H_0 : \mu_1 = \mu_4$	$p_3 = .005$	$p_{[6]}$
4	$H_0 : \mu_2 = \mu_3$	$p_4 = .620$	$p_{[1]}$
5	$H_0 : \mu_2 = \mu_4$	$p_5 = .130$	$p_{[2]}$
6	$H_0 : \mu_3 = \mu_4$	$p_6 = .014$	$p_{[4]}$

Problems

17. You perform five tests and get the p-values .049, .048, .045, .047, and .042. Based on the Bonferroni inequality, which would be rejected with FWE equal to .05?

18. Referring to the previous problem, which would be rejected with Rom's procedure?

19. What do the last two exercises illustrate?

20. Five tests are performed aimed at comparing the medians of dependent groups. The p-values are .24, .001, .005, .12, .04. Which should be rejected when using Rom's method if FWE is to be .05?

11.4 Some modern advances and insights

As in previous chapters, if the goal is to compare means, without being sensitive to other ways the groups might differ, the methods covered here can be unsatisfactory. Again, concerns arise when groups differ in terms of their variances and skewness. And as always, outliers can result in relatively low power. An advantage of the methods that allow unequal variances is that with sufficiently large sample sizes, they will provide adequate inferences about the means when dealing with non-normal distributions, but as usual, it is generally unclear just how large the sample sizes must be to achieve this goal. This issue is complicated because adequate sample sizes are a function of several factors, including the extent to which the groups differ in terms of their variances and skewness. However, if these methods reject, it is reasonable to conclude the groups differ in some manner, consistent with chapters 9 and 10. If the goal is to compute confidence intervals for the difference between the means, it seems some type of bootstrap-t method is a relatively good choice, but this does not eliminate all practical concerns.[2]

When comparing the medians of independent groups, currently the one method that performs well, even when tied values occur, is based on a simple extension of the percentile bootstrap method for comparing medians that was described in chapter 9 (Wilcox, 2006). Briefly, for each pair of groups, determine a p-value. Then use Rom's method to control the probability of at least one Type I error.

Again, consistent with previous chapters, a compromise amount of trimming can have practical value in terms of both Type I errors and power. A simple method, when dealing with independent groups, is to apply Yuen's test for each pair of groups under study. Then, based on the resulting degrees of freedom, use table 10 to determine an appropriate critical value. Very effective methods, based in part on the percentile bootstrap technique, are available as well.

Example 1

Imagine that six hypotheses are to be tested with the goal of having FWE equal to .05. So when referring to table 10, $C = 6$. In each case, 20% trimmed means

2. For computational details and appropriate software, see Wilcox, 2003, section 12.7.1.

are compared with Yuen's method, and for the first hypothesis the test statistic is $T_y = 2.4$, with the degrees of freedom are $\nu = 24$. Then from table 10, the crictical value is $t = 2.85$, and because $2.4 < 2.85$, you fail to reject.

As for comparing dependent groups based on medians or 20% trimmed means, a simple approach is to apply the methods in chapter 9 to each pair of groups of interest. Then, Rom's method can be used to control FWE. As usual, bootstrap methods can be used as well.

Finally, this chapter outlines how to perform multiple comparisons in a two-way design using an extension of Welch's test. It is noted that a simple extension of Yuen's test can be used to compare 20% trimmed means, but no details are given here.

A Summary of Some Key Points

- Methods designed to control the probability of at least one Type I error, which assume normality and equal variances, such as the Tukey-Kramer technique, appear to perform fairly well when all groups have identical distributions. That is, they perform well in terms of Type I errors when testing the hypothesis that groups have identical distributions. Some departures from normality and unequal variances can be tolerated, but the extent to which this is true is a complicated function of the sample sizes, the type of departure from normality encountered, and the degree to which the variances differ.
- In some cases, methods aimed at comparing means are satisfactory, but for reasons explained in previous chapters, the methods in this chapter run the risk of relatively poor power.
- Among the methods covered here, Dunnett's T3 and the Games-Howell method are best for comparing means, but there are practical reasons for considering more advanced techniques (not described in this book) based on a bootstrap-*t* procedure. But despite any practical advantages these advanced methods offer, not all practical problems are addressed when the goal is to compare means.
- Methods based on 20% trimmed means perform well over a relatively broad range of situations in terms of both Type I errors and power. If the results are similar to those obtained with means, this suggests that methods based on the sample means are probably satisfactory in terms of making inferences about the population means. When they differ, perhaps methods based on means remain satisfactory, but there is some uncertainty about the extent to which this is true.
- Methods based on medians can be useful and effective. An advantage of using a percentile bootstrap method is that regardless of whether tied values occur, excellent control over the probability of at least one Type I error can be achieved, except perhaps for very small sample sizes.
- Chapter 9 pointed out that different methods for comparing any two groups can give us a different perspective on how they differ. And the choice of method can make a practical difference in terms of how much power will be achieved. Of course, this issue remains relevant here and in fact becomes more complex. For example, comparing the medians of groups 1 and 2 might be ideal in terms of power, but when comparing groups 1 and 3, perhaps comparing means or 20% trimmed means is more effective.

- Chapter 9 noted that if we apply many methods when comparing two groups, a technical issue arises. That issue is controlling the the probability of at least one Type I error. If, for example, we compare two groups with means, then 20% trimmed means, and then medians, we could control the probability of at least one Type I error using the Bonferroni method or Rom's technique. Note, however, that such an adjustment will lower power.

12

CATEGORICAL DATA

This chapter covers some basic methods for analyzing categorical data. Categorical data simply means that observations belong to two or more groups. The simplest case is the binomial distribution, which was introduced in chapter 4. For this special situation, there are only two categories usually called success and failure. Contingency tables, also introduced in chapter 4, are used to summarize categorical data and represent a generalization of the binomial to more than two groups. The main goal here is to describe some additional methods for analyzing such data.

12.1 One-way contingency tables

One-way contingency tables represent a generalization of the binomial distribution, introduced in chapter 4, where the outcome associated with a single variable has two or more categories. Consider, for example, a multiple-choice test item having five choices. One choice is correct and the other four, called distractors, are incorrect. Among students who choose the wrong answer, are choices made at random? The variable of interest is the response among students choosing one of the four distractors. Let p_1, p_2, p_3 and p_4 be the probability that, among students who failed to choose the correct answer, they picked distractor 1, 2, 3, or 4, respectively. Here we let n represent the total number of students choosing a wrong answer. Then, if students pick a wrong response at random, it should be the case that these four probabilities are equal. In symbols, the hypothesis

$$H_0: p_1 = p_2 = p_3 = p_4 = 1/4$$

should be true.

More generally, assume that k mutually exclusive outcomes are possible and that the goal is to test the hypothesis that the probabilities associated with these k categories are equal. Said more formally, the goal is to test

$$H_0: p_1 = p_2 = \cdots = p_k = \frac{1}{k}. \tag{12.1}$$

In case it helps, note that for the special case $k = 2$, there are only two categories and the situation reduces to the binomial distribution. In particular, the hypothesis given by equation (12.1) is tantamount to testing H_0: $p = 1/2$, where p is the probability of a success.

Let n_1 be the number of participants who belong to category 1, let n_2 be the number of participants who belong to category 2, and so on. So the total number of participants is

$$n = \sum n_k.$$

The classic test of the hypothesis given by equation (12.1) was derived by Karl Pearson and is based on the test statistic

$$X^2 = \frac{\sum \left(n_j - \frac{n}{k}\right)^2}{n/k}. \tag{12.2}$$

When the null hypothesis is true, and the sample size, n, is sufficiently large, X^2 will have, approximately, what is called a chi-squared distribution with $\nu = k - 1$ degrees of freedom. Figure 12.1 shows chi-squared distributions with degrees of freedom $\nu = 2$ and $\nu = 4$. If the Type I error probability is to be α, the hypothesis is rejected if X^2 exceeds c, the $1 - \alpha$ quantile, which is given in table 3 in appendix B.

A word of explanation, when reading table 3 in appendix B, might help. The first column, headed by ν indicates the degrees of freedom. Notice the terms in the top row. The subscripts denote the quantiles. For example, under $\chi^2_{.10}$ and in the first row where $\nu = 1$, you will see the entry .01579. This means that for a chi-squared distribution with 1 degree of freedom, the probability of getting a value less than or equal to .01579 is .1. Similarly, under $\chi^2_{.975}$ and in the row corresponding to $\nu = 10$ is the entry 20.4637. That is, the probability of getting a value less than or equal to 20.4637, when sampling from a chi-squared distribution with 10 degrees of freedom, is .975. In the context of testing some hypothesis with a Type I error probability of $\alpha = .025$, the critical value, corresponding to 10 degrees of freedom, would be 20.4637.

Example 1

Consider a six-sided die with each side having between one and six spots. That is, one side has a single spot, another has two spots, and so on. A gambling

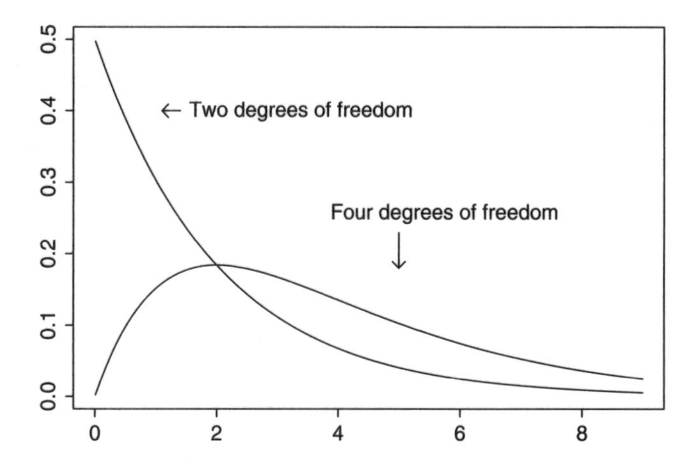

Figure 12.1 Shown are two chi-squared distributions. One has degrees of freedom 2 and the other has degrees of freedom 4.

casino assumes that when an individual tosses the die, all six sides have the same probability of occurring. A gambler believes she can toss the die in particular fashion so that this is not true; the belief is that some sides have a higher probability than others. To find out whether the sides have different probabilities of occurring, the gambler tosses the die 102 times and gets the following results: $n_1 = 10$, $n_2 = 15$, $n_3 = 30$, $n_4 = 25$, $n_5 = 7$, and $n_6 = 15$, so $n = 102$. That is, for 10 of the tosses 1 spot occurred, for 15 of the tosses, two spots resulted, and so forth. To test the hypothesis that all six possibilities have the same probability, first note that the degrees of freedom are $\nu = k - 1 = 6 - 1 = 5$. If the Type I error probability is to be .05, then from table 3 in appendix B, the critical value is $c = 11.0707$. Because $n = 102$, $n/k = 102/6 = 17$, so

$$X^2 = \frac{(10-17)^2 + (15-17)^2 + (30-17)^2 + (25-17)^2 + (7-17)^2 + (15-17)^2}{17}$$

$$= 22.94.$$

Because 22.94 is greater than the critical value, 11.0707, reject and conclude that the probabilities are not equal.

The method just described can be generalized to situations where the goal is to test the hypothesis that the probabilities associated with the k categories have specified values other than $1/k$. For instance, imagine that a certain medication sometimes causes a rash. For present purposes, assume the severity of the rash is rated on a four point scale where a 1 means no rash and a 4 means the rash is severe. Among a very large number of individuals, it is found that the proportion of individuals belonging to these four groups is .5, .3, .2, and .1, respectively. A modification of the medication is under consideration and one goal is to determine whether this alters the probabilities associated with the four possible outcomes. That is, the goal is to test H_0: $p_1 = .5, p_2 = .3, p_3 = .2, p_4 = .1$. More generally, given a set of k probabilities $p_{01}, \cdots, p_{0k}$, the goal is to test

$$H_0 : p_1 = p_{01}, \cdots, p_k = p_{0k}. \qquad (12.3)$$

Among n observations, if the null hypothesis is true, the expected number of individuals falling into category 1 is np_{01}. In essence, if we focus on whether or not an individual belongs to category 1, we are dealing with a binomial random variable. That is, among the n participants, the probability of belonging to category 1 is p_{01} and the probability of not belonging to category 1 is $1 - p_{01}$. And from the basic features of a binomial distribution covered in chapter 4, the expected number of individuals falling into category 1 is np_{01}. In a similar manner, the expected number of individuals falling into category 2 is np_{02}, still assuming that the null hypothesis is true, and so on. The test statistics is

$$X^2 = \sum \frac{(n_j - np_{0j})^2}{np_{0j}}. \qquad (12.4)$$

Written in words, the test statistics is

$$X^2 = \sum \frac{(\text{observed} - \text{expected})^2}{\text{expected}}. \qquad (12.5)$$

Again the hypothesis is rejected if $X^2 \geq c$, where c is the $1 - \alpha$ quantile of a chi-squared distribution with $k - 1$ degrees of freedom.

Example 2

Continuing the illustration dealing with the likelihood of getting a rash, suppose that among 88 adults taking the medication, the number of individuals corresponding to the four ratings for a rash are $n_1 = 40$, $n_2 = 30$, $n_3 = 15$, and $n_4 = 3$. This says, for example, that 40 of the 88 participants did not get a rash and 3 got the severest form. As previously noted, the goal is to test H_0: $p_1 = .5$, $p_2 = .3$, $p_3 = .2$, $p_4 = .1$. So the expected values corresponding to the four types of rash, when the null hypothesis is true, are 44, 26.4, 17.6, and 8.8, respectively. (That is, $n = 88$, and so $np_1 = 44$, $np_2 = 26.4$, $np_3 = 17.6$, and $np_4 = 8.8$.) The resulting test statistic is

$$X^2 = \frac{(40-44)^2}{44} + \frac{(30-26.4)^2}{26.4} + \frac{(15-17.6)^2}{17.6} + \frac{(3-8.8)^2}{8.8} = 5.08.$$

With $\alpha = .05$, the critical value is 7.8148, so the null hypothesis is not rejected.

Example 3

In 1988, the Department of Education conducted a survey of 17-year-old students to determine how much time they spend doing homework. Each student was classified into one of five categories: (1) none was assigned, (2) did not do it, (3) spent less than 1 hour, (4) 1–2 hours, or (5) Spent more than 2 hours doing homework. The percentage of students in each of these categories was found to be 20.8%, 13.4%, 27.8%, 26% and 12%, respectively. A local school board wants to know how their students compare based on 100 randomly sampled students. That is, the goal is to test

$$H_0: p_1 = .208, \quad p_2 = .134, \quad p_3 = .278, \quad p_4 = .260 \quad \text{and} \quad p_5 = .120.$$

Imagine that 100 randomly sampled students, the number of hours they study are as summarized in table 12.1. It can be seen that $X^2 = 8.7$. If the Type I error is to .05, the critical value is 9.49, so fail to reject.

Gaining perspective: A closer look at the the Chi-squared distribution

At first glance, it might appear that normality is not assumed when applying the chi-squared test just described. But this is not true. It can be shown that the test statistic, X^2, is based on a sum of squared terms, where the individual terms are approximately normal when the sample size is sufficiently large. Moreover, the formal definition of a chi-squared distribution is based on the sum of squared standard normal random variables.

Table 12.1 Hypothetical data on homework survey

Not assigned	Not done	< 1 hour	1–2 hours	> 2 hours
15	10	25	30	20

The case where there are two categories

Some comments are useful about the special case where there are $k=2$ categories. Now we have the binomial probability function where p is the probability of success. We saw in chapter 6 how to a compute a confidence interval for p and this method can be used to test the hypothesis that the probability of success is equal to some specified value, say p_0. If the confidence interval contains p_0, fail to reject

$$H_0: p = p_0;$$

otherwise, reject. Put another way, let $\hat{p}$ be the proportion of observed successes among n trials, let

$$Z = \frac{\hat{p} - p_0}{\sqrt{\hat{p}(1-\hat{p})/n}},$$

and let c be the $1-\alpha/2$ quantile of a standard normal distribution, which is read from table 1 in appendix B. Then reject the null hypothesis if

$$|Z| \geq c.$$

Example 4

Someone claims that for a randomly sampled stock, the probability that its value will increase after six months is .4. As a check on this claim, 20 stocks are randomly sampled and it is found that five gain value. So $n=20$, $p_0=.4$ and the estimated probability of success is $\hat{p}=5/20=.25$. To test the claim, with a Type I error probability of .05, compute

$$Z = \frac{.25 - .4}{\sqrt{.25(.75)/20}} = -1.549.$$

The critical value is $c=1.96$, which is greater than $|Z|$, so fail to reject.

To underscore the connection between the chi-squared distribution and the standard normal, note that an alternative approach to the last example is to reject if $Z^2 \geq c^2$. It can be seen that this corresponds to assuming that Z^2 has a chi-squared distribution with one degree of freedom, meaning that c^2 can be read from table 3 in appendix B.

Problems

1. For a one-way contingency table having four categories, the frequencies corresponding to each category are: 23, 14, 8, 32. Test the hypothesis that each category has the same probability. Use $\alpha=.05$.
2. For a one-way contingency table, the following frequencies are observed: 23, 34, 43, 53, 16. Using $\alpha=.01$, test the hypothesis that all five categories have the same probability.
3. A game show allows contestants to pick one of six boxes, one of which contains a large sum of money. To reduce the expected winnings among the contestants, the organizers speculate that contestants will not pick at random, but that some boxes are more likely to be chosen over others. To find out, a study is done where

potential contestants get to choose a box, in resulting the following frequencies: 6, 20, 30, 35, 10, 5. Test the hypothesis that each box has the same probability of being chosen. Use $\alpha = .01$

4. Imagine that 54 individuals are asked whether they agree, disagree or have no opinion that persons with a college degree feel more satisfied with their lives. The observed frequencies are 9, 30, and 15. Test $H_0 : p_1 = p_2 = p_3$ with $\alpha = .05$.

5. It is speculated that the probabilities associated with four categories are .1, .3, .5, and .1. The observed frequencies are 10, 40, 50, and 5. Test this speculation using $\alpha = .05$.

6. It is speculated that the probabilities associated with five categories are .2, .3, .3, .1 and .1. The observed frequencies are 20, 50, 40, 10 and 15. Test this hypothesis using $\alpha = .05$.

7. Someone claims that the proportion of adults getting low, medium, and high amounts of exercise is .5, .3, and .2, respectively. To check this claim you sample 100 individuals and find that 40 get low amounts of exercise, 50 get medium amounts, and 10 get high amounts. Test the claim with $\alpha = .05$.

8. A geneticist postulates that in the progeny of a certain dihybrid cross, the four phenotypes should be present in the ratio 9:3:3:1. So if p_1, p_2, p_3, and p_4 are the probabilities associated with these four phenotypes, the issue is whether there is empirical evidence indicating that H_0: $p_1 = 9/16, p_2 = 3/16, p_3 = 3/16, p_4 = 1/16$ should be rejected. The observed frequencies corresponding to these four phenotypes, among 800 members of the progeny generation, are 439, 168, 133, and 60. Test H_0 using $\alpha = .05$.

9. Does the likelihood of a particular crime vary depending on the day of the week? To find out, the number of crimes for Monday through Sunday were recorded and found to be 38, 31, 40, 39, 40, 44, and 48. Test the hypothesis that the likelihood of a crime is the same for each day of the week, using $\alpha = .05$.

12.2 Two-way contingency tables

This section considers the more general case of a two-way contingency table. There is a vast literature on how to analyze two-way contingency tables (e.g., Agresti, 1990, 1996; Andersen, 1997; Lloyd, 1996; and Powers and Xie, 1999), but only the basics are covered here.

We begin with a problem that arises in a variety of settings. To be concrete, imagine a survey of 1,600 randomly sampled adults who, at two different times, are asked whether they approve of a particular political leader. At issue is whether the approval rating at time 1 differs from the approval rating at time 2. For illustrative purposes, suppose the results are as indicated in table 12.2. For example, of the 1,600 respondents, 794 respond that they approve at both time 1 and 2. The number who approve at time 1, ignoring time 2, is 944, and the number who disapprove at time 2 is 720. Consequently, the estimated probability of getting an approval rating at time 1 is $944/1,600 = .59$.

It is convenient to write this contingency table in a more generic form as shown in table 12.3. So p_{11} represents the probability that a randomly sampled individual

Table 12.2 Approval rating of a political leader

	Time 2		
Time 1	Approve	Disapprove	Total
Approve	794	150	944
Disapprove	86	570	656
Total	880	720	1600

Table 12.3 Probabilities associated with a two-way contingency table

	Time 2		
Time 1	Approve	Disapprove	Total
Approve	p_{11}	p_{12}	$p_{1+} = p_{11} + p_{12}$
Disapprove	p_{21}	p_{22}	$p_{2+} = p_{21} + p_{22}$
Total	$p_{+1} = p_{11} + p_{21}$	$p_{+2} = p_{12} + p_{22}$	p_{++}

Table 12.4 Notation for observed frequencies

	Time 2		
Time 1	Approve	Disapprove	Total
Approve	n_{11}	n_{12}	$n_{1+} = n_{11} + n_{12}$
Disapprove	n_{21}	n_{22}	$n_{2+} = n_{21} + n_{22}$
Total	$n_{+1} = n_{11} + n_{21}$	$n_{+2} = n_{12} + n_{22}$	n

approves at both time 1 and time 2, p_{1+} is the probability of approving at time 1, ignoring time 2, and p_{+1} is the probability of approving at time 2, ignoring time 1. The frequencies associated with the possible outcomes are denoted as shown in table 12.4. So, for example, n_{11} is the number of individuals who approve at both time 1 and time 2. The estimates of the probabilities in table 12.3, namely p_{11}, p_{12}, p_{21}, and p_{22}, are

$$\hat{p}_{11} = \frac{n_{11}}{n}, \hat{p}_{12} = \frac{n_{12}}{n}, \hat{p}_{21} = \frac{n_{21}}{n}, \hat{p}_{22} = \frac{n_{22}}{n},$$

respectively. For example, based on the data in table 12.2, the estimate of the probability $\hat{p}_{11}$ is $\hat{p}_{11} = n_{11}/n = 794/1600 = .49625$. In words, the estimated probability of approving at both time 1 and 2 is .496. In a similar manner the marginal probabilities $\hat{p}_{1+}, \hat{p}_{+1}, \hat{p}_{2+}$, and $\hat{p}_{+2}$ are estimated with

$$\hat{p}_{1+} = \frac{n_{1+}}{n}, \hat{p}_{+1} = \frac{n_{+1}}{n}, \hat{p}_{2+} = \frac{n_{2+}}{n}, \hat{p}_{+2} = \frac{n_{+2}}{n},$$

respectively.

Now consider the issue of whether the approval rating has changed from time 1 to time 2. The approval rating at time 1 is p_{1+}, at time 2 it is p_{+1}, and so the change in the approval rating is

$$d = p_{1+} - p_{+1}.$$

This difference is estimated with

$$\begin{aligned}\hat{d} &= \hat{p}_{1+} - \hat{p}_{+1} \\ &= \frac{n_{11} + n_{12}}{n} - \frac{n_{11} + n_{21}}{n} \\ &= \frac{n_{12} - n_{21}}{n}.\end{aligned}$$

In formal terms, the goal is to test

$$H_0 : d = 0, \tag{12.6}$$

the hypothesis that the difference between the two approval ratings is zero. The standard approach is to proceed along the lines outlined in chapter 7. That is, we first need to find a method for estimating the squared standard error of d, where the squared standard error of d roughly refers to the variance of d over many studies. Using methods not covered in this book, it can be shown that an appropriate estimate is

$$s_d^2 = \frac{1}{n}\{\hat{p}_{1+}(1 - \hat{p}_{1+}) + \hat{p}_{+1}(1 - \hat{p}_{+1}) - 2(\hat{p}_{11}\hat{p}_{22} - \hat{p}_{12}\hat{p}_{21})\}.$$

So if the null hypothesis is true, the central limit theorem says that with a sufficiently large sample size, $\hat{d}$ will have, approximately, a normal distribution with mean 0 and variance s_d^2. This means that an appropriate test statistic is

$$Z = \frac{\hat{d}}{s_d},$$

which will have, approximately, a standard normal distribution when the null hypothesis is true and the sample size is reasonably large. The computations are simplified by noting that this last equation can be written as

$$Z = \frac{n_{12} - n_{21}}{\sqrt{n_{12} + n_{21}}},$$

Proceeding along the lines in chapter 7, if the Type I error probability is to be α, reject the hypothesis of no change if $|Z| \geq c$, where c is the $1 - \alpha/2$ quantile of a standard normal distribution read from table 1 in appendix B. A $1 - \alpha$ confidence interval for d is

$$\hat{d} \pm c s_d.$$

Example 1

For the data in table 12.2,

$$\hat{d} = \hat{p}_{1+} - \hat{p}_{+1} = 0.59 - 0.55 = 0.04$$

meaning that the change in approval rating is estimated to be 0.04. The estimated squared standard error of $\hat{d}$ is

$$s_d^2 = \frac{1}{1600}\{.59(1-.59)+.55(1-.55)-2(.496(.356)-.094(.054))\}$$
$$=0.0000915,$$

the estimated standard error is $s_d = \sqrt{0.0000915} = .0096$ so the test statistic is

$$Z = \frac{.04}{.0096} = 4.17.$$

With $\alpha = .05$, $c = 1.96$, and because $|Z| > 1.96$, reject and conclude that the approval rating has changed. To illustrate the alternative method for computing the statistic Z, which is called *McNemar's test*, we see that $n_{12} = 150$, $n_{21} = 86$, so

$$Z = \frac{150-86}{150+86} = 4.17,$$

which agrees with value previously obtained. The .95 confidence interval for the change in the approval rating is

$$0.04 \pm 1.96(0.0096) = (0.021, 0.059).$$

So the data suggest that, with a reasonably high probability, the approval rating has changed by at least 0.021 and as much as 0.059.

A test for independence

The next goal is to describe a classic technique for detecting dependence in a contingency table. Imagine that you want to investigate personality style versus blood pressure. Suppose each participant is classified as having personality type A or B and that each is labeled as having or not having high blood pressure. Some hypothetical results are shown in table 12.5. The goal is to test the hypothesis that personality type and high blood pressure are independent. The notion of independence was formally introduced in chapter 4, but a quick review, in the context of the problem at hand, might help.

If the probability of a randomly sampled participant having a type A personality is $p_{1+} = 0.4$, and the probability of having high blood pressure is $p_{+1} = .2$, then based on the product rule in chapter 4, independence implies that the probability of having a type A personality *and* high blood pressure is

$$p_{11} = p_{1+} \times p_{+1} = .4 \times .2 = .08.$$

Table 12.5 Hypothetical results on personality versus blood pressure

	Blood Pressure		
Personality	High	Not High	Total
A	8	67	75
B	5	20	25
Total	13	87	100

If, for example, $p_{11} = .0799999$, they are dependent although in some sense they are close to being independent. Similarly, independence implies that

$$p_{12} = p_{1+} \times p_{+2}$$

$$p_{21} = p_{2+} \times p_{+1}$$

$$p_{22} = p_{2+} \times p_{+2}.$$

In table 12.5, $n_{11} = 8$ is the number of participants among the 100 sampled who have both a type A personality and high blood pressure. Similarly, $n_{12} = 67$, $n_{21} = 5$, and $n_{22} = 20$. The hypothesis of independence can be tested with

$$X^2 = \frac{n(n_{11}n_{22} - n_{12}n_{21})^2}{n_{1+}n_{2+}n_{+1}n_{+2}}. \tag{12.7}$$

When the null hypothesis of independence is true, X^2 has, approximately, a chi-squared distribution with 1 degree of freedom.

Example 2

For the data in table 12.5,

$$X^2 = \frac{100[8(20) - 67(5)]^2}{75(25)(13)(87)} = 1.4.$$

With $\nu = 1$ degree of freedom and $\alpha = .05$, the critical value is 3.84, and because $1.4 < 3.84$, the hypothesis of independence is not rejected.

Generally, the chi-squared test of independence performs reasonably well in terms of Type I errors (e.g., Hosmane, 1986), but difficulties can arise, particularly when the number of observations in any of the cells is relatively small. For instance, if any of the n_{ij} values is less than or equal to 5, problems might occur in terms of Type I errors. There are a variety of methods for improving upon the chi-squared test, but details are not given here. (See the final section of this chapter for a list of books dedicated to categorical data.)

It is noted that the notion of a contingency table is readily generalized to situations where the variables can have more than two outcomes. A common notation is to let R represent the number of possible values for the first variable, and to let C represent the number of possible values for the second. Put another way, R indicates the number of rows in a contingency table and C represents the number of columns. Consider, for example, two raters who rate the same figure skaters on a three-point scale. The outcomes based on a rating of 100 skaters might look something like what is shown in table 12.6. So here, $R = C = 3$. This table says that there were 13 instances where rater A gave a score of 2 and simultaneously rater B gave a score of 3. In this more general context, p_{ij} represents the probability of an observation belonging to the ith row of the first variable (rater A) and the jth row of the second (rater B). And n_{ij} represents the number of times, among a sample of n observations, an outcome belongs to the ith row and jth column. The other notation introduced in the previous section is generalized in an obvious way. For example, $n_{i+} = n_{i1} + \cdots n_{iC}$ represents the total number of observations belonging to the ith row. In table 12.6, $n_{2+} = 11 + 6 + 13 = 30$ is the number of skaters who got a rating of 2 from rater B.

Table 12.6 Ratings of 100 figure skaters

	Rater B			
Rater A	1	2	3	Total
1	20	12	8	40
2	11	6	13	30
3	19	2	9	30
Total	50	20	30	100

For the more general case where a contingency table has R rows and C columns, independence corresponds to a situation where for the ith row and jth column,

$$p_{ij} = p_{i+}p_{+j},$$

where

$$p_{i+} = \sum_{j=1}^{C} p_{ij},$$

and

$$p_{+j} = \sum_{i=1}^{R} p_{ij}.$$

(This follows from the product rule introduced in chapter 4.) In words, for any row, say the ith, and any column, say the jth, if we multiply the marginal probabilities, independence means that the result must be equal to probability of being in the ith row and jth column. If for any row and column, this is not true, then there is dependence.

Example 3

Imagine that the probabilities for the two raters considered in table 12.6 are

	Rater B		
Rater A	1	2	3
1	.18	.08	.06
2	.14	.05	.19
3	.15	.01	.14

Then the probability that rater A gives a rating of 1 is $p_{1+} = .18 + .08 + .06 = .32$, the probability that rater B gives a rating of 2 is $p_{+2} = .08 + .05 + .01 = .14$, and the probability that rater B gives a rating of 3 is $p_{+3} = .39$. We see that $p_{1+}p_{+1} = .32 \times .47 = .1507$. The probability that simultaneously rater A gives a rating of 1 and rater B gives a rating of 1 is $p_{11} = .18$. Because $p_{11} = .18 \neq .1507$, the raters are dependent. As usual, however, we do not know the true probabilities and so we must rely on observations to make a decision about whether there is dependence.

For the more general contingency table considered here, the hypothesis of independence can be tested with

$$X^2 = \sum_{i=1}^{R} \sum_{j=1}^{C} \frac{n(n_{ij} - \frac{n_{i+}n_{+j}}{n})^2}{n_{i+}n_{+j}}. \quad (12.8)$$

The degrees of freedom are

$$\nu = (R-1)(C-1).$$

If X^2 exceeds the $1-\alpha$ quantile of a chi-square distribution with ν degrees of freedom, which is read from table 3 in appendix B, you reject and conclude that there is dependence.

Example 4

For the data in table 12.6, it can be seen that $X^2 = 9.9$, the degrees of freedom are 4, and with $\alpha = .05$, the critical value is 9.49. So the hypothesis of independence would be rejected. That is, when it comes to judging skaters, there is an association between the ratings given by these two raters.

Measures of association

Simply rejecting the hypothesis of independence does not tell us how strong the association happens to be. That is, to what extent is the association important? Measuring the strength of an association between two categorical variables turns out to be a nontrivial task. This section covers some of the more basic strategies aimed at accomplishing this goal. As in previous chapters, different methods provide different perspectives. What is required is a good understanding of what these measures tell us so that a judicious choice can be made when trying to understand data.

It helps to begin with what might seem like a reasonable approach, but which turns out to be unsatisfactory: use the p-value associated with the chi-squared test for independence. Generally, p-values are unsatisfactory when it comes to characterizing the extent to which groups differ and variables are related, and the situation at hand is no exception.

Another approach that has received serious consideration is to use some function of the test statistic for independence, X^2. A well-known choice is the *phi coefficient* given by

$$\phi = \frac{X}{\sqrt{n}},$$

but this measure, plus all other functions of X^2, have been found to have little value as measures of association (e.g., Fleiss, 1981; Goodman and Kruskal, 1954).

The probability of agreement

A simple yet potentially useful measure of association is the probability of agreement. In the example dealing with rating skaters, the probability of agreement refers to the probability that the two raters give the same rating to a skater. In this particular case, the probability of agreement is just $p = p_{11} + p_{22} + p_{33}$. That is, the probability of agreement is the probability that both raters give a rating of 1, or a rating of 2, or a rating of 3.

For the more general case where there are R rows and R columns, the probability of agreement is

$$p = p_{11} + \cdots + p_{RR}.$$

Notice that when working with the probability of agreement, in essence, we are dealing with a binomial probability function. That is, among n observations, either there is agreement or there is not. The total number of times we get agreement is

$$n_a = n_{11} + \cdots + n_{RR},$$

and the estimate of p is just $\hat{p} = n_a/n$. Moreover, a confidence interval for the probability of agreement can be computed as described in chapter 6. In particular, if the goal is to compute a $1-\alpha$ confidence interval for p, let c be the $1-\alpha/2$ quantile of a standard normal distribution, which is read from table 1 in appendix B, in which case an approximate $1-\alpha$ confidence interval for p is

$$\hat{p} \pm c\sqrt{\frac{\hat{p}(1-\hat{p})}{n}}. \qquad (12.9)$$

Equation (12.9) is the classic, routinely taught method, but, as noted in chapter 6, it is generally regarded as begin rather inaccurate relative to other methods that might be used. A simple improvement is the Agresti-Coull method, which in the present context is applied as follows. Let

$$\hat{n} = n + C^2,$$

$$\hat{n_a} = n_a + \frac{c^2}{2},$$

and

$$\hat{p} = \frac{n_a}{\hat{n}}.$$

Then the $1-\alpha$ confidence interval for p is

$$\hat{p} \pm c\sqrt{\frac{\hat{p}(1-\hat{p})}{\hat{n}}}.$$

Example 5

For the skating data in table 12.6, the number of times there is agreement between the two judges is $n_a = 20 + 6 + 9 = 35$ and the total number of ratings is $n = 100$. Consequently, the estimated probability of agreement is $\hat{p} = 35/100 = .35$. To compute a .95 confidence interval for p, we see from table 1 in appendix B that $c = 1.96$, and so an approximate .95 confidence interval for p is given by

$$.35 \pm 1.96\sqrt{.35(.65)/100} = (.26,\ .44).$$

Using the Agresti-Coull method, $\hat{n_a} = 35 + 1.96^2/2 = 36.92$, $\hat{n} = 100 + 1.96^2 = 103.842$, $\hat{p} = .3555$ and the .95 confidence interval is

$$.355 \pm 1.96\sqrt{.355(.6445)/103.842} = (.263,\ .447).$$

Odds and odds ratio

Some of the measures of association that have proven to be useful in applied work are based in part on conditional probabilities. First, however, some additional notation is required. The notation $p_{1|i}$ refers to the conditional probability of a participants belonging to category 1 of the second factor (the second column in contingency table) given that the individual is a member of the ith row or level of the first. In the personality versus blood pressure illustration, $p_{1|1}$ is the probability not of having high blood pressure, given that someone has a Type A personality. Similarly, $p_{2|1}$ is the probability of having blood pressure, given that the participants has a Type A personality, and $p_{1|2}$ is the probability of high blood pressure, given that the participant has a Type B personality. In general, $p_{j|i}$ refers to the conditional probability of being in the jth column of the second factor, given that the participant is a member of the ith row. From chapter 3,

$$p_{j|i} = \frac{p_{ij}}{p_{i+}},$$

where for the situation under consideration,

$$p_{i+} = p_{i1} + p_{i2}.$$

As previously indicated, the term p_{i+} is called a *marginal probability.*

To simplify matters, attention is focused on a two-way contingency table having two rows and two columns only. Associated with each row of such contingency tables is a quantity called its *odds*. First note that for now 1, there are two conditional probabilities associated with the two columns, namely $p_{1|1}$ and $p_{2|1}$. In symbols, the odds for row 1 is

$$\Omega_1 = \frac{p_{1|1}}{p_{2|1}},$$

while the odds for row 2 is

$$\Omega_2 = \frac{p_{1|2}}{p_{2|2}}.$$

(The symbol Ω is an upper case Greek omega.)

Example 6

Consider the personality versus blood pressure data shown in table 12.5. The estimated probabilities are summarized in table 12.7. For instance, the estimated probability that, simultaneously, a randomly sampled adult has a Type A personality and high blood pressure is

$$\hat{p}_{11} = \frac{n_{11}}{n} = \frac{8}{100} = 0.08.$$

Table 12.7 Estimated probabilities for personality versus blood pressure

	Blood Pressure		
Personality	High	Not High	Total
A	.08	.67	.75
B	.05	.20	.25
Total	.13	.87	1.00

Consequently, the estimated probability of having high blood pressure, given that someone has a Type A personality, is

$$\hat{p}_{1|1} = \frac{0.08}{0.08 + 0.67} = 0.1067.$$

Similarly, the probability of not having high blood pressure, given that you have a Type A personality, is

$$\hat{p}_{2|1} = \frac{0.67}{0.08 + 0.67} = 0.8933.$$

The estimate of the odds for this row is

$$\begin{aligned} \hat{\Omega}_1 &= \frac{\hat{p}_{1|1}}{\hat{p}_{2|1}} \\ &= \frac{0.1067}{0.8933} \\ &= 0.12. \end{aligned}$$

That is, given that someone has a Type A personality, the probability of having high blood pressure is estimated to be about 12% of the probability that the person's blood pressure is not high. In symbols, your estimate is that $p_{1|1} = .12 \times p_{2|1}$ since

$$\hat{p}_{1|1} = \hat{\Omega}_1 \times \hat{p}_{2|1}.$$

Put another way, among Type A personalities, the probability of not having high blood pressure is about $1/0.12 = 8.4$ times as high as the probability that blood pressure is high. As for Type B personalities, the odds is estimated to be

$$\hat{\Omega}_2 = 0.25.$$

This says that if you have a Type B personality, the chance of having hypertension is estimated to be about a fourth of the probability that blood pressure is not high. Notice that you can measure the relative risk of hypertension, based on personality type, by comparing the two odds just estimated. Typically a comparison is made with their ratios. This means you use what is called the *odds ratio*, which is estimated with

$$\begin{aligned} \hat{\theta} &= \frac{\hat{\Omega}_1}{\hat{\Omega}_2} \\ &= \frac{0.12}{0.25} \\ &= 0.48. \end{aligned}$$

This says that among Type A personalities, the relative risk of having hypertension is about half what it is for individuals who are Type B personalities.

Table 12.8 Mortality rates per 100,000 person-years from lung cancer and coronary artery disease for smokers and nonsmokers of cigarettes

	Smokers	Nonsmokers	Difference
Cancer of the lung	48.33	4.49	43.84
Coronary artery disease	294.67	169.54	125.13

In terms of population probabilities, the odds ratio can be written as

$$\theta = \frac{p_{11}p_{22}}{p_{12}p_{21}},$$

and for this reason, θ is often called the *cross-product ratio*. A simpler way of writing the estimate of θ is

$$\hat{\theta} = \frac{n_{11}n_{22}}{n_{12}n_{21}}.$$

Under independence, it can be shown that $\theta = 1$. If $\theta > 1$, then participants in row 1 (Type A personality in the illustration) are more likely to belong to the first category of the second factor (high blood pressure) than are participants in row 2. If $\theta < 1$, then the reverse is true. That is, participants are less likely to belong to the first category of the second factor than are participants in row 2.

All measures of association are open to the criticism that they reduce your data down to a point where important features can become obscured. This criticism applies to the odds ratio, as noted by Berkson (1958) and discussed by Fleiss (1981). Table 12.8 shows the data analyzed by Berkson on mortality and smoking. It can be seen that the estimated odds ratio is

$$\hat{\theta} = \frac{10.8}{1.7} = 6.35,$$

and this might suggest that cigarette smoking has a greater effect on lung cancer than on coronary artery disease. Berkson pointed out that it is *only* the difference in mortality that permits a valid assessment of the effect of smoking on a cause of death. The difference for coronary artery disease is considerably larger than it is for smoking, as indicated in the last column of table 12.8,indicating that smoking is more serious in terms of coronary artery disease. The problem with the odds ratio in this example is that it throws away all the information on the number of deaths due to either cause.

Problems

10. For a random sample of 200 adults, each adult was classified as having a high or low income, and each was asked whether they are optimistic about the future. The results were

	Yes	No	Total
High	35	42	77
Low	80	43	123
Total	115	85	200

What is the estimated probability that a randomly sampled adult has a high income and is optimistic? Compute a .95 confidence interval for the true probability.

11. Referring to the previous exercise, compute a .95 confidence interval for $\delta = p_{1+} - p_{+1}$. Also test $H_0 : p_{12} = p_{21}$ using Z with a Type I error probability of $\alpha = .05$.

12. In problem 10, would you reject the hypothesis that income and outlook are independent? Use $\alpha = .05$. Would you use the ϕ coefficient to measure the association between income and outlook? Why?

13. In problem 10, estimate the odds ratio and interpret the results.

14. You observe

	Income (father)			
Income (daughter)	High	Medium	Low	Total
High	30	50	20	100
Medium	50	70	30	150
Low	10	20	40	70
Total	90	140	90	320

Estimate the proportion of agreement and compute a .95 confidence interval.

12.3 Some modern advances and insights

There are many methods for analyzing categorical data beyond those covered here (e.g., Agresti, 1990, 1996; Andersen, 1997; Lloyd, 1996; and Powers and Xie, 1999; Simonoff, 2003). One topic of special interest is loglinear models (e.g., Agresti, 1990; Fienberg, 1980); they can be useful when studying associations in three-way and higher contingency tables. Another area of interest is a regression problem where the outcome is binary. For example, what is the probability of getting a heart attack during the next year if an individual's cholesterol level is 250? One approach to this problem is to use what is called logistic regression. For books dedicated to this topic, see, for example, Kleinbaum (1994) or Hosmer and Lemeshow (1989).

13

RANK-BASED AND NONPARAMETRIC METHODS

Nonparametric and rank-based methods provide yet another approach to comparing groups and studying associations. Generally, they provide a different and potentially useful perspective regarding how groups compare and how variables are related. This chapter introduces some classic, well-known methods and some recent advances and insights are covered as well.

13.1 Comparing independent groups

This section describes methods for comparing independent groups. We begin with a classic technique, outline its practical problems, and then mention modern methods for dealing with these issues.

Wilcoxon-Mann-Whitney test

The classic rank-based method for comparing two independent groups is called the Wilcoxon-Mann-Whitney test. It was originally derived by Wilcoxon (1945), and later it was realized that Wilcoxon's method was the same as a procedure proposed by Mann and Whitney (1947).

Some care is needed when describing the goal of the Wilcoxon-Mann-Whitney test. To begin, imagine that a single observation is randomly sampled from each group. Let p be the probability that the randomly sampled observation from the first group is less than the randomly sampled observation from the second. In symbols, if X_{11} is the first observation from the first group and X_{12} is the first observation from the second group,

$$p = P(X_{11} < X_{12}),$$

which is a natural way of characterizing how the groups differ. The Wilcoxon-Mann-Whitney test is based on a direct estimate of p with the intended goal of testing

$$H_0 : p = .5. \qquad (13.1)$$

In words, if the groups do not differ in any manner, then there should be a 50% chance that an observation from the first group is less than an observation from the second.

Again, imagine that a single observation is randomly sampled from the first group, but now, n_2 observations are randomly sampled from the second. For illustrative purposes, imagine that the value from the first group is 8, $n_2 = 10$, and that the values from the second group are

$$2, 4, 6, 9, 13, 15, 18, 22, 25, 29.$$

So we see that 8 is less than 7 of the observations in the second group. Based on this data, a natural estimate of p is just the proportion of times the single observation in the first group is less than the values in the second. Here we write this estimate as p_1, where the subscript 1 means that the first observation in the first group is being used. In the illustration, $p_1 = 7/10$.

Now suppose that an additional observation is randomly sampled from the first group. For illustrative purposes, suppose the value is 23. Then, continuing the illustration, we see that 23 is less than two of the observations in the second group, and so now the estimate of p would be $p_2 = 2/10$. Continuing in this fashion when there are n_1 observations in the first group, we get n_1 estimates of p, which we label as $p_1, \ldots, p_{n_1}$. A natural way of combining these n_1 estimates into a single estimate of p is to average them, this is what is done in practice, and the result is labeled

$$\hat{p} = \frac{1}{n_1} \sum p_i. \tag{13.2}$$

Example 1

Consider the following observations:

$$\text{Group 1: } 30, 60, 28, 38, 42, 69$$

$$\text{Group 2: } 19, 21, 27, 73, 71, 25, 59, 61.$$

The first value in the first group, 30, is less than four of the eight observations in the second group, so p_1 is 4/8. In a similar manner, the second observation in the first group is 60, it is less than three of the values in the second group, so $p_2 = 3/8$. Continuing in this fashion, $p_3 = 4/8$, $p_4 = 4/8$, $p_5 = 4/8$ and $p_6 = 2/8$. Consequently, the estimate of p, the probability that a value from the first group is less than a value from the second, is

$$\hat{p} = \frac{1}{6}\left(\frac{4}{8} + \frac{3}{8} + \frac{4}{8} + \frac{4}{8} + \frac{4}{8} + \frac{2}{8}\right) = .4375.$$

An alternative method for estimating p, that is equivalent to the method just described, explains why the Wilcoxon-Mann-Whitney is called a rank-based technique. The method begins by combining the observations into a single group, writing them ascending order, and then assigning ranks. This means that the smallest value among all of the observations gets a rank of 1, the next smallest gets a rank of 2, and the largest observation gets a rank of $n_1 + n_2$. Next, the sum of the ranks associated with the second group is computed, which we label S. Letting

$$U = S - \frac{n_2(n_2 + 1)}{2},$$

the estimate of p is

$$\hat{p} = \frac{U}{n_1 n_2}.$$

The quantity U is called the Wilcoxon-Mann-Whitney U statistic.

Example 2

Continuing the last example, if we pool the observations into a single groupd and write them in ascending order, we get

Pooled Data:	**19**	**21**	**25**	**27**	28	30	38	42	**59**	60	**61**	69	**71**	**73**
Ranks:	**1**	**2**	**3**	**4**	5	6	7	8	**9**	10	**11**	12	**13**	**14.**

For convenience, the values corresponding to the second group and their corresponding ranks are written in boldface. The sum of the ranks associated with the second group is

$$S = 1 + 2 + 3 + 4 + 9 + 11 + 13 + 14 = 57,$$

so

$$U = 57 - \frac{8(9)}{2} = 21,$$

and

$$\hat{p} = \frac{21}{6(8)} = .4375,$$

consistent with the previous example.

Now consider the problem of testing H_0: $p = .5$. The usual description of the test statistic is in terms of U. If $p = .5$, then U estimates $n_1 n_2/2$. The test statistic is based in part on an estimate of VAR(U), the variance of U over many studies, which is the squared standard error of U. If we assume there are no tied (duplicated) values and that the groups have *identical distributions*, then VAR(U) is given by

$$\sigma_u^2 = \frac{n_1 n_2 (n_1 + n_2 + 1)}{12}.$$

This means that the null hypothesis can be tested with

$$Z = \frac{U - \frac{n_1 n_2}{2}}{\sigma_u}, \tag{13.3}$$

which has, approximately, a standard normal distribution when the assumptions are met and H_0 is true. In particular, reject if

$$|Z| \geq c,$$

where c is the $1 - \alpha/2$ quantile of a standard normal distribution, which is read from table 1 in appendix B. To make sure that it is clear that the method is based on an estimate of p, it is noted that the test statistic Z can be written as

$$Z = \frac{\hat{p} - .5}{\sigma_u / \sqrt{n_1 n_2}}.$$

Example 3

Continuing the last illustration, $n_1 = 6$, $n_2 = 8$, so $\sigma_u^2 = 6(8)(6+8+1)/12 = 60$ Consequently,

$$Z = \frac{21 - 24}{\sqrt{60}} = -0.387.$$

With $\alpha = .05$, the critical value is 1.96, $|Z|$ is less than 1.96, so fail to reject.

When does the Wilcoxon-Mann-Whitney test perform well?

A positive feature of the Wilcoxon-Mann-Whitney test is that when groups have identical distributions, it performs reasonably well in terms of controlling the probability of a Type I error. In practical terms, like Student's t test, when it rejects, it is reasonable to conclude that the groups differ in some manner. But when distributions differ, if the primary goal is to make inferences about p, practical concerns arise. A fundamental reason why is that now, under general conditions, the wrong standard for U is being used. This can create problems in terms of both Type I error probabilities and power. Consequently, although the Wilcoxon-Mann-Whitney test is based on an estimate of p, a more accurate description is that it tests the hypothesis of identical distributions. Indeed, if tied values are impossible, and the goal is to test the hypothesis of identical distributions, the probability of a Type I error can be controlled exactly under random sampling using techniques not covered in this book.

The Wilcoxon-Mann-Whitney test and medians

Sometimes the Wilcoxon-Mann-Whitney test is described as a method for comparing the medians of two groups. There are restrictive conditions where this is true, but when these conditions are not met, the Wilcoxon-Mann-Whitney test performs poorly as a method for comparing medians (e.g., Fung, 1980). For example, there are situations where power decreases as the difference between the population medians increases, and confidence intervals for the difference between the medians cannot be computed (Kendall and Stuart, 1973; Hettmansperger, 1984).

Dealing with tied values

There are two issues that should be mentioned regarding tied values. The first is that the formulation of the null hypothesis must be altered. To begin, note that there are three possible outcomes when sampling a single observation from each group: (1) the observation from the first group is greater than the observation from the second, (2) the observations have identical values, or (3) the observation from the first group is less than the observation from the second. For convenience, let p_0 be the probability that the two values are identical and let

$$P = p + .5p_0.$$

Then the goal is to test

$$H_0 : P = .5.$$

But again, although the Wilcoxon-Mann-Whitney test is based on an estimate of P, a more accurate description is that the hypothesis of identical distributions is being tested.

The second issue has to do with how ranks are computed. The standard strategy is to use what are called the midranks, which involves averaging the ranks corresponding to the duplicated values. To illustrate what this means, consider the values 45, 12, 32, 64, 13, and 25. There are no tied values and the smallest value has a rank 1, the next smallest has rank 2, and so on, as previously explained. A common notation for the rank corresponding to the ith observation is R_i. So in the example, the first observation is $X_1 = 45$ and its rank is $R_1 = 5$. Similarly, $X_2 = 12$ and its rank is $R_2 = 1$.

Now consider a situation where there are tied values: 45, 12, 13, 64, 13, and 25. Putting these values in ascending order yields 12, 13, 13, 25, 45, 64. So the value 12 gets a rank of 1, but there are two identical values having a rank of 2 and 3. The *midrank* is simply the average of the ranks among the tied values. Here, this means that the rank assigned to the two values equal to 13 would be $(2 + 3)/2 = 2.5$, the average of their corresponding ranks. So the ranks for all six values would be 1, 2.5, 2.5, 4, 5, 6.

Generalizing, consider

$$7,\ 7.5,\ 7.5,\ 8,\ 8,\ 8.5,\ 9,\ 11,\ 11,\ 11.$$

There are 10 values, so if there were no tied values, their ranks would be 1, 2, 3, 4, 5, 6, 7, 8, 9, and 10. But because there are two values equal to 7.5, their ranks are averaged yielding a rank of 2.5 for each. There are two values equal to 8, their original ranks were 4 and 5, so their final ranks (their midranks) are both 4.5. There are three values equal to 11, their original ranks were 8, 9 and 10, the average of these ranks is 9, so their midranks are all equal to 9. So the ranks for the 10 observations are

$$1,\ 2.5,\ 2.5,\ 4.5,\ 4.5,\ 6,\ 7,\ 9,\ 9,\ 9.$$

The Kolmogorov-Smirnov test

Yet another way of testing the hypothesis that two independent groups have identical distributions is with the Kolmogorov-Smirnov test. Unlike Student's t test and the Wilcoxon-Mann-Whitney test, the Kolmogorov-Smirnov test is designed to be sensitive to any differences among the quantiles. For example, if the medians differ, this test is capable of detecting this difference. If the medians are equal, but say the quartiles differ, the Kolmogorov-Smirnov test is designed to detect this difference as well. A criticism of the Kolmogorov-Smirnov test is that when tied values can occur, its power can be relatively poor. But despite this, there are situations where an extension of the method (not described here) can provide a deeper and more detailed sense of how groups compare that goes beyond any of the methods covered in this book that are based on a single measure of location. And regardless of any practical problems caused by tied values, there are situations where its power can be relatively high in comparison with other techniques.

To apply it, let $\hat{F}_1(x)$ be the proportion of observations in group one that are less than or equal to x, and let $\hat{F}_2(x)$ be the corresponding proportion for group two. For convenience, let

$$Y_1 = X_{11},\ Y_2 = X_{21}, \ldots, Y_{n_1} = X_{n_1 1}$$

and

$$Y_{n_1+1} = X_{12},\ Y_{n_1+2} = X_{22}, \ldots, Y_{n_1+n_2} = X_{n_2,2}.$$

That is, the n_1 observations in group 1 are labeled $Y_1, \ldots, Y_{n_1}$ and the n_2 observations in group 2 are labeled $Y_{n_1+1}, \ldots, Y_{n_1+n_2}$. Let

$$V_i = |\hat{F}_1(Y_i) - \hat{F}_2(Y_i)|,$$

$i = 1, \ldots, n_1 + n_2$. So V_1, for example, is computed by determining the proportion of values in group 1 that are less than or equal to Y_1, the proportion of values in group 2 that are less than or equal to Y_1, and then taking the absolute value of the difference between these two proportions. The Kolmogorov-Smirnov test statistic is

$$KS = \max\{V_1, \ldots, V_{n_1+n_2}\}, \tag{13.4}$$

the largest V value. For large sample sizes, an approximate critical value when $\alpha = .05$ is

$$c = 1.36\sqrt{\frac{n_1 + n_2}{n_1 n_2}}.$$

The Kolmogorov-Smirnov test rejects the hypothesis of identical distributions if $KS \geq c$.[1]

Example 4

The Kolmogorov-Smirnov test is illustrated with the following data.

Group 1: 7.6, 8.4, 8.6, 8.7, 9.3, 9.9, 10.1, 10.6, 11.2
Group 2: 5.2, 5.7, 5.9, 6.5, 6.8, 8.2, 9.1, 9.8, 10.8,
11.3, 11.5, 12.3, 12.5, 13.4, 14.6.

The sample sizes are 9 and 15. In symbols, $n_1 = 9$ and $n_2 = 15$. Focus on the first observation in the first group, which is 7.6. Then $\hat{F}_1(7.6) = 1/9$ because only one of the nine values in group 1 is less than or equal to 7.6. In a similar manner, $\hat{F}_1(8.4) = 2/9$ and $\hat{F}_1(8.6) = 3/9$. Notice that for the second group, the smallest observation is 5.2. So for the second group, $\hat{F}_2(5.2) = 1/15$, but for the first group, $\hat{F}_1(5.2) = 0/9$. That is, for the first group, there are no values

1. The software S-PLUS and R come with a function for applying the Kolmogorov-Smirnov test. But in general, and particularly when tied values occur, a better choice is the R or S-PLUS function ks, stored in the files mentioned in chapter 1; it provides exact control over the probability of a Type I error assuming random sampling only. For more details, see Wilcox (2005).

less than or equal to 5.2, so the estimated probability of getting a value less than or equal to 5.2 is 0. The test statistic *KS* can be computed by proceeding as follows:

Group 1	Group 2	$\lvert\hat{F}_1(Y) - \hat{F}_2(Y)\rvert$	Group 1	Group 2	$\lvert\hat{F}_1(Y) - \hat{F}_2(Y)\rvert$
	5.2	$\lvert 0/9 - 1/15\rvert = 1/15$		9.8	$\lvert 5/9 - 8/15\rvert = 1/45$
	5.7	$\lvert 0/9 - 2/15\rvert = 2/15$	9.9		$\lvert 6/9 - 8/15\rvert = 2/15$
	5.9	$\lvert 0/9 - 3/15\rvert = 1/5$	10.1		$\lvert 7/9 - 8/15\rvert = 11/45$
	6.5	$\lvert 0/9 - 4/15\rvert = 4/15$	10.6		$\lvert 8/9 - 8/15\rvert = 16/45$
	6.8	$\lvert 0/9 - 5/15\rvert = 1/3$		10.8	$\lvert 8/9 - 9/15\rvert = 13/45$
7.6		$\lvert 1/9 - 5/15\rvert = 2/9$	11.2		$\lvert 9/9 - 9/15\rvert = 2/5$
	8.2	$\lvert 1/9 - 6/15\rvert = 13/45$		11.3	$\lvert 9/9 - 10/15\rvert = 1/3$
8.4		$\lvert 2/9 - 6/15\rvert = 8/45$		11.5	$\lvert 9/9 - 11/15\rvert = 4/15$
8.6		$\lvert 3/9 - 6/15\rvert = 1/15$		12.3	$\lvert 9/9 - 12/15\rvert = 1/5$
8.7		$\lvert 4/9 - 6/15\rvert = 2/45$		12.5	$\lvert 9/9 - 13/15\rvert = 2/15$
	9.1	$\lvert 4/9 - 7/15\rvert = 1/45$		13.4	$\lvert 9/9 - 14/15\rvert = 1/15$
9.3		$\lvert 5/9 - 7/15\rvert = 4/45$		24.6	$\lvert 9/9 - 15/15\rvert = 0$

The largest absolute difference just computed (among columns 3 and 6) is $KS = 2/5 = .4$. The approximate .05 critical value is $c = .573$, so fail to reject.

At first glance, the Kolmogorov-Smirnov test might seem relatively uninteresting because when it rejects, it is unclear in what sense the groups differ. An important extension of the method does provide an indication of which quantiles differ (Doksum and Sievers, 1976), but the details are too involved to give here. Suffice it to say that the Kolmogorov-Smirnov test can yield interesting perspectives on how groups differ and by how much.[2]

Comparing more than two groups: The Kruskall-Wallis test

The best-known rank-based method for more than two independent groups is the Kruskall-Wallis test. The original goal of this method was to extend the Wilcoxon-Mann-Whitney test to more than two groups. Recall that the Wilcoxon-Mann-Whitney test is based on an estimate of p, the probability that a randomly sampled observation from the first groups is less than a randomly sampled observation from the second, is equal to .5. The Kruskall-Wallis test was intended to test the hypothesis that $p = .5$ for any two groups. But like the Wilcoxon-Mann-Whitney test, a more accurate description of the method is that it tests the hypothesis that all groups have identical distributions.

As usual, we let J represent the number of groups, n_j represent the number of observations in the jth group, and N is the total number of observations. In symbols, $N = \sum n_j$. The method begins by pooling all N observations and assigning ranks. That is, the smallest observation among all N values gets a rank of 1, the next smallest a rank of 2, and so on. If X_{ij} is the ith observation in the jth group, we let R_{ij} be its rank among the pooled data. When there are tied values, midranks are used. Next, sum the ranks for

2. For illustrations, see Wilcox (2003, 2005).

each group. In symbols, compute

$$R_j = \sum_{i=1}^{n_j} R_{ij},$$

$(j = 1, \ldots, J)$. Letting

$$S^2 = \frac{1}{N-1}\left(\sum_{j=1}^{J}\sum_{i=1}^{n_j} R_{ij}^2 - \frac{N(N+1)^2}{4}\right),$$

the test statistic is

$$T = \frac{1}{S^2}\left(-\frac{N(N+1)^2}{4} + \sum \frac{R_j^2}{n_j}\right).$$

If there are no ties, S^2 simplifies to

$$S^2 = \frac{N(N+1)}{12},$$

and T becomes

$$T = -3(N+1) + \frac{12}{N(N+1)} \sum \frac{R_j^2}{n_j}.$$

The hypothesis of identical distributions is rejected if $T \geq c$, where c is some appropriate critical value. For small sample sizes, exact critical values are available from Iman, Quade, and Alexander (1975). For large sample sizes, the critical value is approximately equal to the $1-\alpha$ quantile of a chi-squared distribution with $J-1$ degrees of freedom.

Example 5

Table 13.1 shows data for three groups and the corresponding ranks. For example, after pooling all $N = 10$ values, $X_{11} = 40$ has a rank of $R_{11} = 1$, the value 56 has a rank of 6, and so forth. The sums of the ranks corresponding to each group are $R_1 = 1+6+2 = 9$, $R_2 = 3+7+8 = 18$ and $R_3 = 9+10+5+4 = 28$. The number of groups is $J = 3$, so the degrees of freedom are $\nu = 2$, and from table 3 in appendix B, the critical value is approximately $c = 5.99$ with $\alpha = .05$. Because there are no ties among the N observations,

$$T = -3(10+1) + \frac{12}{10 \times 11}\left(\frac{9^2}{3} + \frac{18^2}{3} + \frac{28^2}{4}\right) = 3.109.$$

Because $3.109 < 5.99$, fail to reject. That is, you are unable to detect a difference among the distributions.

Table 13.1 Hypothetical data illustrating the Kruskall-Wallis test

	Group 1		Group 2		Group 3	
i	X_{i1}	R_{i1}	X_{i2}	R_{i2}	X_{i3}	R_{i3}
1	40	1	45	3	61	9
2	56	6	58	7	65	10
3	42	2	60	8	55	5
4					47	4

Problems

1. For the values 1, 1, 1, 1, 2, 2, 2, 3, 3, 4, 5, 5, 6, 6, 6, compute the midranks.

2. For two independent groups, you observe

$$\text{Group 1: } 8, 10, 28, 36, 22, 18, 12$$

$$\text{Group 2: } 11, 9, 23, 37, 25, 43, 39, 57.$$

Compare these two groups with the Wilcoxon-Mann-Whitney test using $\alpha = .05$.

3. For the data in the previous exercise, apply the Kolmogorov-Smirnov test, again using $\alpha = .05$.

4. Comment on using the Wilcoxon-Mann-Whitney test when the goal is to compare medians.

5. When the Wilcoxon-Mann-Whitney rejects, how should this be interpreted?

6. You want to compare the effects of two different cold medicines on reaction time. Suppose you measure the decrease in reaction times for one group of participants who take one capsule of drug A, and you do the same for a different group of participants who take drug B. The results are

$$\text{A: } 1.96, 2.24, 1.71, 2.41, 1.62, 1.93$$

$$\text{B: } 2.11, 2.43, 2.07, 2.71, 2.50, 2.84, 2.88.$$

Compare these two groups with the Wilcoxon-Mann-Whitney test using $\alpha = .05$. What is your estimate of the probability that a randomly sampled subject receiving drug A will have less of a reduction in reaction time than a randomly sampled subject receiving drug B?

7. For the data in the previous exercise, apply the Kolmogorov-Smirnov test, again using $\alpha = .05$. (It is interesting to note that if an exact critical value is used, rather than the approximate critical value described in the text, the opposite conclusion is reached regarding whether to reject.)

8. When there are many tied values, speculate on whether the Wilcoxon-Mann-Whitney test will have more power than the Kolmogorov-Smirnov test.

9. Two methods for reducing shoulder pain after laparoscopic surgery were compared by Jorgensen et al. (1995). The data were

$$\text{Group 1: } 1, 2, 1, 1, 1, 1, 1, 1, 1, 1, 2, 4, 1, 1$$

$$\text{Group 2: } 3, 3, 4, 3, 1, 2, 3, 1, 1, 5, 4.$$

Compare these groups using the Wilcoxon-Mann-Whitney test using $\alpha = .05$.

10. Imagine two groups of cancer patients are compared, the first group having a rapidly progressing form of the disease and the other having a slowly progressing form. At issue is whether psychological factors are related to the progression of cancer. The outcome measure is one where highly negative scores indicated a

tendency to present the appearance of serenity in the presence of stress. The results are

$$\text{Group 1: } -25, -24, -22, -22, -21, -18, -18, -18, -18, -17, -16, -14, -14, -13, -13, -13, -13, -9, -8, -7, -5, 1, 3, 7, 7$$

$$\text{Group 2: } -21, -18, -16, -16, -16, -14, -13, -13, -12, -11, -11, -11, -9, -9, -9, -9, -7, -6, -3, -2, 3, 10.$$

Compare these groups with Wilcoxon-Mann-Whitney test using $\alpha = .05$.

11. Repeat the previous problem, only use the Kolmogorov-Smirnov test.
12. For three independent groups, you observe

$$\text{Group 1: } 4, 6, 7, 8, 9, 15, 12, 19$$

$$\text{Group 2: } 16, 18, 2, 21, 29, 30, 24, 27$$

$$\text{Group 3: } 20, 22, 26, 31, 32, 38, 39, 41.$$

Perform the Kruskal-Wallis test with $\alpha = .05$.

13. In the previous exercise, the largest observation is 41, which is in the third group. The sample means can be seen to be 10, 20.9, and 31.1, respectively, and the ANOVA F test in chapter 10 rejects with $\alpha = .05$. If the largest observation is made even larger, eventually the F test will not reject, even though the sample mean of the third group increases as well. Is the same true when using the Kruskal-Wallis test?

13.2 Comparing two dependent groups

The sign test

A simple method for comparing dependent groups is the so-called sign test. In essence, it is based on making inferences about the probability of success associated with a binomial distribution.

Imagine that n pairs of observations have been randomly sampled. The pairs of observations might reflect blood pressure before and after taking some medication, the attitudes of a married couple regarding abortion, and so on. Following the notation introduced in chapter 9, these pairs of observations are denoted by

$$(X_{11}, X_{12})$$
$$\vdots$$
$$(X_{n1}, X_{n2}).$$

Primarily for convenience, it is momentarily assumed that tied values never occur. That is, for any pair of observations, their difference is not equal to zero. Let p be the probability that for a randomly sampled pair of observations, the observation from group 1 is less than the observation from group 2. More formally,

$$p = P(X_{i1} < X_{i2}).$$

Letting

$$D_1 = X_{11} - X_{12}$$
$$\vdots$$
$$D_n = X_{n1} - X_{n2},$$

an estimate of p is simply the proportion of D_i values that are less than zero. If we let V indicate the number of D_i values less than zero, then and estimate of p is

$$\hat{p} = \frac{V}{n}. \tag{13.5}$$

In the terminology of chapter 4, V represents the number of successes, where 'success' refers to the event that, for a pair of observations, the first is less than the second. From chapter 12, the hypothesis

$$H_0 : p = .5$$

can be tested by computing

$$Z = \frac{\hat{p} - .5}{\sqrt{\hat{p}(1 - \hat{p})/n}}$$

and rejecting if $|Z| \geq c$, where c is the $1 - \alpha/2$ quantile of a standard normal distribution read from table 1 in appendix B.

As for situations where ties can occur, a simple strategy—one that is routinely used—is to ignore or discard these cases. So if among N pairs of observations, there are n pairs of observations that do not have identical values, then an estimate of p is

$$\hat{p} = \frac{V}{n}, \tag{13.6}$$

where again V is the number of times the first observation is less than the second.

Example 1

Consider the values

Group 1: $22, 12, 34, 43, 54, 46, 33, 19, 10, 54, 66$

Group 2: $21, 34, 19, 34, 43, 22, 33, 27, 29, 19, 11.$

The differences are:

$$1, -22, 15, 9, 11, 24, 0, -8, -19, 35, 55.$$

Removing the one case where the difference is zero leaves $n = 10$ values:

$$1, -22, 15, 9, 11, 24, -8, -19, 35, 55.$$

Because the number of negative values is $V = 3$, the estimate of p is $\hat{p} = 3/10 = .3$, and $|Z| = 1.38$. If the Type I error probability is to be $\alpha = .05$, $c = 1.96$, so fail to reject the hypothesis that the probability of a negative difference is .5.

Wilcoxon signed rank test

Another classic method for comparing two dependent groups is the Wilcoxon signed rank test. The null hypothesis is that the two dependent groups have identical distributions. The method begins by forming difference scores as was done in conjunction with the paired T-test in chapter 9 or the the sign test just described. If any difference scores are equal to zero, they are discarded, and we let n indicate that number of differences remaining. In symbols, these differences are again denoted by

$$D_1, \ldots, D_n.$$

Next, rank the $|D_i|$ values and let U_i denote the result for $|D_i|$. So, for example, if the D_i values are 6, −2, 12, 23, −8, then $U_1 = 2$ because after taking absolute values, 6 has a rank of 2. Similarly, $U_2 = 1$ because after taking absolute values, the second value, −2, has a rank of 1. Next set

$$R_i = U_i,$$

if $D_i > 0$; otherwise

$$R_i = -U_i.$$

Positive numbers are said to have a sign of 1, negative numbers a sign of −1, so R_i is the value of the rank corresponding to $|D_i|$ multiplied by the sign of D_i. The test statistic is

$$W = \frac{\sum R_i}{\sqrt{\sum R_i^2}}.$$

If there are no ties, this last equation simplifies to

$$W = \frac{\sqrt{6}\sum R_i}{\sqrt{n(n+1)(2n+1)}}.$$

Decision rule: reject if $|W| \geq c$, where c is the $1 - \alpha/2$ quantile of a standard normal distribution read from table 1 in appendix B.

Example 2

Imagine a study aimed at reducing feelings of depression. The Wilcoxon signed rank test is illustrated with the following values for depression taken at two different times:

$$\text{Time 1: } 45, 12, 34, 56, 78, 12, 43, 65, 76, 66$$

$$\text{Time 2: } 35, 10, 22, 66, 28, 10, 45, 56, 43, 65.$$

The differences are

$$10, 2, 12, -10, 50, 2, -2, 9, 33, 1.$$

The ranks of the absolute values are

$$6.5, 3.0, 8.0, 6.5, 10.0, 3.0, 3.0, 5.0, 9.0, 1.0.$$

Multiplying these ranks by the sign of the differences yields

$$6.5, 3.0, 8.0, -6.5, 10.0, 3.0, -3.0, 5.0, 9.0, 1.0.$$

and a little arithmetic yields $W = 1.84$. With $\alpha = .05$, the crtical value is $c = 1.96$, so fail to reject.

Problems

14. For two dependent groups you get

Group 1: 10, 14, 15, 18, 20, 29, 30, 40

Group 2: 40, 8, 15, 20, 10, 8, 2, 3.

Compare the two groups with the sign test and the Wilcoxon signed rank test with $\alpha = .05$.

15. For two dependent groups you get

Group 1: 86 71 77 68 91 72 77 91 70 71 88 87

Group 2: 88 77 76 64 96 72 65 90 65 80 81 72.

Apply the Wilcoxon signed rank test with $\alpha = .05$.

13.3 Rank-based correlations

As in chapter 8, imagine that we have n randomly sampled pairs of observations. For example, for every individual, we might measure blood pressure and cholesterol levels. Or we might measure levels of anxiety and the amount of an antidepressant that is being taken. As in chapter 8, these pairs of points are labeled $(X_1, Y_1), \ldots, (X_n, Y_n)$. So, X_1 might be blood pressure of the first participant and Y_1 might be her cholesterol level. There are two classic alternatives to Pearson's correlation that can be used, among other things, to establish that two variables are dependent. They are called Kendall's tau and Spearmans' rho.

Kendall's tau

Kendall's tau is based on the following idea. Consider two pairs of observations, which are labeled as (X_1, Y_1) and (X_2, Y_2). These two pairs of numbers are said to be concordant if Y increases as X increases, or if Y decreases as X decreases. If two pairs of observations are not concordant, they are said to be discordant.

Put another way, Kendall's tau reflects how close the relationship is to being monotone. A monotone relationship is one that consistently increases or decreases, but the increase does not necessarily follow a straight line. For each pair of points, Kendall's tau is concerned with whether the slope between these two points is positive (concordant) or negative (discordant). If in general, as one variable gets large, the other tends to increase as well, then Kendall's tau will be positive. In a similar manner, if in general, as one variable gets large the other tends to get smaller, then Kendall's tau will be negative. Based on all pairs of points, Kendall's tau is the difference between the positive slopes and the negative slopes divided by the total number of slopes. Consequently, its value lies between -1 and 1. Said another way, among all pairs of points, Kendall's tau is just the average number that are concordant minus the average number that are discordant. If two measures are independent, then this difference should be approximately equal to zero. If all of the slopes are positive, meaning that as the first variable increases, the second always increaes as well, Kendall's tau is equal to 1. And if the second variable always decreases, Kendall's tau is equal to -1.

Example 1

The pairs of points $(X_1, Y_1) = (12, 32)$ and $(X_2, Y_2) = (14, 42)$ are concordant because X increases from 12 to 14 and the corresponding Y values increase as well. That is, the slope between these two points is positive. The pairs of points $(X_1, Y_1) = (10, 28)$ and $(X_2, Y_1) = (14, 24)$ have a negative slope and are discordant.

To describe how to compute tau in a more formal manner, let $K_{ij} = 1$ if the ith and jth pairs of observations are concordant, otherwise $K_{ij} = -1$. Next, sum all of the K_{ij} for which $i < j$, which is denoted by

$$\sum_{i<j} K_{ij}.$$

Then Kendall's tau is given by

$$\hat{\tau} = \frac{2\sum_{i<j} K_{ij}}{n(n-1)}, \qquad (13.7)$$

where τ is a lower case Greek tau. In words, Kendall's tau is just the average of the K_{ij} values. (In the notation used here, the number of K_{ij} values is $n(n-1)/2$.) As previously indicated, $\hat{\tau}$ has a value between -1 and 1. If $\hat{\tau}$ is positive, there is a tendency for Y to increase with X—possibly in a nonlinear fashion—and if $\hat{\tau}$ is negative, the reverse is true.

Example 2

Consider the values

$$(X_1, Y_1) = (2, 1)$$
$$(X_2, Y_2) = (6, 5)$$
$$(X_3, Y_3) = (8, 7)$$
$$(X_4, Y_4) = (3, 12).$$

We see that for the first two pairs of points ($i = 1$ and $j = 2$), X increases from 2 to 6, Y increases from 1 to 5, so they are concordant. That is $K_{12} = 1$. More generally, we see that

i	j	K_{ij}
1	2	1
1	3	1
1	4	1
2	3	1
2	4	−1
3	4	−1

Then Kendall's tau is just the average of these six values, namely, $\hat{\tau} = .333$. To illustrate the notation, the sum of these values is $\sum_{i<j} K_{ij} = 2$, $n = 4$, so again

$$\hat{\tau} = \frac{2(2)}{4(3)} = .333.$$

The population analog of $\hat{\tau}$ is labeled τ and can be shown to be zero when X and Y are independent. That is, τ is the value of Kendall's tau if all individuals could be measured, and if there is no association, $\tau = 0$. So if there is empirical evidence that $\tau \neq 0$, this indicates that X and Y are dependent. *If* X and Y are independent, and if tied values never occur, it can be shown that the variance of $\hat{\tau}$ over many studies, VAR($\hat{\tau}$), which is the squared standard error of $\hat{\tau}$, is given by

$$\sigma_{\tau}^2 = \frac{2(2n+5)}{9n(n-1)}.$$

To test

$$H_0 : \tau = 0,$$

compute

$$Z = \frac{\hat{\tau}}{\sigma_{\tau}},$$

and reject if

$$|Z| \geq c,$$

where c is the $1 - \alpha/2$ quantile of a standard normal distribution, which can be read from table 1 in appendix B.

Example 3

Continuing the last example, there are four pairs of points ($n = 4$), $2(2n+5) = 26$, $9n(n-1) = 108$, so $\sigma_{\tau}^2 = 26/108 = .2407$, and $Z = .333/\sqrt{.2407} = .68$. With $\alpha = .05$, $c = 1.96$, so fail to reject.

Kendall's tau provides protection against outliers among the X values, ignoring Y, as well as outliers among the Y values, ignoring X. Imagine that $\hat{\tau} = .4$ and that the largest X value is 40. If this largest X is increased to one million, $\hat{\tau}$ is not altered, it remains equal to .4. Nevertheless, a few unusual points, properly placed, can have a tremendous influence on the value of Kendall's tau.

Example 4

Consider the points in figure 13.1 and notice that the two points in the lower right corner appear to be unusual compared to the others, and indeed they are unusual based on how the data were generated. If we ignore these two outliers, Kendall's tau is $\hat{\tau} = .37$, and we reject the hypothesis that $\tau = 0$ with the Type I error set at $\alpha = .05$. (The p-value is .023.) But if we include these two outliers, Kendall's tau drops to $\hat{\tau} = .13$ and the p-value increases to .41, so we no longer reject, even though all but two of the values were generated in a manner where X and Y are dependent.

Spearman's rho

Spearman's rho, labeled r_s, is just Pearson's correlation based on the ranks associated with X versus the ranks associated with Y. Under independence, the population analog

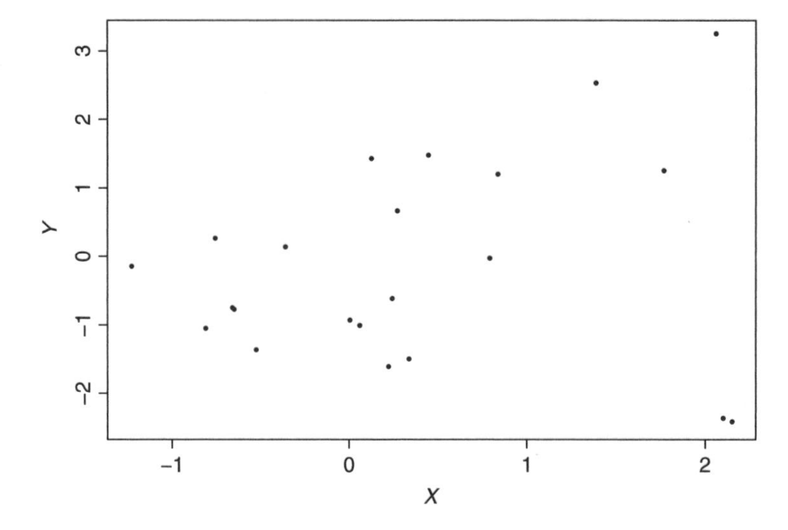

Figure 13.1 This scatterplot illustrates that outliers, properly placed, can have a large influence on both Kendall's tau Spearman's rho.

of r_s, ρ_s, is zero. Also, like Kendall's tau, Spearman's rho is exactly equal to one if there is a monotonic increasing relationship between X and Y. That is, Y always increases as X gets large. And $\rho_s = -1$ if the association is monotonic decreasing instead.

The usual approach to testing

$$H_0 : \rho_s = 0,$$

is based on

$$T = r_s \frac{\sqrt{n-2}}{\sqrt{1-r_s^2}}.$$

When there is independence, T has, approximately, a Student's t-distribution with $\nu = n-2$ degrees of freedom. So reject and conclude there is an association if $|T| \geq t$, where t is the $1-\alpha/2$ quantile of a Student's t-distribution with $n-2$ degrees of freedom.

Like Kendall's tau, Spearman's rho provides protection against outliers among the X values, ignoring Y, as well as outliers among the Y values, ignoring X, yet unusual points can have a big impact on its value.

Example 5

In the last example it was illustrated that Kendall's tau can be influenced by a few unusual values. We repeat this illustration with Spearman's rho. Ignoring the two unusual values in the lower-right corner of figure 13.1, Spearman's rho is $r_s = .54$ and the hypothesis of independence is rejected with $\alpha = .05$. (The p-value is .01.) But when we include the two unusual values, $r_s = .16$ and the p-value is .48. So now we fail to detect any association.

Problems

16. For the following pairs of observations, test $H_0 : \tau = 0$ with $\alpha = .05$.

Time 1: 10 16 15 20

Time 2: 25 8 18 9.

17. Repeat the previous exercise, only use Spearman's rho instead.

13.4 Some modern advances and insights

The methods described in this chapter perform well, in terms of controlling the probability of a Type I error, when comparing groups with identical distributions or when dealing with associations where the variables are independent. But in terms of detecting and describing true differences between groups or true associations between two variables, many new and improved rank-based methods are now available.

The Brunner-Munzel method

Recall that the Wilcoxon-Mann-Whitney test is based on an estimate of P. When there are no tied values, P is the probability that a randomly sampled observation from the first group is less than a randomly sampled observation from the second. But if the goal is to test H_0: $P = .5$, the method is unsatisfactory under general conditions because it uses the wrong standard error when the distributions differ. There are two methods that appear to correct this problem to a reasonable degree. One was derived by Brunner and Munzel (2000). As usual, let X_{ij} be the ith observation from the jth group ($i = 1, \ldots, n_j; j = 1, 2$). To apply the Brunner-Munzel method, first pool all $N = n_1 + n_2$ observations and assign ranks. In the event there are tied values, ranks are averaged as previously illustrated. (That is, midranks are used.) As previously described and illustrated, the results for the jth group are labeled R_{ij}, $i = 1, \ldots, n_j$. That is, R_{ij} is the rank corresponding to X_{ij} among the pooled values. Let $\bar{R}_1$ be the average of the ranks corresponding to group one and $\bar{R}_2$ is the average for group 2. So

$$\bar{R}_1 = \frac{1}{n_1} \sum_{i=1}^{n_1} R_{i1}$$

and

$$\bar{R}_2 = \frac{1}{n_2} \sum_{i=1}^{n_2} R_{i2}.$$

Next, for the first group, rank the observations ignoring group 2 and label the results $V_{11}, \ldots V_{n_1 1}$. Do the same for group 2 (ignoring group 1) and label the ranks $V_{12}, \ldots V_{n_2 2}$. The remaining calculations are shown in box 13.1.[3]

3. The function bpm, which belongs to the library of R or S-PLUS functions mentioned in chapter 1, performs the calculations. When using SPSS, access to this function can be obtained via the software zumastat, also mentioned in chapter 1.

BOX 13.1 The Brunner-Munzel method for two independent groups.

Compute

$$S_j^2 = \frac{1}{n_j - 1}\sum_{i=1}^{n_j}\left(R_{ij} - V_{ij} - \bar{R}_j + \frac{n_j+1}{2}\right)^2,$$

$$s_j^2 = \frac{S_j^2}{(N-n_j)^2}$$

$$s_e = \sqrt{N}\sqrt{\frac{s_1^2}{n_1} + \frac{s_2^2}{n_2}},$$

$$U_1 = \left(\frac{S_1^2}{N-n_1} + \frac{S_2^2}{N-n_2}\right)^2$$

and

$$U_2 = \frac{1}{n_1-1}\left(\frac{S_1^2}{N-n_1}\right)^2 + \frac{1}{n_2-1}\left(\frac{S_2^2}{N-n_2}\right)^2.$$

The test statistic is

$$W = \frac{\bar{R}_2 - \bar{R}_1}{\sqrt{N}s_e},$$

and the degrees of freedom are

$$\hat{\nu} = \frac{U_1}{U_2}.$$

Decision Rule: Reject $H_0: P = .5$ if $|W| \geq t$, where t is the $1-\alpha/2$ quantile of a Student's t-distribution with $\hat{\nu}$ degrees of freedom. An estimate of P is

$$\hat{P} = \frac{1}{n_1}\left(R_2 - \frac{n_2+1}{2}\right) = \frac{1}{N}(\bar{R}_2 - \bar{R}_1) + \frac{1}{2}.$$

An approximate $1-\alpha$ confidence interval for P is

$$\hat{P} \pm ts_e.$$

Example 1

Table 13.2 reports data from a study of hangover symptoms among sons of alcoholics versus a control. Note that there are many tied values among these data. In the second group, for example, 14 of the 20 values are zero. Welch's test for means has a p-value of .14, Yuen's test has a p-value of .076, the Brunner-Munzel method has a p-value of .042, and its .95 confidence interval for P is (.167, .494). The main point is that if we compare groups with the goal that the Type I error probability be .05 level, the choice of method makes a difference in whether or not we reject. But it should be noted that a criticism of using multiple methods for comparing groups is that if the groups do not differ in

Table 13.2 The effect of alcohol

Group 1:	0	32	9	0	2	0	41	0	0	0
	6	18	3	3	0	11	11	2	0	11
Group 2:	0	0	0	0	0	0	0	0	1	8
	0	3	0	0	32	12	2	0	0	0

any manner, and if each test is performed at the .05 level, the probability of making a Type I error among all of the methods used will be greater than .05. A simple method for dealing with this is to perform each test at the $.05/C$ level, where C is the number of tests used to compare the groups. That is, use the Bonferroni method described in chapter 11. Here, the number of tests used to compare the two groups is $C = 3$. The lowest p-value is .042, this is larger than .05/3, so now we would fail to reject.

Although no details are given here, it should be noted that an alternative to the Brunner-Munzel method is recommended by Cliff (1996).[4] It appears that with very small sample sizes, it can be a little more satisfactory than the Brunner-Munzel method in terms of Type I errors (Neuhäuser, et al., 2007). For extensions of the Brunner-Munzel method to more than two groups, as well as more modern rank-based methods for comparing dependent groups see Brunner, et al. (2002) as well as Wilcox (2003, 2005). Reiczigel, et al. (2005) studied a bootstrap method for making inferences about p, and it appears to have an advantage over the Brunner-Munzel method when sample sizes are small and tied values never occur. However, with tied values it can perform poorly in situations where the Brunner-Munzel method performs reasonably well. Currently, Cliff's method appears to be the best choice for general use.

The Brunner-Dette-Munk method

The Kruskall-Wallis test performs relatively well, in terms of controlling the probability of a Type I error, when the null hypothesis of identical distributions is true, but concerns about relatively low power arise when distributions differ. An alternative rank-based method, which deserves consideration, was derived by Brunner, Dette and Munk (1997). The explicit goal is to test the hypothesis that all J groups have identical distributions. But a rough characterization of the method is that it is designed to be sensitive to differences among the average ranks. The method pools the data and assigns ranks as was done in the Kruskall-Wallis test. The remaining calculations are too involved to give here, but complete details and software can be found in Brunner, et al. (2002) and Wilcox (2003, 2005).

Finally, a classic technique for comparing more than two dependent groups, based on ranks, is called *Friedman's test*. Like the other classic techniques covered in this chapter, it performs well, in terms of Type I errors, when comparing groups that have identical distributions, but when groups differ, several methods are now available that are aimed at providing higher power. Details can be found in Brunner, et al. (2002) as well as Wilcox (2003, 2005).

4. In the library of S-PLUS and R functions mentioned in chapter 1, the function cid performs the calculations and the function cidv2 can be used to compute a p-value.

A Summary of Some Key Points

- Generally, nonparametric methods provide alternative ways of describing and detecting differences among groups and associations among variables.
- Rank-based methods are sometimes described as methods for comparing medians, but under general conditions they are highly unsatisfactory for this purpose. More generally, if the goal is to compare measures of location, such as the mean, median or trimmed mean, methods in this chapter are unsatisfactory.
- An advantage of rank-based methods is that they are insensitive to outliers when comparing groups. That is, power might be high relative to methods for comparing means. Kendall's tau and Spearman's rho provide some protection against outliers, but they are not completely satisfactory in this regard; methods covered in more advanced courses are more satisfactory.

13.5 Some final comments on comparing groups

Multiple methods for comparing two or more groups have been described at various points in this book. And there are additional methods, not covered here, that have practical value. How should one choose a method from among the many that are available? There is no agreement on how best to proceed. Using data from 24 dissertations, Wu (2002) compared the power of a variety methods. No single method was always best and in some cases methods based on means performed well, but in general, methods based on means had the poorest power. Methods based on a 20% trimmed mean did not always compete well with other techniques, but it was the most likely approach to provide the best power. But it should be kept in mind that different methods tell us different things about how groups differ and by how much. For instance, we saw examples where despite any negative features associated with means, comparing means can be argued to be more meaningful than comparing medians or 20% trimmed means. Simultaneously, there are situations where means provide a dubious summary of what the typical response happens to be. That is, some thought is required when choosing a method. Multiple methods could be used, but now there is the issue of controlling the probability of at least one Type I error. Methods are available for dealing with this issue, some of which were described in chapter 11, but a concern is that now power might be poor versus using a single technique. One possibility is to use 20% trimmed means to make a decision about whether groups differ, and then use alternative techniques to gain perspective on the nature of any differences that might exist. Another possibility is choose a single method, and if no differences are found try some alternative techniques. If many methods are used and each is performed with the Type I error probability set equal to .05, again there is the issue of controlling the probability of at least one Type I error. Suppose five methods are used to compare the groups, with a method based on medians yielding the lowest p-value, which is .04. It can be argued that the evidence is weak that the groups differ, but that there are indications that perhaps the medians do indeed differ. So one possibility is to perform a new study where groups are compared using medians only. Repeating the study might require a considerable effort, but this seems preferable to always using only one method for comparing groups, and if no differences are found, throwing the data away. That is, it is undesirable to routinely miss a true difference due to the method used to compare the groups.

APPENDIX A: SOLUTIONS TO SELECTED EXERCISE PROBLEMS

Chapter 2

1. (a) 22, (b) 2, (c) 20, (d) 484, (e) 27, (f) −41, (g) 2, (h) 220, (i) 12, (j) 54.
2. (a) $\sum X_i/i$, (b) $\sum U_i^i$, (c) $(\sum Y_i)^4$.
3. (a) $\sum X_i/i$.
4. (a) $\bar{X} = -.2, M = 0$, (b)$\bar{X} = 186.1667, M = 6.5$.
5. $\bar{X} = 83.93, M = 80.5$.
6. 338.1.
7. (a) 30.6, (b) 120.6, (c) 1020.6.
8. 24.5 in all three cases.
9. One
10. About half.
11. $q1 = -6.58$ $q2 = 7.4$ $(j = 3, h = .41667)$.
12. $q1 = -6$, $q2 = 3$, $(j = 4, h = .16667)$.
13. About a fourth.
14. Clearly, $X_{(1)}$ has a value between $X_{(1)}$ and $X_{(n)}$.
15. If we multiply all of the values by any constant c, then in particular $X_{(1)}$ becomes $cX_{(1)}$.
16. Range = 18, $s^2 = 32$, $s = 5.66$.
17. Note that $n\bar{X} = \sum(X_i)$, So $\sum(X_i - \bar{X}) = (\sum X_i) - n\bar{X} = 0$.
18. $s = .37$.
19. $s = 11.58$.
20. 20 is an outlier.
21. Again, 20 is an outlier.
22. Yes.
26. Sometimes, even with two or more outliers, the classic rule might catch all of the outliers, depending on where they are, but the boxplot rule is better at avoiding masking.
27. $.2 \times n = .2 \times 21 = 4.2$, so $g = 4$, $\bar{X}_t = 80.08$.
30. 2.
31. 5. The mean is least resistant and the median is the most resistant.
33. About 20%, or more than g.x
34. $\bar{X} = 229.2$, $\bar{X}_t = 220.8$ $M = 221$.
35. 24, 24, 25, 32, 35, 36, 36, 42, 42.
36. $s_w^2 = 51.36$.
37. Smaller, $s^2 = 81$, $s_w^2 = 51.4$.
38. Yes.
40. $s_w^2 = 1,375.6$.
41. $\bar{X}_t = 82$, $s_w^2 = 69.2$.

Chapter 3

1. mean is 2.85, variance is 1.94697, sd = 1.395.
2. mean is 2.52, variance is 1.2245, sd = 1.11.
3. mean is 3, variance is 1.5, sd = 1.22.
4. mean is 18.387, variance is 85.04, sd = 9.22.
5. mean is 11.1, variance is 42.3, sd = 6.5.
8. No.
9. Yes.
10. The boxplot rule can declare values to be outliers that do not appear to be outliers based on a histogram.
12. 34.6.
13. There would be only one stem.
14. median = 80, quartiles = 50 and 121, IQR = 121 − 50 = 71, largest value not declared an outlier is 215.
15. Values less than −56.5 or greater than 227.5 are declared outliers.
18. When the population histogram is symmetric and outliers are rare.
19. In some cases, 100 is sufficient but in others a much larger sample size is needed.
20. Generally, the boxplot is better than a histogram.
21. Not necessarily. Situations are encountered where the sample histogram is a poor indication of the population histogram.

Chapter 4

1. No
2. No
3. Yes
4. .5
5. 0.
6. 0.5
7. 0.8
8. 0.7
10. mean = 2.3, variance is .81 and the standard deviaitoni is .9
11. mean = 3.2, variance is 1.76 and the standard deviation is 1.3.
12. $\mu - \sigma = 3.2 - 1.33 = 1.87$ and $\mu + \sigma = 3.2 + 1.33 = 4.53$. But the only possible values between 1.87 and 4.53 are 2, 3 and 4. So the answer is $p(2) + p(3) + p(4) = .7$.
13. mean = 2, standard deviation is .63.
14. Increase.
15. mean = 3, variance is 1.6.
16. smaller
17. larger
18. (a) .3, (b) .03/.3, (c) .09/.3, (d) .108/.18.
19. Yes.
20. (a) 1253/3398, (b) 757/1828, (c) 757/1253, (d) no, (e) 1831/3398
21. Yes, this can only happen if the conditional probabilities change when you are told X.
22. Yes.
23. Yes.
24. (a) 0.006, (b) 0.3823, (c) 0.1673, (d) 0.367, (e) 0.7874
25. (a) 0.00475, (b) 0.29687, (c) 0.4845, (d) 0.6864.
26. 0.1859.
27. 0.29848
28. 10.8 and 4.32
29. 4.4, 3.52.

30. .7,0.0105
31. .3, .007.
32. (a) 0.3222, (b) 0.6778, (c) 0.302.
33. Two heads and a tail has probability .44, versus three heads with probability .343.
34. .5
35. (a) $.75^5$, (b) $.25^5$, (c) $1 - .25^5 - 5(.75)(.25)^4$.
36. (a) 0.5858, (b) 0.7323, (c) 0.5754, (d) 0.7265.
37. 10, 6, .4, 0.0096.
38. (a) 0.0668, (b) 0.0062, (c) 0.0062, (d) .683.
39. (a) 0.691, (b) 0.894, (c) .77.
40. (a) .31, (b) .885, c) 0.018, d) .221.
41. (a) -2.33, (b) 1.93, (c) -0.174, (d) .3,
42. (a) 1.43, (b) -0.01, (c) 1.7, (d) 1.28.
43. (a) .133, (b) .71, (c) .133, (d) .733
44. (a) .588, (b) .63, (c) .71, (d) .95.
45. (a) .1587, (b) .382, (c) .383, (d) .683.
46. $c = 1.96$
47. 1.28.
48. .16.
49. 84.45.
50. 1−.91.
51. .87.
52. .001.
53. .68.
54. .95.
55. .115.
56. .043.
57. Yes
59. No, could be much larger
60. No.
61. Yes.
62. Yes.

Chapter 5

1. (a) 0.885, (b) $1 - 0.885 = .115$, (c) 0.7705.
2. 0.3151.
3. 0. It is impossible to get $\hat{p} = .05$ when $n = 10$.
4. Note that with $n = 25$, there are only two ways of getting $\hat{p} \leq .05$: when the number of successes is 0 or 1. And the probability of getting 0 or 1 successes is 0.2712.
6. .4
7. $.4(.6)/30 = .008$
8. (a) .0228, (b) .1587, (c) 0.9545.
9. (a) .1587, (b) .023, (c) $.977 - .023 = 0.954$.
10. .023.
11. .6822.
12. 0.866.
13. $\sqrt{160.78/9} = 4.23$.
14. (a) .055, (b) .788, (c) .992 (d) $.788 - .055$.
15. (a) .047, (b) .952 (c) 1−.047, (d) $.952 - .047$.
16. Skewed, heavy-tailed distribution.
17. Symmetric, light-tailed.

19. No tied values.
21. There are tied values.
22. No.
23. There is an outlier. $s_M = 3.60$, $s/\sqrt{n} = 17.4$.
25. 69.4.
26. There is an outlier.
27. The sample mean has the smallest standard error under normality. So if there is an ideal estimator, it must be the mean, but under non-normality it can perform poorly.
28. If the distribution is heavy-tailed, meaning outliers are common, a trimmed mean or median can be more accurate on average.

Chapter 6

2. $1.28, 1.75, 2.33$.
3. $45 \pm 1.96(5/5) = (43.04, 46.96)$.
4. $45 \pm 2.576(5/5)$.
5. No, .95 CI $= (1141.8, 1158.2)$.
6. (a) $(52.55, 77.45)$, (b) $(180.82, 189.18)$, (c) $(10.68, 27.32)$.
7. Length $= 2c\sigma/\sqrt{n}$. When n is doubled, length $= 2c\sigma/\sqrt{2n}$. Therefore, the ratio of the lengths is $1/\sqrt{2}$. That is, the length is decreased by a factor of $1/\sqrt{2}$. For $4n$, decreased by a factor of 1/2.
8. $\bar{X} = 19.97$. so .95 CI $= (19.9602, 19.9798)$. b) No. c) Just because the confidence contains 20, this does not necessarily mean that $\mu = 20$.
9. .99 CI $= (1.896363, 2.303637)$.
10. .90 CI $= (1.62644, 1.77356)$.
11. (a) 2.085963, (b) 2.84534, (c) 1.725.
12. (a) $(19.56179, 32.43821)$, (b) $(122.0542, 141.9458)$, (c) $(47, 57)$.
13. (a) $(16.75081, 35.24919)$, (b) $(118.3376, 145.6624)$, (c) $(45.28735, 58.71265)$.
14. $(161.5030, 734.7075)$.
15. $(10.69766, 23.50234)$.
16. $(28.12, 39.88)$.
17. $(109.512, 414.488)$.
18. $1 - .022 = .978$.
19. .964.
20. $n = 19$ and $k = 3$. .9993.
21. $n = 15$, $\hat{p} = 5/15$, CI $= (0.09476973, 0.5718969)$ using equation (6.10). Agresti-Coull gives (0.15, 0.585).
22. (a) .0064. (b) .0039, (c) .0016, (d) .00083.
23. $(0.0412, 0.1588)$. Agresti-Coull = (0.05, 0.176).
24. $(0.26187550.3181245)$.
25. $(0.04528041, 0.07471959)$.
27. $(0.0009958988, 0.0013299610)$.
28. $(0.1856043, 0.3143957)$.
29. $(0.00264542, 0.03735458)$.
30. $(0, .00004)$.
31. $.18 \pm 1.96\sqrt{.18(.82)/1000}$
33. A large sample size might be needed.
35. (a). $52 \pm 2.13\sqrt{12}/(.6\sqrt{24})$. (b). $10 \pm 2.07\sqrt{30}/(.6\sqrt{36})$. (c). $16 \pm 2.07\sqrt{9}/(.6\sqrt{12})$.
37. (293.6, 595.9).
38. Under normality, the ideal estimator is the mean. So if an ideal exists, it must be the mean, but under general conditions, it performs poorly.
39. For trimmed mean, .95 confidence interval is (34.8, 54.8). for the mean, it is (30.7, 58.3).
40. There is an outlier.

Chapter 7

1. $Z = -1.265$, Fail to reject.
2. Fail to reject.
3. (74.9, 81.1).
4. .103.
5. .206.
6. $Z = -14$, reject.
7. Reject
8. (118.6, 121.4)
9. Yes, because $\bar{X}$ is consistent with H_0.
10. $Z = 10$, reject.
11. $Z = 2.12$. Reject.
19. Increase α.
20. (a) $T = 1$, fail to reject. (b) $T = .5$, fail to reject. (c) $T = 2.5$, reject, (c)
22. (a) $T = .8$, fail to reject. (b) $T = .4$, fail to reject. (c) $T = 2$, reject.
24. $T = .39$, fail to reject.
25. $T = -2.61$. Reject
26. $T = 2$, $c = 2.6$, fail to reject.
27. $T = .75$, $c = -1.86$, fail to reject.
28. $T = -1.6$, $-c = -2.26$, fail to reject.
29. $T = 2.3$, $c = 2.54$, fail to reject.
30. (a) $T = .6\sqrt{20}(44 - 42)/9 = 0.596$, $c = 2.2$, fail to reject. (b) $T = 0.2981$, fail to reject.
32. $n = 10$, so the degrees of freedom are 5, $c = 2.571$, $T = -3.1$, reject.
33. The degrees of freedom are 14, $c = 2.977$, $T = 0.129$, fail to reject.

Chapter 8

3. Least squares minimizes the sum of the squared residuals. So for any choice for the slope and intercept, the sum of the squared residuals will be at most 47.
4. In equation (8.5), $C = (n - 1)s_x^2 = 132$. So the estimated slope is $144/132 = 1.09$.
5. $b_1 = -0.0355$, $b_0 = 39.93$.
6. $b_1 = .0039$, $b_0 = .485$.
8. $b_1 = 0.0754$, $b_0 = 1.253$.
9. $\hat{Y} = -0.0355(600) + 39.93 = 18.63$, but daily calories of 600 is greater than any value used to compute the slope and intercept. That is, extrapolation is being used.
17. $(-2.65, -0.35)$.
18. $(-9.5436, -4.4564)$.
20. $b_1 = 3$, $b_0 = -.5$.
21. (a) $b_1 = 3$, $b_0 = -1$.(b) $T = 2.11$, critical value is $t = -0.099$, fail to reject. (c) $(0.602, 5.398)$.
22. (a) $b_1 = .25$, $b_0 = 7$. (b) $(-0.76, 1.26)$.
23. $(2.04, 4.16)$ indicating that the slope is probably greater than 2.
24. $(2.78, 9.22)$.
25. (a) $r = .8$. $T = 6.67$, reject. (b) $r = .5$, $T = 1.29$, fail to reject.
26. $r = .5$, $T = 2.8$, $t = 1.7$, reject. This indicates dependence.
27. (a). Yes. (b) Yes, outliers can mask a negative association. (c) Plot the data.
28. Nothing, this does not change r.
29. The absolute value of the slope gets larger.
30. The residuals are larger, meaning the the correlation will get smaller.
31. The slope and intercept were chosen so as to minimize the second sum. $\bar{Y}$ is the regression line with $b_1 = 0$, and so the first sum must be bigger than the second.
32. There are various ways there might be dependence that is not detected by r.

33. (a) Many factors affect r. Outliers can result in a large r but a poor fit. (b). $\hat{Y}$ does much better than $\bar{Y}$, in terms of minimizing the sum of the squared residuals, but both perform poorly.
34. No. You need to look at the data.
35. The confidence interval can be relatively long and is potentially inaccurate.
36. Again, the confidence interval can be relatively long and is potentially inaccurate.

Chapter 9

3. $T = 9.3$, reject.
4. $W = 10.5$, reject.
5. Welch's test might have more power.
6. $T = 3.795$. reject
7. $WT = 3.795$. reject.
8. When the sample variances are approximately equal, the choice between T and W makes little difference.
9. $\nu = 31.86066$, $t2.037283$, $\mathrm{CI} = (1.4, 8.6)$.
10. $\mathrm{CI} = (1.49, 8.61)$.
11. No, fail to reject.
12. $W = 1.95$, $t = 2.06$, fail to reject.
13. The actual probability coverage could differ substantially from .95.
14. $Z = .62$, fail to reject.
15. $Z = -.26$, $\mathrm{CI} = (-0.17, .13)$.
16. $Z = -.4$, $\mathrm{CI} = (-.13, .086)$.
17. $T = -3.32$, $\nu = 49$, reject, $\mathrm{CI} = (-13.358907, -3.277457)$.
19. Difference between the means is -8.3 versus -7.5. Also, there are no outliers suggesting that the standard error of the median will be larger than the standard error of the means, which turns out to be the case.
20. $Z = 2.23$.
21. $T = 5.25$. Reject.
22. (1.88, 4.12)
23. $T = 2.91$, reject, $\mathrm{CI} = (1.28, 7.43)$.
24. No. The method used in problem 22 is not appropriate for testing the hypothesis of equal medians. It tests the hypothesis that the median of the difference scores is zero, which is not necessarily the same.
29. .021

Chapter 10

1. $\mathrm{MSWG} = (6.214 + 3.982 + 2.214)/3 = 4.14$.
2. $\mathrm{MSBG} = 25.04$, $F = 6.05$, critical values is $f = 3.47$.
5. $\mathrm{MSWG} = 9.5$
7. No.
8. Don't know when power is high enough to detect situations where unequal variances is a practical issue.
11. MSBG estimated $2 + 25 = 27$. MSWG estimates 2, the common variance. Because the null hypothesis is false, MSBG estimates a larger quantity, on average.
14. The distributions differ, suggesting that in particular the means differ.
15. Low power due to outliers, violating the equal variance assumptions, differences in skewness, small sample sizes.
16. No.
17. Unclear whether the test has enough power to detect a departure from normality that has practical importance.

18. Increase the variances.
19. There is a main effect for A and B, no interaction.
20. There is a main effect for A and B and an interaction.
21. $N = 90$, Factor A, $F = 400/50 = 8$, $\nu_1 = 1$, $\nu_2 = 90 - 6 = 84$, $f = 6.95$, reject. Factor B, $F = 4$, $\nu_1 = 2$, $f = 4.87$, fail to reject. Interaction, $F = 4$, $\nu_1 = 2$, $f = 4.87$, fail to reject.
22. $N = 200$, Factor A, $F = 6$, $\nu_1 = 3$, $\nu_2 = 200 - 20 = 180$, $f = 2.65$, reject. Factor B, $F = 4$, $\nu_1 = 4$, $f = 2.42$, reject. Interaction, $F = 3$, $\nu_1 = 12$, $f = 1.81$, reject.

Chapter 11

1. The probability of at least one Type I error can be unacceptably high.
2. $\text{MSWG} = 11.6$, $T = |15 - 10|/\sqrt{11.6(1/20 + 1/20)} = 4.64$ $\nu = 100 - 5 = 95$, reject.
3. $T = |15 - 10|/\sqrt{11.6(1/20 + 1/20)/2} = 6.565$, $q = 3.9$, reject.
4. $\text{MSWG} = 8$. $T = |20 - 12|/\sqrt{8(1/10 + 1/10)} = 6.325$ $\nu = 50 - 5 = 45$, reject.
5. $T = |20 - 12|/\sqrt{8(1/10 + 1/10)/2} = 8.94$, $q = 4.01$, reject.
8. $W = (15 - 10)/\sqrt{4/20 + 9/20} = 6.2$, $\hat{\nu} = 33$, $c = 2.99$, reject.
9. $(15 - 10)/\sqrt{.5(4/20 + 9/20)} = 8.77$, reject.
10. $W = (20 - 12)/\sqrt{5/10 + 6/10} = 7.63$, reject.
11. $|20 - 12|/\sqrt{(5/10 + 6/10)/2} = 10.787$, $q = 4.01$, reject.
12. $|16 - 22|/\sqrt{64 + 5} = -0.72$, fail to reject.
13. Tied values can result in poor control over Type I errors.
14. $W = (24 + 14 - 36 - 24)/\sqrt{6 + 8 + 8 + 5} = -4.23$, $\nu = 23$, $t = 2.07$.
15. $\nu = 23$. Critical value equals 2.57.
16. $W = -.38$, $\nu = 23.4$, $c = 2.57$, fail to reject.
17. None
18. All of them.
19. Rom's method has as much or more power than the Bonferroni.
20. Tests with p-values .001 and .005 would be rejected.

Chapter 12

1. $X^2 = 17.2$, critical value $c = 7.8$.
2. $X^2 = 26.2$, critical value $c = 13.3$.
3. $X^2 = 46$, critical value $c = 15.1$.
4. $X^2 = 13$, critical value $c = 5.99$.
5. $X^2 = 5.3$, critical value $c = 7.8$.
6. $X^2 = 5.1$, critical value $c = 9.5$.
7. $X^2 = 20.3$, critical value $c = 5.99$.
8. $X^2 = 6.36$, critical value $c = 7.81$.
9. $X^2 = 4.15$, critical value $c = 12.59$.
10. $\hat{p}_{11} = 35/200$. $\text{CI} = (0.12, 0.23)$.
11. $d = -.19$, $\text{CI} = (-0.29, -0.085)$. $Z = -3.4$, reject.
12. $X^2 = 7.4$, reject. No, phi coefficient is known to be an unsatisfactory measure of association.
13. $\hat{\theta} = .448$. Individuals with high incomes are about half as likely to be optimistic about the future.
14. $\hat{p} = .4375$, .95 $\text{CI} = (.385, .494)$.

Chapter 13

1. 2.5, 2.5, 2.5, 2.5, 6.0, 6.0, 6.0, 8.5, 8.5, 10.0, 11.5, 11.5, 14.0, 14.0, 14.0.
2. $Z = 1.5$, fail to reject.
3. $KS = .5$, fail to reject.
4. It is unsatisfactory under general conditions.
5. The distributions differ in some manner.
6. $Z = 2.43$, reject. The estimate of p is .9.
7. $KS = .71$, critical value is .76, fail to reject.
8. A speculation is that the Kolmogorov-Smirnov test will have less power.
9. $Z = 2.44$, reject.
10. $Z = 1.64$, fail to reject.
11. $KS = .32$, fail to reject.
12. $T = 14.06$, reject.
13. No.
14. Sign test, $\hat{p} = .29$, reject. Wilcoxon signed rank test, $W = 2.28$, reject.
15. $W = .76$, fail to reject.
16. $\hat{\tau} = -0.667$, $Z = -1.36$, fail to reject.
17. $r_s = -0.8$, fail to reject.

Appendix B: Tables

Table 1 Standard normal distribution

z	$P(Z \leq z)$	z	$P(Z \leq z)$	z	$P(Z \leq z)$	z	$P(Z \leq z)$
−3.00	0.0013	−2.99	0.0014	−2.98	0.0014	−2.97	0.0015
−2.96	0.0015	−2.95	0.0016	−2.94	0.0016	−2.93	0.0017
−2.92	0.0018	−2.91	0.0018	−2.90	0.0019	−2.89	0.0019
−2.88	0.0020	−2.87	0.0021	−2.86	0.0021	−2.85	0.0022
−2.84	0.0023	−2.83	0.0023	−2.82	0.0024	−2.81	0.0025
−2.80	0.0026	−2.79	0.0026	−2.78	0.0027	−2.77	0.0028
−2.76	0.0029	−2.75	0.0030	−2.74	0.0031	−2.73	0.0032
−2.72	0.0033	−2.71	0.0034	−2.70	0.0035	−2.69	0.0036
−2.68	0.0037	−2.67	0.0038	−2.66	0.0039	−2.65	0.0040
−2.64	0.0041	−2.63	0.0043	−2.62	0.0044	−2.61	0.0045
−2.60	0.0047	−2.59	0.0048	−2.58	0.0049	−2.57	0.0051
−2.56	0.0052	−2.55	0.0054	−2.54	0.0055	−2.53	0.0057
−2.52	0.0059	−2.51	0.0060	−2.50	0.0062	−2.49	0.0064
−2.48	0.0066	−2.47	0.0068	−2.46	0.0069	−2.45	0.0071
−2.44	0.0073	−2.43	0.0075	−2.42	0.0078	−2.41	0.0080
−2.40	0.0082	−2.39	0.0084	−2.38	0.0087	−2.37	0.0089
−2.36	0.0091	−2.35	0.0094	−2.34	0.0096	−2.33	0.0099
−2.32	0.0102	−2.31	0.0104	−2.30	0.0107	−2.29	0.0110
−2.28	0.0113	−2.27	0.0116	−2.26	0.0119	−2.25	0.0122
−2.24	0.0125	−2.23	0.0129	−2.22	0.0132	−2.21	0.0136
−2.20	0.0139	−2.19	0.0143	−2.18	0.0146	−2.17	0.0150
−2.16	0.0154	−2.15	0.0158	−2.14	0.0162	−2.13	0.0166
−2.12	0.0170	−2.11	0.0174	−2.10	0.0179	−2.09	0.0183
−2.08	0.0188	−2.07	0.0192	−2.06	0.0197	−2.05	0.0202
−2.04	0.0207	−2.03	0.0212	−2.02	0.0217	−2.01	0.0222
−2.00	0.0228	−1.99	0.0233	−1.98	0.0239	−1.97	0.0244
−1.96	0.0250	−1.95	0.0256	−1.94	0.0262	−1.93	0.0268
−1.92	0.0274	−1.91	0.0281	−1.90	0.0287	−1.89	0.0294
−1.88	0.0301	−1.87	0.0307	−1.86	0.0314	−1.85	0.0322
−1.84	0.0329	−1.83	0.0336	−1.82	0.0344	−1.81	0.0351
−1.80	0.0359	−1.79	0.0367	−1.78	0.0375	−1.77	0.0384
−1.76	0.0392	−1.75	0.0401	−1.74	0.0409	−1.73	0.0418
−1.72	0.0427	−1.71	0.0436	−1.70	0.0446	−1.69	0.0455
−1.68	0.0465	−1.67	0.0475	−1.66	0.0485	−1.65	0.0495
−1.64	0.0505	−1.63	0.0516	−1.62	0.0526	−1.61	0.0537
−1.60	0.0548	−1.59	0.0559	−1.58	0.0571	−1.57	0.0582
−1.56	0.0594	−1.55	0.0606	−1.54	0.0618	−1.53	0.0630
−1.52	0.0643	−1.51	0.0655	−1.50	0.0668	−1.49	0.0681
−1.48	0.0694	−1.47	0.0708	−1.46	0.0721	−1.45	0.0735
−1.44	0.0749	−1.43	0.0764	−1.42	0.0778	−1.41	0.0793
−1.40	0.0808	−1.39	0.0823	−1.38	0.0838	−1.37	0.0853
−1.36	0.0869	−1.35	0.0885	−1.34	0.0901	−1.33	0.0918
−1.32	0.0934	−1.31	0.0951	−1.30	0.0968	−1.29	0.0985
−1.28	0.1003	−1.27	0.1020	−1.26	0.1038	−1.25	0.1056
−1.24	0.1075	−1.23	0.1093	−1.22	0.1112	−1.21	0.1131
−1.20	0.1151	−1.19	0.1170	−1.18	0.1190	−1.17	0.1210
−1.16	0.1230	−1.15	0.1251	−1.14	0.1271	−1.13	0.1292
−1.12	0.1314	−1.11	0.1335	−1.10	0.1357	−1.09	0.1379
−1.08	0.1401	−1.07	0.1423	−1.06	0.1446	−1.05	0.1469
−1.04	0.1492	−1.03	0.1515	−1.02	0.1539	−1.01	0.1562
−1.00	0.1587	−0.99	0.1611	−0.98	0.1635	−0.97	0.1662
−0.96	0.1685	−0.95	0.1711	−0.94	0.1736	−0.93	0.1762

Table 1 (continued)

z	$P(Z \leq z)$	z	$P(Z \leq z)$	z	$P(Z \leq z)$	z	$P(Z \leq z)$
−0.92	0.1788	−0.91	0.1814	−0.90	0.1841	−0.89	0.1867
−0.88	0.1894	−0.87	0.1922	−0.86	0.1949	−0.85	0.1977
−0.84	0.2005	−0.83	0.2033	−0.82	0.2061	−0.81	0.2090
−0.80	0.2119	−0.79	0.2148	−0.78	0.2177	−0.77	0.2207
−0.76	0.2236	−0.75	0.2266	−0.74	0.2297	−0.73	0.2327
−0.72	0.2358	−0.71	0.2389	−0.70	0.2420	−0.69	0.2451
−0.68	0.2483	−0.67	0.2514	−0.66	0.2546	−0.65	0.2578
−0.64	0.2611	−0.63	0.2643	−0.62	0.2676	−0.61	0.2709
−0.60	0.2743	−0.59	0.2776	−0.58	0.2810	−0.57	0.2843
−0.56	0.2877	−0.55	0.2912	−0.54	0.2946	−0.53	0.2981
−0.52	0.3015	−0.51	0.3050	−0.50	0.3085	−0.49	0.3121
−0.48	0.3156	−0.47	0.3192	−0.46	0.3228	−0.45	0.3264
−0.44	0.3300	−0.43	0.3336	−0.42	0.3372	−0.41	0.3409
−0.40	0.3446	−0.39	0.3483	−0.38	0.3520	−0.37	0.3557
−0.36	0.3594	−0.35	0.3632	−0.34	0.3669	−0.33	0.3707
−0.32	0.3745	−0.31	0.3783	−0.30	0.3821	−0.29	0.3859
−0.28	0.3897	−0.27	0.3936	−0.26	0.3974	−0.25	0.4013
−0.24	0.4052	−0.23	0.4090	−0.22	0.4129	−0.21	0.4168
−0.20	0.4207	−0.19	0.4247	−0.18	0.4286	−0.17	0.4325
−0.16	0.4364	−0.15	0.4404	−0.14	0.4443	−0.13	0.4483
−0.12	0.4522	−0.11	0.4562	−0.10	0.4602	−0.09	0.4641
−0.08	0.4681	−0.07	0.4721	−0.06	0.4761	−0.05	0.4801
−0.04	0.4840	−0.03	0.4880	−0.02	0.4920	−0.01	0.4960
0.01	0.5040	0.02	0.5080	0.03	0.5120	0.04	0.5160
0.05	0.5199	0.06	0.5239	0.07	0.5279	0.08	0.5319
0.09	0.5359	0.10	0.5398	0.11	0.5438	0.12	0.5478
0.13	0.5517	0.14	0.5557	0.15	0.5596	0.16	0.5636
0.17	0.5675	0.18	0.5714	0.19	0.5753	0.20	0.5793
0.21	0.5832	0.22	0.5871	0.23	0.5910	0.24	0.5948
0.25	0.5987	0.26	0.6026	0.27	0.6064	0.28	0.6103
0.29	0.6141	0.30	0.6179	0.31	0.6217	0.32	0.6255
0.33	0.6293	0.34	0.6331	0.35	0.6368	0.36	0.6406
0.37	0.6443	0.38	0.6480	0.39	0.6517	0.40	0.6554
0.41	0.6591	0.42	0.6628	0.43	0.6664	0.44	0.6700
0.45	0.6736	0.46	0.6772	0.47	0.6808	0.48	0.6844
0.49	0.6879	0.50	0.6915	0.51	0.6950	0.52	0.6985
0.53	0.7019	0.54	0.7054	0.55	0.7088	0.56	0.7123
0.57	0.7157	0.58	0.7190	0.59	0.7224	0.60	0.7257
0.61	0.7291	0.62	0.7324	0.63	0.7357	0.64	0.7389
0.65	0.7422	0.66	0.7454	0.67	0.7486	0.68	0.7517
0.69	0.7549	0.70	0.7580	0.71	0.7611	0.72	0.7642
0.73	0.7673	0.74	0.7703	0.75	0.7734	0.76	0.7764
0.77	0.7793	0.78	0.7823	0.79	0.7852	0.80	0.7881
0.81	0.7910	0.82	0.7939	0.83	0.7967	0.84	0.7995
0.85	0.8023	0.86	0.8051	0.87	0.8078	0.88	0.8106
0.89	0.8133	0.90	0.8159	0.91	0.8186	0.92	0.8212
0.93	0.8238	0.94	0.8264	0.95	0.8289	0.96	0.8315
0.97	0.8340	0.98	0.8365	0.99	0.8389	1.00	0.8413
1.01	0.8438	1.02	0.8461	1.03	0.8485	1.04	0.8508
1.05	0.8531	1.06	0.8554	1.07	0.8577	1.08	0.8599
1.09	0.8621	1.10	0.8643	1.11	0.8665	1.12	0.8686
1.13	0.8708	1.14	0.8729	1.15	0.8749	1.16	0.8770

Continued

Table 1 Standard normal distribution (continued)

z	$P(Z \leq z)$	z	$P(Z \leq z)$	z	$P(Z \leq z)$	z	$P(Z \leq z)$
1.17	0.8790	1.18	0.8810	1.19	0.8830	1.20	0.8849
1.21	0.8869	1.22	0.8888	1.23	0.8907	1.24	0.8925
1.25	0.8944	1.26	0.8962	1.27	0.8980	1.28	0.8997
1.29	0.9015	1.30	0.9032	1.31	0.9049	1.32	0.9066
1.33	0.9082	1.34	0.9099	1.35	0.9115	1.36	0.9131
1.37	0.9147	1.38	0.9162	1.39	0.9177	1.40	0.9192
1.41	0.9207	1.42	0.9222	1.43	0.9236	1.44	0.9251
1.45	0.9265	1.46	0.9279	1.47	0.9292	1.48	0.9306
1.49	0.9319	1.50	0.9332	1.51	0.9345	1.52	0.9357
1.53	0.9370	1.54	0.9382	1.55	0.9394	1.56	0.9406
1.57	0.9418	1.58	0.9429	1.59	0.9441	1.60	0.9452
1.61	0.9463	1.62	0.9474	1.63	0.9484	1.64	0.9495
1.65	0.9505	1.66	0.9515	1.67	0.9525	1.68	0.9535
1.69	0.9545	1.70	0.9554	1.71	0.9564	1.72	0.9573
1.73	0.9582	1.74	0.9591	1.75	0.9599	1.76	0.9608
1.77	0.9616	1.78	0.9625	1.79	0.9633	1.80	0.9641
1.81	0.9649	1.82	0.9656	1.83	0.9664	1.84	0.9671
1.85	0.9678	1.86	0.9686	1.87	0.9693	1.88	0.9699
1.89	0.9706	1.90	0.9713	1.91	0.9719	1.92	0.9726
1.93	0.9732	1.94	0.9738	1.95	0.9744	1.96	0.9750
1.97	0.9756	1.98	0.9761	1.99	0.9767	2.00	0.9772
2.01	0.9778	2.02	0.9783	2.03	0.9788	2.04	0.9793
2.05	0.9798	2.06	0.9803	2.07	0.9808	2.08	0.9812
2.09	0.9817	2.10	0.9821	2.11	0.9826	2.12	0.9830
2.13	0.9834	2.14	0.9838	2.15	0.9842	2.16	0.9846
2.17	0.9850	2.18	0.9854	2.19	0.9857	2.20	0.9861
2.21	0.9864	2.22	0.9868	2.23	0.9871	2.24	0.9875
2.25	0.9878	2.26	0.9881	2.27	0.9884	2.28	0.9887
2.29	0.9890	2.30	0.9893	2.31	0.9896	2.32	0.9898
2.33	0.9901	2.34	0.9904	2.35	0.9906	2.36	0.9909
2.37	0.9911	2.38	0.9913	2.39	0.9916	2.40	0.9918
2.41	0.9920	2.42	0.9922	2.43	0.9925	2.44	0.9927
2.45	0.9929	2.46	0.9931	2.47	0.9932	2.48	0.9934
2.49	0.9936	2.50	0.9938	2.51	0.9940	2.52	0.9941
2.53	0.9943	2.54	0.9945	2.55	0.9946	2.56	0.9948
2.57	0.9949	2.58	0.9951	2.59	0.9952	2.60	0.9953
2.61	0.9955	2.62	0.9956	2.63	0.9957	2.64	0.9959
2.65	0.9960	2.66	0.9961	2.67	0.9962	2.68	0.9963
2.69	0.9964	2.70	0.9965	2.71	0.9966	2.72	0.9967
2.73	0.9968	2.74	0.9969	2.75	0.9970	2.76	0.9971
2.77	0.9972	2.78	0.9973	2.79	0.9974	2.80	0.9974
2.81	0.9975	2.82	0.9976	2.83	0.9977	2.84	0.9977
2.85	0.9978	2.86	0.9979	2.87	0.9979	2.88	0.9980
2.89	0.9981	2.90	0.9981	2.91	0.9982	2.92	0.9982
2.93	0.9983	2.94	0.9984	2.95	0.9984	2.96	0.9985
2.97	0.9985	2.98	0.9986	2.99	0.9986	3.00	0.9987

Note: This table was computed with IMSL subroutine ANORIN.

Table 2 Binomial probability function (values of entries are $P(X \leq k)$)

$n = 5$

	p										
k	.05	.1	.2	.3	.4	.5	.6	.7	.8	.9	.95
0	0.774	0.590	0.328	0.168	0.078	0.031	0.010	0.002	0.000	0.000	0.000
1	0.977	0.919	0.737	0.528	0.337	0.188	0.087	0.031	0.007	0.000	0.000
2	0.999	0.991	0.942	0.837	0.683	0.500	0.317	0.163	0.058	0.009	0.001
3	1.000	1.000	0.993	0.969	0.913	0.813	0.663	0.472	0.263	0.081	0.023
4	1.000	1.000	1.000	0.998	0.990	0.969	0.922	0.832	0.672	0.410	0.226

$n = 6$

	p										
k	.05	.1	.2	.3	.4	.5	.6	.7	.8	.9	.95
0	0.735	0.531	0.262	0.118	0.047	0.016	0.004	0.001	0.000	0.000	0.000
1	0.967	0.886	0.655	0.420	0.233	0.109	0.041	0.011	0.002	0.000	0.000
2	0.998	0.984	0.901	0.744	0.544	0.344	0.179	0.070	0.017	0.001	0.000
3	1.000	0.999	0.983	0.930	0.821	0.656	0.456	0.256	0.099	0.016	0.002
4	1.000	1.000	0.998	0.989	0.959	0.891	0.767	0.580	0.345	0.114	0.033
5	1.000	1.000	1.000	0.999	0.996	0.984	0.953	0.882	0.738	0.469	0.265

$n = 7$

	p										
k	.05	.1	.2	.3	.4	.5	.6	.7	.8	.9	.95
0	0.698	0.478	0.210	0.082	0.028	0.008	0.002	0.000	0.000	0.000	0.000
1	0.956	0.850	0.577	0.329	0.159	0.062	0.019	0.004	0.000	0.000	0.000
2	0.996	0.974	0.852	0.647	0.420	0.227	0.096	0.029	0.005	0.000	0.000
3	1.000	0.997	0.967	0.874	0.710	0.500	0.290	0.126	0.033	0.003	0.000
4	1.000	1.000	0.995	0.971	0.904	0.773	0.580	0.353	0.148	0.026	0.004
5	1.000	1.000	1.000	0.996	0.981	0.938	0.841	0.671	0.423	0.150	0.044
6	1.000	1.000	1.000	1.000	0.998	0.992	0.972	0.918	0.790	0.522	0.302

$n = 8$

	p										
k	.05	.1	.2	.3	.4	.5	.6	.7	.8	.9	.95
0	0.663	0.430	0.168	0.058	0.017	0.004	0.001	0.000	0.000	0.000	0.000
1	0.943	0.813	0.503	0.255	0.106	0.035	0.009	0.001	0.000	0.000	0.000
2	0.994	0.962	0.797	0.552	0.315	0.145	0.050	0.011	0.001	0.000	0.000
3	1.000	0.995	0.944	0.806	0.594	0.363	0.174	0.058	0.010	0.000	0.000
4	1.000	1.000	0.990	0.942	0.826	0.637	0.406	0.194	0.056	0.005	0.000
5	1.000	1.000	0.999	0.989	0.950	0.855	0.685	0.448	0.203	0.038	0.006
6	1.000	1.000	1.000	0.999	0.991	0.965	0.894	0.745	0.497	0.187	0.057
7	1.000	1.000	1.000	1.000	0.999	0.996	0.983	0.942	0.832	0.570	0.337

$n = 9$

	p										
k	.05	.1	.2	.3	.4	.5	.6	.7	.8	.9	.95
0	0.630	0.387	0.134	0.040	0.010	0.002	0.000	0.000	0.000	0.000	0.000
1	0.929	0.775	0.436	0.196	0.071	0.020	0.004	0.000	0.000	0.000	0.000
2	0.992	0.947	0.738	0.463	0.232	0.090	0.025	0.004	0.000	0.000	0.000
3	0.999	0.992	0.914	0.730	0.483	0.254	0.099	0.025	0.003	0.000	0.000
4	1.000	0.999	0.980	0.901	0.733	0.500	0.267	0.099	0.020	0.001	0.000
5	1.000	1.000	0.997	0.975	0.901	0.746	0.517	0.270	0.086	0.008	0.001
6	1.000	1.000	1.000	0.996	0.975	0.910	0.768	0.537	0.262	0.053	0.008
7	1.000	1.000	1.000	1.000	0.996	0.980	0.929	0.804	0.564	0.225	0.071
8	1.000	1.000	1.000	1.000	1.000	0.998	0.990	0.960	0.866	0.613	0.370

$n = 10$

	p										
k	.05	.1	.2	.3	.4	.5	.6	.7	.8	.9	.95
0	0.599	0.349	0.107	0.028	0.006	0.001	0.000	0.000	0.000	0.000	0.000
1	0.914	0.736	0.376	0.149	0.046	0.011	0.002	0.000	0.000	0.000	0.000
2	0.988	0.930	0.678	0.383	0.167	0.055	0.012	0.002	0.000	0.000	0.000
3	0.999	0.987	0.879	0.650	0.382	0.172	0.055	0.011	0.001	0.000	0.000
4	1.000	0.998	0.967	0.850	0.633	0.377	0.166	0.047	0.006	0.000	0.000
5	1.000	1.000	0.994	0.953	0.834	0.623	0.367	0.150	0.033	0.002	0.000
6	1.000	1.000	0.999	0.989	0.945	0.828	0.618	0.350	0.121	0.013	0.001
7	1.000	1.000	1.000	0.998	0.988	0.945	0.833	0.617	0.322	0.070	0.012
8	1.000	1.000	1.000	1.000	0.998	0.989	0.954	0.851	0.624	0.264	0.086
9	1.000	1.000	1.000	1.000	1.000	0.999	0.994	0.972	0.893	0.651	0.401

$n = 15$

	p										
k	.05	.1	.2	.3	.4	.5	.6	.7	.8	.9	.95
0	0.463	0.206	0.035	0.005	0.000	0.000	0.000	0.000	0.000	0.000	0.000
1	0.829	0.549	0.167	0.035	0.005	0.000	0.000	0.000	0.000	0.000	0.000
2	0.964	0.816	0.398	0.127	0.027	0.004	0.000	0.000	0.000	0.000	0.000
3	0.995	0.944	0.648	0.297	0.091	0.018	0.002	0.000	0.000	0.000	0.000
4	0.999	0.987	0.836	0.515	0.217	0.059	0.009	0.001	0.000	0.000	0.000
5	1.000	0.998	0.939	0.722	0.403	0.151	0.034	0.004	0.000	0.000	0.000
6	1.000	1.000	0.982	0.869	0.610	0.304	0.095	0.015	0.001	0.000	0.000
7	1.000	1.000	0.996	0.950	0.787	0.500	0.213	0.050	0.004	0.000	0.000
8	1.000	1.000	0.999	0.985	0.905	0.696	0.390	0.131	0.018	0.000	0.000
9	1.000	1.000	1.000	0.996	0.966	0.849	0.597	0.278	0.061	0.002	0.000
10	1.000	1.000	1.000	0.999	0.991	0.941	0.783	0.485	0.164	0.013	0.001
11	1.000	1.000	1.000	1.000	0.998	0.982	0.909	0.703	0.352	0.056	0.005
12	1.000	1.000	1.000	1.000	1.000	0.996	0.973	0.873	0.602	0.184	0.036
13	1.000	1.000	1.000	1.000	1.000	1.000	0.995	0.965	0.833	0.451	0.171
14	1.000	1.000	1.000	1.000	1.000	1.000	1.000	0.995	0.965	0.794	0.537

$n = 20$

	p										
k	.05	.1	.2	.3	.4	.5	.6	.7	.8	.9	.95
0	0.358	0.122	0.012	0.001	0.000	0.000	0.000	0.000	0.000	0.000	0.000
1	0.736	0.392	0.069	0.008	0.001	0.000	0.000	0.000	0.000	0.000	0.000
2	0.925	0.677	0.206	0.035	0.004	0.000	0.000	0.000	0.000	0.000	0.000
3	0.984	0.867	0.411	0.107	0.016	0.001	0.000	0.000	0.000	0.000	0.000
4	0.997	0.957	0.630	0.238	0.051	0.006	0.000	0.000	0.000	0.000	0.000
5	1.000	0.989	0.804	0.416	0.126	0.021	0.002	0.000	0.000	0.000	0.000
6	1.000	0.998	0.913	0.608	0.250	0.058	0.006	0.000	0.000	0.000	0.000
7	1.000	1.000	0.968	0.772	0.416	0.132	0.021	0.001	0.000	0.000	0.000
8	1.000	1.000	0.990	0.887	0.596	0.252	0.057	0.005	0.000	0.000	0.000
9	1.000	1.000	0.997	0.952	0.755	0.412	0.128	0.017	0.001	0.000	0.000
10	1.000	1.000	0.999	0.983	0.872	0.588	0.245	0.048	0.003	0.000	0.000
11	1.000	1.000	1.000	0.995	0.943	0.748	0.404	0.113	0.010	0.000	0.000
12	1.000	1.000	1.000	0.999	0.979	0.868	0.584	0.228	0.032	0.000	0.000
13	1.000	1.000	1.000	1.000	0.994	0.942	0.750	0.392	0.087	0.002	0.000
14	1.000	1.000	1.000	1.000	0.998	0.979	0.874	0.584	0.196	0.011	0.000
15	1.000	1.000	1.000	1.000	1.000	0.994	0.949	0.762	0.370	0.043	0.003
16	1.000	1.000	1.000	1.000	1.000	0.999	0.984	0.893	0.589	0.133	0.016
17	1.000	1.000	1.000	1.000	1.000	1.000	0.996	0.965	0.794	0.323	0.075
18	1.000	1.000	1.000	1.000	1.000	1.000	0.999	0.992	0.931	0.608	0.264
19	1.000	1.000	1.000	1.000	1.000	1.000	1.000	0.999	0.988	0.878	0.642

$n = 25$

	p										
k	.05	.1	.2	.3	.4	.5	.6	.7	.8	.9	.95
0	0.277	0.072	0.004	0.000	0.000	0.000	0.000	0.000	0.000	0.000	0.000
1	0.642	0.271	0.027	0.002	0.000	0.000	0.000	0.000	0.000	0.000	0.000
2	0.873	0.537	0.098	0.009	0.000	0.000	0.000	0.000	0.000	0.000	0.000
3	0.966	0.764	0.234	0.033	0.002	0.000	0.000	0.000	0.000	0.000	0.000
4	0.993	0.902	0.421	0.090	0.009	0.000	0.000	0.000	0.000	0.000	0.000
5	0.999	0.967	0.617	0.193	0.029	0.002	0.000	0.000	0.000	0.000	0.000
6	1.000	0.991	0.780	0.341	0.074	0.007	0.000	0.000	0.000	0.000	0.000
7	1.000	0.998	0.891	0.512	0.154	0.022	0.001	0.000	0.000	0.000	0.000
8	1.000	1.000	0.953	0.677	0.274	0.054	0.004	0.000	0.000	0.000	0.000
9	1.000	1.000	0.983	0.811	0.425	0.115	0.013	0.000	0.000	0.000	0.000
10	1.000	1.000	0.994	0.902	0.586	0.212	0.034	0.002	0.000	0.000	0.000
11	1.000	1.000	0.998	0.956	0.732	0.345	0.078	0.006	0.000	0.000	0.000
12	1.000	1.000	1.000	0.983	0.846	0.500	0.154	0.017	0.000	0.000	0.000
13	1.000	1.000	1.000	0.994	0.922	0.655	0.268	0.044	0.002	0.000	0.000
14	1.000	1.000	1.000	0.998	0.966	0.788	0.414	0.098	0.006	0.000	0.000
15	1.000	1.000	1.000	1.000	0.987	0.885	0.575	0.189	0.017	0.000	0.000
16	1.000	1.000	1.000	1.000	0.996	0.946	0.726	0.323	0.047	0.000	0.000
17	1.000	1.000	1.000	1.000	0.999	0.978	0.846	0.488	0.109	0.002	0.000
18	1.000	1.000	1.000	1.000	1.000	0.993	0.926	0.659	0.220	0.009	0.000
19	1.000	1.000	1.000	1.000	1.000	0.998	0.971	0.807	0.383	0.033	0.001
20	1.000	1.000	1.000	1.000	1.000	1.000	0.991	0.910	0.579	0.098	0.007
21	1.000	1.000	1.000	1.000	1.000	1.000	0.998	0.967	0.766	0.236	0.034
22	1.000	1.000	1.000	1.000	1.000	1.000	1.000	0.991	0.902	0.463	0.127
23	1.000	1.000	1.000	1.000	1.000	1.000	1.000	0.998	0.973	0.729	0.358
24	1.000	1.000	1.000	1.000	1.000	1.000	1.000	1.000	0.996	0.928	0.723

Table 3 Percentage points of the Chi-square distribution

ν	$\chi^2_{.005}$	$\chi^2_{.01}$	$\chi^2_{.025}$	$\chi^2_{.05}$	$\chi^2_{.10}$
1	0.0000393	0.0001571	0.0009821	0.0039321	0.0157908
2	0.0100251	0.0201007	0.0506357	0.1025866	0.2107213
3	0.0717217	0.1148317	0.2157952	0.3518462	0.5843744
4	0.2069889	0.2971095	0.4844186	0.7107224	1.0636234
5	0.4117419	0.5542979	0.8312111	1.1454763	1.6103077
6	0.6757274	0.8720903	1.2373447	1.6353836	2.2041321
7	0.9892554	1.2390423	1.6898699	2.1673594	2.8331099
8	1.3444128	1.6464968	2.1797333	2.7326374	3.4895401
9	1.7349329	2.0879011	2.7003908	3.3251143	4.1681604
10	2.1558590	2.5582132	3.2469759	3.9403019	4.8651857
11	2.6032248	3.0534868	3.8157606	4.5748196	5.5777788
12	3.0738316	3.5705872	4.4037895	5.2260313	6.3037949
13	3.5650368	4.1069279	5.0087538	5.8918715	7.0415068
14	4.0746784	4.6604300	5.6287327	6.5706167	7.7895403
15	4.6009169	5.2293501	6.2621403	7.2609539	8.5467529
16	5.1422071	5.8122101	6.9076681	7.9616566	9.3122330
17	5.6972256	6.4077673	7.5641880	8.6717682	10.0851974
18	6.2648115	7.0149183	8.2307510	9.3904572	10.8649368
19	6.8439512	7.6327391	8.9065247	10.1170273	11.6509628
20	7.4338474	8.2603989	9.5907822	10.8508148	12.4426041
21	8.0336685	8.8972015	10.2829285	11.5913391	13.2396393
22	8.6427155	9.5425110	10.9823456	12.3380432	14.0414886
23	9.2604370	10.1957169	11.6885223	13.0905151	14.8479385
24	9.8862610	10.8563690	12.4011765	13.8484344	15.6587067
25	10.5196533	11.5239716	13.1197433	14.6114349	16.4734497
26	11.1602631	12.1981506	13.8439331	15.3792038	17.2919159
27	11.8076019	12.8785095	14.5734024	16.1513977	18.1138763
28	12.4613495	13.5647125	15.3078613	16.9278717	18.9392395
29	13.1211624	14.2564697	16.0470886	17.7083893	19.7678223
30	13.7867584	14.9534760	16.7907562	18.4926147	20.5992126
40	20.7065582	22.1642761	24.4330750	26.5083008	29.0503540
50	27.9775238	29.7001038	32.3561096	34.7638702	37.6881561
60	35.5294037	37.4848328	40.4817810	43.1865082	46.4583282
70	43.2462311	45.4230499	48.7503967	51.7388763	55.3331146
80	51.1447754	53.5226593	57.1465912	60.3912201	64.2818604
90	59.1706543	61.7376862	65.6405029	69.1258850	73.2949219
100	67.3031921	70.0493622	74.2162018	77.9293976	82.3618469

Table 3 (continued)

ν	$\chi^2_{.900}$	$\chi^2_{.95}$	$\chi^2_{.975}$	$\chi^2_{.99}$	$\chi^2_{.995}$
1	2.7056	3.8415	5.0240	6.6353	7.8818
2	4.6052	5.9916	7.3779	9.2117	10.5987
3	6.2514	7.8148	9.3486	11.3465	12.8409
4	7.7795	9.4879	11.1435	13.2786	14.8643
5	9.2365	11.0707	12.8328	15.0870	16.7534
6	10.6448	12.5919	14.4499	16.8127	18.5490
7	12.0171	14.0676	16.0136	18.4765	20.2803
8	13.3617	15.5075	17.5355	20.0924	21.9579
9	14.6838	16.9191	19.0232	21.6686	23.5938
10	15.9874	18.3075	20.4837	23.2101	25.1898
11	17.2750	19.6754	21.9211	24.7265	26.7568
12	18.5494	21.0263	23.3370	26.2170	28.2995
13	19.8122	22.3627	24.7371	27.6882	29.8194
14	21.0646	23.6862	26.1189	29.1412	31.3193
15	22.3077	24.9970	27.4883	30.5779	32.8013
16	23.5421	26.2961	28.8453	31.9999	34.2672
17	24.7696	27.5871	30.1909	33.4087	35.7184
18	25.9903	28.8692	31.5264	34.8054	37.1564
19	27.2035	30.1434	32.8523	36.1909	38.5823
20	28.4120	31.4104	34.1696	37.5662	39.9968
21	29.6150	32.6705	35.4787	38.9323	41.4012
22	30.8133	33.9244	36.7806	40.2893	42.7958
23	32.0069	35.1725	38.0757	41.6384	44.1812
24	33.1962	36.4151	39.3639	42.9799	45.5587
25	34.3815	37.6525	40.6463	44.3142	46.9280
26	35.5631	38.8852	41.9229	45.6418	48.2899
27	36.7412	40.1134	43.1943	46.9629	49.6449
28	37.9159	41.3371	44.4608	48.2784	50.9933
29	39.0874	42.5571	45.7223	49.5879	52.3357
30	40.2561	43.7730	46.9792	50.8922	53.6721
40	51.8050	55.7586	59.3417	63.6909	66.7660
50	63.1670	67.5047	71.4201	76.1538	79.4899
60	74.3970	79.0820	83.2977	88.3794	91.9516
70	85.5211	90.5283	95.0263	100.4409	104.2434
80	96.5723	101.8770	106.6315	112.3434	116.3484
90	107.5600	113.1425	118.1392	124.1304	128.3245
100	118.4932	124.3395	129.5638	135.8203	140.1940

Note: This table was computed with IMSL subroutine CHIIN.

Table 4 Percentage points of Student's t-distribution

ν	$t_{.9}$	$t_{.95}$	$t_{.975}$	$t_{.99}$	$t_{.995}$	$t_{.999}$
1	3.078	6.314	12.706	31.821	63.6567	318.313
2	1.886	2.920	4.303	6.965	9.925	22.327
3	1.638	2.353	3.183	4.541	5.841	10.215
4	1.533	2.132	2.776	3.747	4.604	7.173
5	1.476	2.015	2.571	3.365	4.032	5.893
6	1.440	1.943	2.447	3.143	3.707	5.208
7	1.415	1.895	2.365	2.998	3.499	4.785
8	1.397	1.856	2.306	2.897	3.355	4.501
9	1.383	1.833	2.262	2.821	3.245	4.297
10	1.372	1.812	2.228	2.764	3.169	4.144
11	1.363	1.796	2.201	2.718	3.106	4.025
12	1.356	1.782	2.179	2.681	3.055	3.930
13	1.350	1.771	2.160	2.650	3.012	3.852
14	1.345	1.761	2.145	2.624	2.976	3.787
15	1.341	1.753	2.131	2.603	2.947	3.733
16	1.337	1.746	2.120	2.583	2.921	3.686
17	1.333	1.740	2.110	2.567	2.898	3.646
18	1.330	1.734	2.101	2.552	2.878	3.610
19	1.328	1.729	2.093	2.539	2.861	3.579
20	1.325	1.725	2.086	2.528	2.845	3.552
20	1.325	1.725	2.086	2.528	2.845	3.552
24	1.318	1.711	2.064	2.492	2.797	3.467
30	1.310	1.697	2.042	2.457	2.750	3.385
40	1.303	1.684	2.021	2.423	2.704	3.307
60	1.296	1.671	2.000	2.390	2.660	3.232
120	1.289	1.658	1.980	2.358	2.617	3.160
∞	1.282	1.645	1.960	2.326	2.576	3.090

Entries were computed with IMSL subroutine TIN.

Table 5 Percentage points of the F distribution, $\alpha = .10$

	ν_1								
ν_2	1	2	3	4	5	6	7	8	9
1	39.86	49.50	53.59	55.83	57.24	58.20	58.91	59.44	59.86
2	8.53	9.00	9.16	9.24	9.29	9.33	9.35	9.37	9.38
3	5.54	5.46	5.39	5.34	5.31	5.28	5.27	5.25	5.24
4	4.54	4.32	4.19	4.11	4.05	4.01	3.98	3.95	3.94
5	4.06	3.78	3.62	3.52	3.45	3.40	3.37	3.34	3.32
6	3.78	3.46	3.29	3.18	3.11	3.05	3.01	2.98	2.96
7	3.59	3.26	3.07	2.96	2.88	2.83	2.79	2.75	2.72
8	3.46	3.11	2.92	2.81	2.73	2.67	2.62	2.59	2.56
9	3.36	3.01	2.81	2.69	2.61	2.55	2.51	2.47	2.44
10	3.29	2.92	2.73	2.61	2.52	2.46	2.41	2.38	2.35
11	3.23	2.86	2.66	2.54	2.45	2.39	2.34	2.30	2.27
12	3.18	2.81	2.61	2.48	2.39	2.33	2.28	2.24	2.21
13	3.14	2.76	2.56	2.43	2.35	2.28	2.23	2.20	2.16
14	3.10	2.73	2.52	2.39	2.31	2.24	2.19	2.15	2.12
15	3.07	2.70	2.49	2.36	2.27	2.21	2.16	2.12	2.09
16	3.05	2.67	2.46	2.33	2.24	2.18	2.13	2.09	2.06
17	3.03	2.64	2.44	2.31	2.22	2.15	2.10	2.06	2.03
18	3.01	2.62	2.42	2.29	2.20	2.13	2.08	2.04	2.00
19	2.99	2.61	2.40	2.27	2.18	2.11	2.06	2.02	1.98
20	2.97	2.59	2.38	2.25	2.16	2.09	2.04	2.00	1.96
21	2.96	2.57	2.36	2.23	2.14	2.08	2.02	1.98	1.95
22	2.95	2.56	2.35	2.22	2.13	2.06	2.01	1.97	1.93
23	2.94	2.55	2.34	2.21	2.11	2.05	1.99	1.95	1.92
24	2.93	2.54	2.33	2.19	2.10	2.04	1.98	1.94	1.91
25	2.92	2.53	2.32	2.18	2.09	2.02	1.97	1.93	1.89
26	2.91	2.52	2.31	2.17	2.08	2.01	1.96	1.92	1.88
27	2.90	2.51	2.30	2.17	2.07	2.00	1.95	1.91	1.87
28	2.89	2.50	2.29	2.16	2.06	2.00	1.94	1.90	1.87
29	2.89	2.50	2.28	2.15	2.06	1.99	1.93	1.89	1.86
30	2.88	2.49	2.28	2.14	2.05	1.98	1.93	1.88	1.85
40	2.84	2.44	2.23	2.09	2.00	1.93	1.87	1.83	1.79
60	2.79	2.39	2.18	2.04	1.95	1.87	1.82	1.77	1.74
120	2.75	2.35	2.13	1.99	1.90	1.82	1.77	1.72	1.68
∞	2.71	2.30	2.08	1.94	1.85	1.77	1.72	.167	1.63

Continued

Table 5 (continued)

ν_2	ν_1 10	12	15	20	24	30	40	60	120	∞
1	60.19	60.70	61.22	61.74	62.00	62.26	62.53	62.79	63.06	63.33
2	9.39	9.41	9.42	9.44	9.45	9.46	9.47	9.47	9.48	9.49
3	5.23	5.22	5.20	5.19	5.18	5.17	5.16	5.15	5.14	5.13
4	3.92	3.90	3.87	3.84	3.83	3.82	3.80	3.79	3.78	3.76
5	3.30	3.27	3.24	3.21	3.19	3.17	3.16	3.14	3.12	3.10
6	2.94	2.90	2.87	2.84	2.82	2.80	2.78	2.76	2.74	2.72
7	2.70	2.67	2.63	2.59	2.58	2.56	2.54	2.51	2.49	2.47
8	2.54	2.50	2.46	2.42	2.40	2.38	2.36	2.34	2.32	2.29
9	2.42	2.38	2.34	2.30	2.28	2.25	2.23	2.21	2.18	2.16
10	2.32	2.28	2.24	2.20	2.18	2.16	2.13	2.11	2.08	2.06
11	2.25	2.21	2.17	2.12	2.10	2.08	2.05	2.03	2.00	1.97
12	2.19	2.15	2.10	2.06	2.04	2.01	1.99	1.96	1.93	1.90
13	2.14	2.10	2.05	2.01	1.98	1.96	1.93	1.90	1.88	1.85
14	2.10	2.05	2.01	1.96	1.94	1.91	1.89	1.86	1.83	1.80
15	2.06	2.02	1.97	1.92	1.90	1.87	1.85	1.82	1.79	1.76
16	2.03	1.99	1.94	1.89	1.87	1.84	1.81	1.78	1.75	1.72
17	2.00	1.96	1.91	1.86	1.84	1.81	1.78	1.75	1.72	1.69
18	1.98	1.93	1.89	1.84	1.81	1.78	1.75	1.72	1.69	1.66
19	1.96	1.91	1.86	1.81	1.79	1.76	1.73	1.70	1.67	1.63
20	1.94	1.89	1.84	1.79	1.77	1.74	1.71	1.68	1.64	1.61
21	1.92	1.87	1.83	1.78	1.75	1.72	1.69	1.66	1.62	1.59
22	1.90	1.86	1.81	1.76	1.73	1.70	1.67	1.64	1.60	1.57
23	1.89	1.84	1.80	1.74	1.72	1.69	1.66	1.62	1.59	1.55
24	1.88	1.83	1.78	1.73	1.70	1.67	1.64	1.61	1.57	1.53
25	1.87	1.82	1.77	1.72	1.69	1.66	1.63	1.59	1.56	1.52
26	1.86	1.81	1.76	1.71	1.68	1.65	1.61	1.58	1.54	1.50
27	1.85	1.80	1.75	1.70	1.67	1.64	1.60	1.57	1.53	1.49
28	1.84	1.79	1.74	1.69	1.66	1.63	1.59	1.56	1.52	1.48
29	1.83	1.78	1.73	1.68	1.65	1.62	1.58	1.55	1.51	1.47
30	1.82	1.77	1.72	1.67	1.64	1.61	1.57	1.54	1.50	1.46
40	1.76	1.71	1.66	1.61	1.57	1.54	1.51	1.47	1.42	1.38
60	1.71	1.66	1.60	1.54	1.51	1.48	1.44	1.40	1.35	1.29
120	1.65	1.60	1.55	1.48	1.45	1.41	1.37	1.32	1.26	1.19
∞	1.60	1.55	1.49	1.42	1.38	1.34	1.30	1.24	1.17	1.00

Note: Entries in this table were computed with IMSL subroutine FIN.

Table 6 Percentage points of the F distribution, $\alpha = .05$

	ν_1								
ν_2	1	2	3	4	5	6	7	8	9
1	161.45	199.50	215.71	224.58	230.16	233.99	236.77	238.88	240.54
2	18.51	19.00	19.16	19.25	19.30	19.33	19.35	19.37	19.38
3	10.13	9.55	9.28	9.12	9.01	8.94	8.89	8.85	8.81
4	7.71	6.94	6.59	6.39	6.26	6.16	6.09	6.04	6.00
5	6.61	5.79	5.41	5.19	5.05	4.95	4.88	4.82	4.77
6	5.99	5.14	4.76	4.53	4.39	4.28	4.21	4.15	4.10
7	5.59	4.74	4.35	4.12	3.97	3.87	3.79	3.73	3.68
8	5.32	4.46	4.07	3.84	3.69	3.58	3.50	3.44	3.39
9	5.12	4.26	3.86	3.63	3.48	3.37	3.29	3.23	3.18
10	4.96	4.10	3.71	3.48	3.33	3.22	3.14	3.07	3.02
11	4.84	3.98	3.59	3.36	3.20	3.09	3.01	2.95	2.90
12	4.75	3.89	3.49	3.26	3.11	3.00	2.91	2.85	2.80
13	4.67	3.81	3.41	3.18	3.03	2.92	2.83	2.77	2.71
14	4.60	3.74	3.34	3.11	2.96	2.85	2.76	2.70	2.65
15	4.54	3.68	3.29	3.06	2.90	2.79	2.71	2.64	2.59
16	4.49	3.63	3.24	3.01	2.85	2.74	2.66	2.59	2.54
17	4.45	3.59	3.20	2.96	2.81	2.70	2.61	2.55	2.49
18	4.41	3.55	3.16	2.93	2.77	2.66	2.58	2.51	2.46
19	4.38	3.52	3.13	2.90	2.74	2.63	2.54	2.48	2.42
20	4.35	3.49	3.10	2.87	2.71	2.60	2.51	2.45	2.39
21	4.32	3.47	3.07	2.84	2.68	2.57	2.49	2.42	2.37
22	4.30	3.44	3.05	2.82	2.66	2.55	2.46	2.40	2.34
23	4.28	3.42	3.03	2.80	2.64	2.53	2.44	2.37	2.32
24	4.26	3.40	3.01	2.78	2.62	2.51	2.42	2.36	2.30
25	4.24	3.39	2.99	2.76	2.60	2.49	2.40	2.34	2.28
26	4.23	3.37	2.98	2.74	2.59	2.47	2.39	2.32	2.27
27	4.21	3.35	2.96	2.73	2.57	2.46	2.37	2.31	2.25
28	4.20	3.34	2.95	2.71	2.56	2.45	2.36	2.29	2.24
29	4.18	3.33	2.93	2.70	2.55	2.43	2.35	2.28	2.22
30	4.17	3.32	2.92	2.69	2.53	2.42	2.33	2.27	2.21
40	4.08	3.23	2.84	2.61	2.45	2.34	2.25	2.18	2.12
60	4.00	3.15	2.76	2.53	2.37	2.25	2.17	2.10	2.04
120	3.92	3.07	2.68	2.45	2.29	2.17	2.09	2.02	1.96
∞	3.84	3.00	2.60	2.37	2.21	2.10	2.01	1.94	1.88

Continued

Table 6 Percentage points of the F distribution, $\alpha = .05$ (continued)

	ν_1									
ν_2	10	12	15	20	24	30	40	60	120	∞
1	241.88	243.91	245.96	248.00	249.04	250.08	251.14	252.19	253.24	254.3
2	19.40	19.41	19.43	19.45	19.45	19.46	19.47	19.48	19.49	19.50
3	8.79	8.74	8.70	8.66	8.64	8.62	8.59	8.57	8.55	8.53
4	5.97	5.91	5.86	5.80	5.77	5.74	5.72	5.69	5.66	5.63
5	4.73	4.68	4.62	4.56	4.53	4.50	4.46	4.43	4.40	4.36
6	4.06	4.00	3.94	3.87	3.84	3.81	3.77	3.74	3.70	3.67
7	3.64	3.57	3.51	3.44	3.41	3.38	3.34	3.30	3.27	3.23
8	3.35	3.28	3.22	3.15	3.12	3.08	3.04	3.00	2.97	2.93
9	3.14	3.07	3.01	2.94	2.90	2.86	2.83	2.79	2.75	2.71
10	2.98	2.91	2.85	2.77	2.74	2.70	2.66	2.62	2.58	2.54
11	2.85	2.79	2.72	2.65	2.61	2.57	2.53	2.49	2.45	2.40
12	2.75	2.69	2.62	2.54	2.51	2.47	2.43	2.38	2.34	2.30
13	2.67	2.60	2.53	2.46	2.42	2.38	2.34	2.30	2.25	2.21
14	2.60	2.53	2.46	2.39	2.35	2.31	2.27	2.22	2.18	2.13
15	2.54	2.48	2.40	2.33	2.29	2.25	2.20	2.16	2.11	2.07
16	2.49	2.42	2.35	2.28	2.24	2.19	2.15	2.11	2.06	2.01
17	2.45	2.38	2.31	2.23	2.19	2.15	2.10	2.06	2.01	1.96
18	2.41	2.34	2.27	2.19	2.15	2.11	2.06	2.02	1.97	1.92
19	2.38	2.31	2.23	2.16	2.11	2.07	2.03	1.98	1.93	1.88
20	2.35	2.28	2.20	2.12	2.08	2.04	1.99	1.95	1.90	1.84
21	2.32	2.25	2.18	2.10	2.05	2.01	1.96	1.92	1.87	1.81
22	2.30	2.23	2.15	2.07	2.03	1.98	1.94	1.89	1.84	1.78
23	2.27	2.20	2.13	2.05	2.00	1.96	1.91	1.86	1.81	1.76
24	2.25	2.18	2.11	2.03	1.98	1.94	1.89	1.84	1.79	1.73
25	2.24	2.16	2.09	2.01	1.96	1.92	1.87	1.82	1.77	1.71
26	2.22	2.15	2.07	1.99	1.95	1.90	1.85	1.80	1.75	1.69
27	2.20	2.13	2.06	1.97	1.93	1.88	1.84	1.79	1.73	1.67
28	2.19	2.12	2.04	1.96	1.91	1.87	1.82	1.77	1.71	1.65
29	2.18	2.10	2.03	1.94	1.90	1.85	1.81	1.75	1.70	1.64
30	2.16	2.09	2.01	1.93	1.89	1.84	1.79	1.74	1.68	1.62
40	2.08	2.00	1.92	1.84	1.79	1.74	1.69	1.64	1.58	1.51
60	1.99	1.92	1.84	1.75	1.70	1.65	1.59	1.53	1.47	1.39
120	1.91	1.83	1.75	1.66	1.61	1.55	1.50	1.43	1.35	1.25
∞	1.83	1.75	1.67	1.57	1.52	1.46	1.39	1.32	1.22	1.00

Note: Entries in this table were computed with IMSL subroutine FIN.

Table 7 Percentage points of the F distribution, $\alpha = .025$

	ν_1								
ν_2	1	2	3	4	5	6	7	8	9
1	647.79	799.50	864.16	899.59	921.85	937.11	948.22	956.66	963.28
2	38.51	39.00	39.17	39.25	39.30	39.33	39.36	39.37	39.39
3	17.44	16.04	15.44	15.10	14.88	14.74	14.63	14.54	14.47
4	12.22	10.65	9.98	9.61	9.36	9.20	9.07	8.98	8.90
5	10.01	8.43	7.76	7.39	7.15	6.98	6.85	6.76	6.68
6	8.81	7.26	6.60	6.23	5.99	5.82	5.70	5.60	5.52
7	8.07	6.54	5.89	5.52	5.29	5.12	5.00	4.90	4.82
8	7.57	6.06	5.42	5.05	4.82	4.65	4.53	4.43	4.36
9	7.21	5.71	5.08	4.72	4.48	4.32	4.20	4.10	4.03
10	6.94	5.46	4.83	4.47	4.24	4.07	3.95	3.85	3.78
11	6.72	5.26	4.63	4.28	4.04	3.88	3.76	3.66	3.59
12	6.55	5.10	4.47	4.12	3.89	3.73	3.61	3.51	3.44
13	6.41	4.97	4.35	4.00	3.77	3.60	3.48	3.39	3.31
14	6.30	4.86	4.24	3.89	3.66	3.50	3.38	3.29	3.21
15	6.20	4.77	4.15	3.80	3.58	3.41	3.29	3.20	3.12
16	6.12	4.69	4.08	3.73	3.50	3.34	3.22	3.12	3.05
17	6.04	4.62	4.01	3.66	3.44	3.28	3.16	3.06	2.98
18	5.98	4.56	3.95	3.61	3.38	3.22	3.10	3.01	2.93
19	5.92	4.51	3.90	3.56	3.33	3.17	3.05	2.96	2.88
20	5.87	4.46	3.86	3.51	3.29	3.13	3.01	2.91	2.84
21	5.83	4.42	3.82	3.48	3.25	3.09	2.97	2.87	2.80
22	5.79	4.38	3.78	3.44	3.22	3.05	2.93	2.84	2.76
23	5.75	4.35	3.75	3.41	3.18	3.02	2.90	2.81	2.73
24	5.72	4.32	3.72	3.38	3.15	2.99	2.87	2.78	2.70
25	5.69	4.29	3.69	3.35	3.13	2.97	2.85	2.75	2.68
26	5.66	4.27	3.67	3.33	3.10	2.94	2.82	2.73	2.65
27	5.63	4.24	3.65	3.31	3.08	2.92	2.80	2.71	2.63
28	5.61	4.22	3.63	3.29	3.06	2.90	2.78	2.69	2.61
29	5.59	4.20	3.61	3.27	3.04	2.88	2.76	2.67	2.59
30	5.57	4.18	3.59	3.25	3.03	2.87	2.75	2.65	2.57
40	5.42	4.05	3.46	3.13	2.90	2.74	2.62	2.53	2.45
60	5.29	3.93	3.34	3.01	2.79	2.63	2.51	2.41	2.33
120	5.15	3.80	3.23	2.89	2.67	2.52	2.39	2.30	2.22
∞	5.02	3.69	3.12	2.79	2.57	2.41	2.29	2.19	2.11

Continued

Table 7 Percentage points of the F distribution, $\alpha = .025$ (continued)

	ν_1									
ν_2	10	12	15	20	24	30	40	60	120	∞
1	968.62	976.71	984.89	993.04	997.20	1,001	1,006	1,010	1,014	1,018
2	39.40	39.41	39.43	39.45	39.46	39.46	39.47	39.48	39.49	39.50
3	14.42	14.33	14.26	14.17	14.13	14.08	14.04	13.99	13.95	13.90
4	8.85	8.75	8.66	8.56	8.51	8.46	8.41	8.36	8.31	8.26
5	6.62	6.53	6.43	6.33	6.28	6.23	6.17	6.12	6.07	6.02
6	5.46	5.37	5.27	5.17	5.12	5.06	5.01	4.96	4.90	4.85
7	4.76	4.67	4.57	4.47	4.41	4.36	4.31	4.25	4.20	4.14
8	4.30	4.20	4.10	4.00	3.95	3.89	3.84	3.78	3.73	3.67
9	3.96	3.87	3.77	3.67	3.61	3.56	3.51	3.45	3.39	3.33
10	3.72	3.62	3.52	3.42	3.37	3.31	3.26	3.20	3.14	3.08
11	3.53	3.43	3.33	3.23	3.17	3.12	3.06	3.00	2.94	2.88
12	3.37	3.28	3.18	3.07	3.02	2.96	2.91	2.85	2.79	2.72
13	3.25	3.15	3.05	2.95	2.89	2.84	2.78	2.72	2.66	2.60
14	3.15	3.05	2.95	2.84	2.79	2.73	2.67	2.61	2.55	2.49
15	3.06	2.96	2.86	2.76	2.70	2.64	2.59	2.52	2.46	2.40
16	2.99	2.89	2.79	2.68	2.63	2.57	2.51	2.45	2.38	2.32
17	2.92	2.82	2.72	2.62	2.56	2.50	2.44	2.38	2.32	2.25
18	2.87	2.77	2.67	2.56	2.50	2.44	2.38	2.32	2.26	2.19
19	2.82	2.72	2.62	2.51	2.45	2.39	2.33	2.27	2.20	2.13
20	2.77	2.68	2.57	2.46	2.41	2.35	2.29	2.22	2.16	2.09
21	2.73	2.64	2.53	2.42	2.37	2.31	2.25	2.18	2.11	2.04
22	2.70	2.60	2.50	2.39	2.33	2.27	2.21	2.14	2.08	2.00
23	2.67	2.57	2.47	2.36	2.30	2.24	2.18	2.11	2.04	1.97
24	2.64	2.54	2.44	2.33	2.27	2.21	2.15	2.08	2.01	1.94
25	2.61	2.51	2.41	2.30	2.24	2.18	2.12	2.05	1.98	1.91
26	2.59	2.49	2.39	2.28	2.22	2.16	2.09	2.03	1.95	1.88
27	2.57	2.47	2.36	2.25	2.19	2.13	2.07	2.00	1.93	1.85
28	2.55	2.45	2.34	2.23	2.17	2.11	2.05	1.98	1.91	1.83
29	2.53	2.43	2.32	2.21	2.15	2.09	2.03	1.96	1.89	1.81
30	2.51	2.41	2.31	2.20	2.14	2.07	2.01	1.94	1.87	1.79
40	2.39	2.29	2.18	2.07	2.01	1.94	1.88	1.80	1.72	1.64
60	2.27	2.17	2.06	1.94	1.88	1.82	1.74	1.67	1.58	1.48
120	2.16	2.05	1.95	1.82	1.76	1.69	1.61	1.53	1.43	1.31
∞	2.05	1.94	1.83	1.71	1.64	1.57	1.48	1.39	1.27	1.00

Note: Entries in this table were computed with IMSL subroutine FIN.

Table 8 Percentage points of the F distribution, $\alpha = .01$

	ν_1								
ν_2	1	2	3	4	5	6	7	8	9
1	4,052	4,999	5,403	5,625	5,764	5,859	5,928	5,982	6,022
2	98.50	99.00	99.17	99.25	99.30	99.33	99.36	99.37	99.39
3	34.12	30.82	29.46	28.71	28.24	27.91	27.67	27.50	27.34
4	21.20	18.00	16.69	15.98	15.52	15.21	14.98	14.80	14.66
5	16.26	13.27	12.06	11.39	10.97	10.67	10.46	10.29	10.16
6	13.75	10.92	9.78	9.15	8.75	8.47	8.26	8.10	7.98
7	12.25	9.55	8.45	7.85	7.46	7.19	6.99	6.84	6.72
8	11.26	8.65	7.59	7.01	6.63	6.37	6.18	6.03	5.91
9	10.56	8.02	6.99	6.42	6.06	5.80	5.61	5.47	5.35
10	10.04	7.56	6.55	5.99	5.64	5.39	5.20	5.06	4.94
11	9.65	7.21	6.22	5.67	5.32	5.07	4.89	4.74	4.63
12	9.33	6.93	5.95	5.41	5.06	4.82	4.64	4.50	4.39
13	9.07	6.70	5.74	5.21	4.86	4.62	4.44	4.30	4.19
14	8.86	6.51	5.56	5.04	4.69	4.46	4.28	4.14	4.03
15	8.68	6.36	5.42	4.89	4.56	4.32	4.14	4.00	3.89
16	8.53	6.23	5.29	4.77	4.44	4.20	4.03	3.89	3.78
17	8.40	6.11	5.18	4.67	4.34	4.10	3.93	3.79	3.68
18	8.29	6.01	5.09	4.58	4.25	4.01	3.84	3.71	3.60
19	8.18	5.93	5.01	4.50	4.17	3.94	3.77	3.63	3.52
20	8.10	5.85	4.94	4.43	4.10	3.87	3.70	3.56	3.46
21	8.02	5.78	4.87	4.37	4.04	3.81	3.64	3.51	3.40
22	7.95	5.72	4.82	4.31	3.99	3.76	3.59	3.45	3.35
23	7.88	5.66	4.76	4.26	3.94	3.71	3.54	3.41	3.30
24	7.82	5.61	4.72	4.22	3.90	3.67	3.50	3.36	3.26
25	7.77	5.57	4.68	4.18	3.85	3.63	3.46	3.32	3.22
26	7.72	5.53	4.64	4.14	3.82	3.59	3.42	3.29	3.18
27	7.68	5.49	4.60	4.11	3.78	3.56	3.39	3.26	3.15
28	7.64	5.45	4.57	4.07	3.75	3.53	3.36	3.23	3.12
29	7.60	5.42	4.54	4.04	3.73	3.50	3.33	3.20	3.09
30	7.56	5.39	4.51	4.02	3.70	3.47	3.30	3.17	3.07
40	7.31	5.18	4.31	3.83	3.51	3.29	3.12	2.99	2.89
60	7.08	4.98	4.13	3.65	3.34	3.12	2.95	2.82	2.72
120	6.85	4.79	3.95	3.48	3.17	2.96	2.79	2.66	2.56
∞	6.63	4.61	3.78	3.32	3.02	2.80	2.64	2.51	2.41

Continued

Table 8 Percentage points of the F distribution, $\alpha = .01$ (continued)

	ν_1									
ν_2	10	12	15	20	24	30	40	60	120	∞
1	6,056	6,106	6,157	6,209	6,235	6,261	6,287	6,313	6,339	6,366
2	99.40	99.42	99.43	99.45	99.46	99.46	99.47	99.48	99.49	99.50
3	27.22	27.03	26.85	26.67	26.60	26.50	26.41	26.32	26.22	26.13
4	14.55	14.37	14.19	14.02	13.94	13.84	13.75	13.65	13.56	13.46
5	10.05	9.89	9.72	9.55	9.46	9.38	9.30	9.20	9.11	9.02
6	7.87	7.72	7.56	7.40	7.31	7.23	7.15	7.06	6.97	6.88
7	6.62	6.47	6.31	6.16	6.07	5.99	5.91	5.82	5.74	5.65
8	5.81	5.67	5.52	5.36	5.28	5.20	5.12	5.03	4.95	4.86
9	5.26	5.11	4.96	4.81	4.73	4.65	4.57	4.48	4.40	4.31
10	4.85	4.71	4.56	4.41	4.33	4.25	4.17	4.08	4.00	3.91
11	4.54	4.40	4.25	4.10	4.02	3.94	3.86	3.78	3.69	3.60
12	4.30	4.16	4.01	3.86	3.78	3.70	3.62	3.54	3.45	3.36
13	4.10	3.96	3.82	3.66	3.59	3.51	3.43	3.34	3.25	3.17
14	3.94	3.80	3.66	3.51	3.43	3.35	3.27	3.18	3.09	3.00
15	3.80	3.67	3.52	3.37	3.29	3.21	3.13	3.05	2.96	2.87
16	3.69	3.55	3.41	3.26	3.18	3.10	3.02	2.93	2.84	2.75
17	3.59	3.46	3.31	3.16	3.08	3.00	2.92	2.83	2.75	2.65
18	3.51	3.37	3.23	3.08	3.00	2.92	2.84	2.75	2.66	2.57
19	3.43	3.30	3.15	3.00	2.92	2.84	2.76	2.67	2.58	2.49
20	3.37	3.23	3.09	2.94	2.86	2.78	2.69	2.61	2.52	2.42
21	3.31	3.17	3.03	2.88	2.80	2.72	2.64	2.55	2.46	2.36
22	3.26	3.12	2.98	2.83	2.75	2.67	2.58	2.50	2.40	2.31
23	3.21	3.07	2.93	2.78	2.70	2.62	2.54	2.45	2.35	2.26
24	3.17	3.03	2.89	2.74	2.66	2.58	2.49	2.40	2.31	2.21
25	3.13	2.99	2.85	2.70	2.62	2.54	2.45	2.36	2.27	2.17
26	3.09	2.96	2.81	2.66	2.58	2.50	2.42	2.33	2.23	2.13
27	3.06	2.93	2.78	2.63	2.55	2.47	2.38	2.29	2.20	2.10
28	3.03	2.90	2.75	2.60	2.52	2.44	2.35	2.26	2.17	2.06
29	3.00	2.87	2.73	2.57	2.49	2.41	2.33	2.23	2.14	2.03
30	2.98	2.84	2.70	2.55	2.47	2.39	2.30	2.21	2.11	2.01
40	2.80	2.66	2.52	2.37	2.29	2.20	2.11	2.02	1.92	1.80
60	2.63	2.50	2.35	2.20	2.12	2.03	1.94	1.84	1.73	1.60
120	2.47	2.34	2.19	2.03	1.95	1.86	1.76	1.66	1.53	1.38
∞	2.32	2.18	2.04	1.88	1.79	1.70	1.59	1.47	1.32	1.00

Note: Entries in this table were computed with IMSL subroutine FIN.

Table 9 Studentized range statistic, q, for $\alpha = .05$

	J (number of groups)									
ν	2	3	4	5	6	7	8	9	10	11
3	4.50	5.91	6.82	7.50	8.04	8.48	8.85	9.18	9.46	9.72
4	3.93	5.04	5.76	6.29	6.71	7.05	7.35	7.60	7.83	8.03
5	3.64	4.60	5.22	5.68	6.04	6.33	6.59	6.81	6.99	7.17
6	3.47	4.34	4.89	5.31	5.63	5.89	6.13	6.32	6.49	6.65
7	3.35	4.17	4.69	5.07	5.36	5.61	5.82	5.99	6.16	6.30
8	3.27	4.05	4.53	4.89	5.17	5.39	5.59	5.77	5.92	6.06
9	3.19	3.95	4.42	4.76	5.03	5.25	5.44	5.59	5.74	5.87
10	3.16	3.88	4.33	4.66	4.92	5.13	5.31	5.47	5.59	5.73
11	3.12	3.82	4.26	4.58	4.83	5.03	5.21	5.36	5.49	5.61
12	3.09	3.78	4.19	4.51	4.76	4.95	5.12	5.27	5.39	5.52
13	3.06	3.73	4.15	4.45	4.69	4.88	5.05	5.19	5.32	5.43
14	3.03	3.70	4.11	4.41	4.64	4.83	4.99	5.13	5.25	5.36
15	3.01	3.67	4.08	4.37	4.59	4.78	4.94	5.08	5.20	5.31
16	3.00	3.65	4.05	4.33	4.56	4.74	4.90	5.03	5.15	5.26
17	2.98	3.63	4.02	4.30	4.52	4.70	4.86	4.99	5.11	5.21
18	2.97	3.61	4.00	4.28	4.49	4.67	4.83	4.96	5.07	5.17
19	2.96	3.59	3.98	4.25	4.47	4.65	4.79	4.93	5.04	5.14
20	2.95	3.58	3.96	4.23	4.45	4.62	4.77	4.90	5.01	5.11
24	2.92	3.53	3.90	4.17	4.37	4.54	4.68	4.81	4.92	5.01
30	2.89	3.49	3.85	4.10	4.30	4.46	4.60	4.72	4.82	4.92
40	2.86	3.44	3.79	4.04	4.23	4.39	4.52	4.63	4.73	4.82
60	2.83	3.40	3.74	3.98	4.16	4.31	4.44	4.55	4.65	4.73
120	2.80	3.36	3.68	3.92	4.10	4.24	4.36	4.47	4.56	4.64
∞	2.77	3.31	3.63	3.86	4.03	4.17	4.29	4.39	4.47	4.55

Table 9 Studentized range statistic, q, for $\alpha = .01$

	J (number of groups)									
ν	2	3	4	5	6	7	8	9	10	11
2	14.0	19.0	22.3	24.7	26.6	28.2	29.5	30.7	31.7	32.6
3	8.26	10.6	12.2	13.3	14.2	15.0	15.6	16.2	16.7	17.8
4	6.51	8.12	9.17	9.96	10.6	11.1	11.5	11.9	12.3	12.6
5	5.71	6.98	7.81	8.43	8.92	9.33	9.67	9.98	10.24	10.48
6	5.25	6.34	7.04	7.56	7.98	8.32	8.62	8.87	9.09	9.30
7	4.95	5.92	6.55	7.01	7.38	7.68	7.94	8.17	8.37	8.55
8	4.75	5.64	6.21	6.63	6.96	7.24	7.48	7.69	7.87	8.03
9	4.59	5.43	5.96	6.35	6.66	6.92	7.14	7.33	7.49	7.65
10	4.49	5.28	5.77	6.14	6.43	6.67	6.88	7.06	7.22	7.36
11	4.39	5.15	5.63	5.98	6.25	6.48	6.68	6.85	6.99	7.13
12	4.32	5.05	5.51	5.84	6.11	6.33	6.51	6.67	6.82	6.94
13	4.26	4.97	5.41	5.73	5.99	6.19	6.38	6.53	6.67	6.79
14	4.21	4.89	5.32	5.63	5.88	6.08	6.26	6.41	6.54	6.66
15	4.17	4.84	5.25	5.56	5.80	5.99	6.16	6.31	6.44	6.55
16	4.13	4.79	5.19	5.49	5.72	5.92	6.08	6.22	6.35	6.46
17	4.10	4.74	5.14	5.43	5.66	5.85	6.01	6.15	6.27	6.38
18	4.07	4.70	5.09	5.38	5.60	5.79	5.94	6.08	6.20	6.31
19	4.05	4.67	5.05	5.33	5.55	5.73	5.89	6.02	6.14	6.25
20	4.02	4.64	5.02	5.29	5.51	5.69	5.84	5.97	6.09	6.19
24	3.96	4.55	4.91	5.17	5.37	5.54	5.69	5.81	5.92	6.02
30	3.89	4.45	4.80	5.05	5.24	5.40	5.54	5.65	5.76	5.85
40	3.82	4.37	4.69	4.93	5.10	5.26	5.39	5.49	5.60	5.69
60	3.76	4.28	4.59	4.82	4.99	5.13	5.25	5.36	5.45	5.53
120	3.70	4.20	4.50	4.71	4.87	5.01	5.12	5.21	5.30	5.37
∞	3.64	4.12	4.40	4.60	4.76	4.88	4.99	5.08	5.16	5.23

Note: The values in this table were computed with the IBM SSP subroutines DQH32 and DQG32

Table 10 Studentized maximum modulus distribution

		C (the number of tests being performed)								
ν	α	2	3	4	5	6	7	8	9	10
2	.05	5.57	6.34	6.89	7.31	7.65	7.93	8.17	8.83	8.57
	.01	12.73	14.44	15.65	16.59	17.35	17.99	18.53	19.01	19.43
3	.05	3.96	4.43	4.76	5.02	5.23	5.41	5.56	5.69	5.81
	.01	7.13	7.91	8.48	8.92	9.28	9.58	9.84	10.06	10.27
4	.05	3.38	3.74	4.01	4.20	4.37	4.50	4.62	4.72	4.82
	.01	5.46	5.99	6.36	6.66	6.89	7.09	7.27	7.43	7.57
5	.05	3.09	3.39	3.62	3.79	3.93	4.04	4.14	4.23	4.31
	.01	4.70	5.11	5.39	5.63	5.81	5.97	6.11	6.23	6.33
6	.05	2.92	3.19	3.39	3.54	3.66	3.77	3.86	3.94	4.01
	.01	4.27	4.61	4.85	5.05	5.20	5.33	5.45	5.55	5.64
7	.05	2.80	3.06	3.24	3.38	3.49	3.59	3.67	3.74	3.80
	.01	3.99	4.29	4.51	4.68	4.81	4.93	5.03	5.12	5.19
8	.05	2.72	2.96	3.13	3.26	3.36	3.45	3.53	3.60	3.66
	.01	3.81	4.08	4.27	4.42	4.55	4.65	4.74	4.82	4.89
9	.05	2.66	2.89	3.05	3.17	3.27	3.36	3.43	3.49	3.55
	.01	3.67	3.92	4.10	4.24	4.35	4.45	4.53	4.61	4.67
10	.05	2.61	2.83	2.98	3.10	3.19	3.28	3.35	3.41	3.47
	.01	3.57	3.80	3.97	4.09	4.20	4.29	4.37	4.44	4.50
11	.05	2.57	2.78	2.93	3.05	3.14	3.22	3.29	3.35	3.40
	.01	3.48	3.71	3.87	3.99	4.09	4.17	4.25	4.31	4.37
12	.05	2.54	2.75	2.89	3.01	3.09	3.17	3.24	3.29	3.35
	.01	3.42	3.63	3.78	3.89	3.99	4.08	4.15	4.21	4.26
14	.05	2.49	2.69	2.83	2.94	3.02	3.09	3.16	3.21	3.26
	.01	3.32	3.52	3.66	3.77	3.85	3.93	3.99	4.05	4.10
16	.05	2.46	2.65	2.78	2.89	2.97	3.04	3.09	3.15	3.19
	.01	3.25	3.43	3.57	3.67	3.75	3.82	3.88	3.94	3.99
18	.05	2.43	2.62	2.75	2.85	2.93	2.99	3.05	3.11	3.15
	.01	3.19	3.37	3.49	3.59	3.68	3.74	3.80	3.85	3.89
20	.05	2.41	2.59	2.72	2.82	2.89	2.96	3.02	3.07	3.11
	.01	3.15	3.32	3.45	3.54	3.62	3.68	3.74	3.79	3.83
24	.05	2.38	2.56	2.68	2.77	2.85	2.91	2.97	3.02	3.06
	.01	3.09	3.25	3.37	3.46	3.53	3.59	3.64	3.69	3.73
30	.05	2.35	2.52	2.64	2.73	2.80	2.87	2.92	2.96	3.01
	.01	3.03	3.18	3.29	3.38	3.45	3.50	3.55	3.59	3.64
40	.05	2.32	2.49	2.60	2.69	2.76	2.82	2.87	2.91	2.95
	.01	2.97	3.12	3.22	3.30	3.37	3.42	3.47	3.51	3.55
60	.05	2.29	2.45	2.56	2.65	2.72	2.77	2.82	2.86	2.90
	.01	2.91	3.06	3.15	3.23	3.29	3.34	3.38	3.42	3.46
∞	.05	2.24	2.39	2.49	2.57	2.63	2.68	2.73	2.77	2.79
	.01	2.81	2.93	3.02	3.09	3.14	3.19	3.23	3.26	3.29

Continued

Table 10 Studentized maximum modulus distribution (continued)

ν	α	11	12	13	14	15	16	17	18	19
2	.05	8.74	8.89	9.03	9.16	9.28	9.39	9.49	9.59	9.68
	.01	19.81	20.15	20.46	20.75	20.99	20.99	20.99	20.99	20.99
3	.05	5.92	6.01	6.10	6.18	6.26	6.33	6.39	6.45	6.51
	.01	10.45	10.61	10.76	10.90	11.03	11.15	11.26	11.37	11.47
4	.05	4.89	4.97	5.04	5.11	5.17	5.22	5.27	5.32	5.37
	.01	7.69	7.80	7.91	8.01	8.09	8.17	8.25	8.32	8.39
5	.05	4.38	4.45	4.51	4.56	4.61	4.66	4.70	4.74	4.78
	.01	6.43	6.52	6.59	6.67	6.74	6.81	6.87	6.93	6.98
6	.05	4.07	4.13	4.18	4.23	4.28	4.32	4.36	4.39	4.43
	.01	5.72	5.79	5.86	5.93	5.99	6.04	6.09	6.14	6.18
7	.05	3.86	3.92	3.96	4.01	4.05	4.09	4.13	4.16	4.19
	.01	5.27	5.33	5.39	5.45	5.50	5.55	5.59	5.64	5.68
8	.05	3.71	3.76	3.81	3.85	3.89	3.93	3.96	3.99	4.02
	.01	4.96	5.02	5.07	5.12	5.17	5.21	5.25	5.29	5.33
9	.05	3.60	3.65	3.69	3.73	3.77	3.80	3.84	3.87	3.89
	.01	4.73	4.79	4.84	4.88	4.92	4.96	5.01	5.04	5.07
10	.05	3.52	3.56	3.60	3.64	3.68	3.71	3.74	3.77	3.79
	.01	4.56	4.61	4.66	4.69	4.74	4.78	4.81	4.84	4.88
11	.05	3.45	3.49	3.53	3.57	3.60	3.63	3.66	3.69	3.72
	.01	4.42	4.47	4.51	4.55	4.59	4.63	4.66	4.69	4.72
12	.05	3.39	3.43	3.47	3.51	3.54	3.57	3.60	3.63	3.65
	.01	4.31	4.36	4.40	4.44	4.48	4.51	4.54	4.57	4.59
14	.05	3.30	3.34	3.38	3.41	3.45	3.48	3.50	3.53	3.55
	.01	4.15	4.19	4.23	4.26	4.29	4.33	4.36	4.39	4.41
16	.05	3.24	3.28	3.31	3.35	3.38	3.40	3.43	3.46	3.48
	.01	4.03	4.07	4.11	4.14	4.17	4.19	4.23	4.25	4.28
18	.05	3.19	3.23	3.26	3.29	3.32	3.35	3.38	3.40	3.42
	.01	3.94	3.98	4.01	4.04	4.07	4.10	4.13	4.15	4.18
20	.05	3.15	3.19	3.22	3.25	3.28	3.31	3.33	3.36	3.38
	.01	3.87	3.91	3.94	3.97	3.99	4.03	4.05	4.07	4.09
24	.05	3.09	3.13	3.16	3.19	3.22	3.25	3.27	3.29	3.31
	.01	3.77	3.80	3.83	3.86	3.89	3.91	3.94	3.96	3.98
30	.05	3.04	3.07	3.11	3.13	3.16	3.18	3.21	3.23	3.25
	.01	3.67	3.70	3.73	3.76	3.78	3.81	3.83	3.85	3.87
40	.05	2.99	3.02	3.05	3.08	3.09	3.12	3.14	3.17	3.18
	.01	3.58	3.61	3.64	3.66	3.68	3.71	3.73	3.75	3.76
60	.05	2.93	2.96	2.99	3.02	3.04	3.06	3.08	3.10	3.12
	.01	3.49	3.51	3.54	3.56	3.59	3.61	3.63	3.64	3.66
∞	.05	2.83	2.86	2.88	2.91	2.93	2.95	2.97	2.98	3.01
	.01	3.32	3.34	3.36	3.38	3.40	3.42	3.44	3.45	3.47

Table 10 Studentized maximum modulus distribution (continued)

ν	α	20	21	22	23	24	25	26	27	28
2	.05	9.77	9.85	9.92	10.00	10.07	10.13	10.20	10.26	10.32
	.01	22.11	22.29	22.46	22.63	22.78	22.93	23.08	23.21	23.35
3	.05	6.57	6.62	6.67	6.71	6.76	6.80	6.84	6.88	6.92
	.01	11.56	11.65	11.74	11.82	11.89	11.97	12.07	12.11	12.17
4	.05	5.41	5.45	5.49	5.52	5.56	5.59	5.63	5.66	5.69
	.01	8.45	8.51	8.57	8.63	8.68	8.73	8.78	8.83	8.87
5	.05	4.82	4.85	4.89	4.92	4.95	4.98	5.00	5.03	5.06
	.01	7.03	7.08	7.13	7.17	7.21	7.25	7.29	7.33	7.36
6	.05	4.46	4.49	4.52	4.55	4.58	4.60	4.63	4.65	4.68
	.01	6.23	6.27	6.31	6.34	6.38	6.41	6.45	6.48	6.51
7	.05	4.22	4.25	4.28	4.31	4.33	4.35	4.38	4.39	4.42
	.01	5.72	5.75	5.79	5.82	5.85	5.88	5.91	5.94	5.96
8	.05	4.05	4.08	4.10	4.13	4.15	4.18	4.19	4.22	4.24
	.01	5.36	5.39	5.43	5.45	5.48	5.51	5.54	5.56	5.59
9	.05	3.92	3.95	3.97	3.99	4.02	4.04	4.06	4.08	4.09
	.01	5.10	5.13	5.16	5.19	5.21	5.24	5.26	5.29	5.31
10	.05	3.82	3.85	3.87	3.89	3.91	3.94	3.95	3.97	3.99
	.01	4.91	4.93	4.96	4.99	5.01	5.03	5.06	5.08	5.09
11	.05	3.74	3.77	3.79	3.81	3.83	3.85	3.87	3.89	3.91
	.01	4.75	4.78	4.80	4.83	4.85	4.87	4.89	4.91	4.93
12	.05	3.68	3.70	3.72	3.74	3.76	3.78	3.80	3.82	3.83
	.01	4.62	4.65	4.67	4.69	4.72	4.74	4.76	4.78	4.79
14	.05	3.58	3.59	3.62	3.64	3.66	3.68	3.69	3.71	3.73
	.01	4.44	4.46	4.48	4.50	4.52	4.54	4.56	4.58	4.59
16	.05	3.50	3.52	3.54	3.56	3.58	3.59	3.61	3.63	3.64
	.01	4.29	4.32	4.34	4.36	4.38	4.39	4.42	4.43	4.45
18	.05	3.44	3.46	3.48	3.50	3.52	3.54	3.55	3.57	3.58
	.01	4.19	4.22	4.24	4.26	4.28	4.29	4.31	4.33	4.34
20	.05	3.39	3.42	3.44	3.46	3.47	3.49	3.50	3.52	3.53
	.01	4.12	4.14	4.16	4.17	4.19	4.21	4.22	4.24	4.25
24	.05	3.33	3.35	3.37	3.39	3.40	3.42	3.43	3.45	3.46
	.01	4.00	4.02	4.04	4.05	4.07	4.09	4.10	4.12	4.13
30	.05	3.27	3.29	3.30	3.32	3.33	3.35	3.36	3.37	3.39
	.01	3.89	3.91	3.92	3.94	3.95	3.97	3.98	4.00	4.01
40	.05	3.20	3.22	3.24	3.25	3.27	3.28	3.29	3.31	3.32
	.01	3.78	3.80	3.81	3.83	3.84	3.85	3.87	3.88	3.89
60	.05	3.14	3.16	3.17	3.19	3.20	3.21	3.23	3.24	3.25
	.01	3.68	3.69	3.71	3.72	3.73	3.75	3.76	3.77	3.78
∞	.05	3.02	3.03	3.04	3.06	3.07	3.08	3.09	3.11	3.12
	.01	3.48	3.49	3.50	3.52	3.53	3.54	3.55	3.56	3.57

Note: This table was computed using the FORTRAN program described in Wilcox (1986b).

References

Agresti, A. (1990). Categorical Data Analysis. New York: Wiley.

Agresti, A. (1996). An Introduction to Categorical Data Analysis. New York: Wiley.

Agresti, A. & Coull, B. A. (1998). Approximate is better than "exact" for interval estimation of binomial proportions. American Statistician 52, 119–126.

Alexander, R. A. & Govern, D. M. (1994). A new and simpler approximation for ANOVA under variance heterogeneity. Journal of Educational Statistics, 19, 91–101.

Algina, J., Keselman, H. J. & Penfield, R. D. (2005). An alternative to Cohen's standardized mean difference effect size: A robust parameter and confidence interval in the two independent groups case. Psychological Methods, 10, 317–328.

Algina, J., Oshima, T. C. & Lin, W.-Y. (1994). Type I error rates for Welch's test and Jamse's second-order test under nonnormality and inequality of variance when there are two groups. Journal of Educational and Behavioral Statistics, 19, 275–291.

Andersen, E. B. (1997). Introduction to the Statistical Analysis of Categorical Data. New York: Springer.

Asiribo, O. & Gurland, J. (1989). Some simple approximate solutions to the Behrens–Fisher problem. Communications in Statistics–Theory and Methods, 18, 1201–1216.

Atkinson, R. L., Atkinson, R. C., Smith, E. E. & Hilgard, E. R. (1985). Introduction to Psychology, 9th Edition. San Diego: Harcourt, Brace, Jovanovich.

Barrett, J. P. (1974). The coefficient of determination—Some limitations. Annals of Statistics, 28, 19–20.

Beal, S. L. (1987). Asymptotic confidence intervals for the difference between two binomial parameters for use with small samples. Biometrics, 43, 941–950.

Belsley, D. A., Kuh, E. & Welsch, R. E. (1980). Regression Diagnostics: Identifying Influential Data and Sources of Collinearity. New York: Wiley.

Berger, R. L. (1996). More powerful tests from confidence interval *p* values. American Statistician, 50, 314–318.

Berkson, J. (1958). Smoking and lung cancer: Some observations on two recent reports. Journal of the American Statistical Association, 53, 28–38.

Bernhardson, C. (1975). Type I error rates when multiple comparison procedures follow a significant F test of ANOVA. Biometrics, 31, 719–724.

Blyth, C. R. (1986). Approximate binomial confidence limits. Journal of the American Statistical Association, 81, 843–855.

Box, G. E. P. (1953). Non-normality and tests on variances. Biometrika, 40, 318–335.

Bradley, J. V. (1978) Robustness? British Journal of Mathematical and Statistical Psychology, 31, 144–152.

Brown, L. D., Cai, T. T. & Das Gupta, A. (2002). Confidence intervals for a binomial proportion and asymptotic expansions. Annals of Statistics 30, 160–201.

Brunner, E., Dette, H. & Munk, A. (1997). Box-type approximations in non-parametric factorial designs. Journal of the American Statistical Association, 92, 1494–1502.

Brunner, E., Domhof, S. & Langer, F. (2002). Nonparametric Analysis of Longitudinal Data in Factorial Experiments. New York: Wiley.
Brunner, E. & Munzel, U. (2000). The nonparametric Behrens-Fisher problem: asymptotic theory and small-sample approximation. Biometrical Journal 42, 17–25.
Carroll, R. J. & Ruppert, D. (1988). Transformation and Weighting in Regression. New York: Chapman and Hall.
Chen, S. & Chen, H. J. (1998). Single-stage analysis of variance under heteroscedasticity. Communications in Statistics–Simulation and Computation 27, 641–666.
Cliff, N. (1996). Ordinal Methods for Behavioral Data Analysis. Mahwah, N. J.: Erlbaum.
Cochran, W. G. & Cox, G. M. (1950). Experimental Design. New York: Wiley.
Coe, P. R. & Tamhane, A. C. (1993). Small sample confidence intervals for the difference, ratio, and odds ratio of two success probabilities. Communications in Statistics—Simulation and Computation, 22, 925–938.
Cohen, J. (1994). The earth is round ($p < .05$). American Psychologist, 49, 997–1003.
Coleman, J. S. (1964). Introduction to Mathmematical Sociology. New York: Free Press.
Cook, R. D. & Weisberg, S. (1992). Residuals and Influence in Regression. New York: Chapman and Hall.
Cressie, N. A. C. & Whitford, H. J. (1986). How to use the two sample t-test. Biometrical Journal, 28, 131–148.
Cumming, G. & Maillardet, R. (2006). Confidence intervals and replication: Where will the next mean fall?. Psychological Methods, 11, 217–227.
De Veaux, R. D. & Hand, D. J. (2005). How to lie with bad data. Statistical Science, 20, 231–238.
Doksum, K. A. & Sievers, G. L. (1976). Plotting with confidence: graphical comparisons of two populations. Biometrika, 63, 421–434.
Dunnett, C. W. (1980a). Pairwise multiple comparisons in the unequal variance case. Journal of the American Statistical Association, 75, 796–800.
Dunnett, C. W. (1980b). Pairwise multiple comparisons in the homogeneous variance, unequal sample size case. Journal of the American Statistical Association, 75, 796–800.
Emerson, J. D. & Hoaglin, D. C. (1983). Stem-and-leaf displays. In D. C. Hoaglin, F. Mosteller & J. W. Tukey (Eds.) Understanding Robust and Exploratory Data Analysis, pp. 7–32. New York: Wiley.
Fienberg, S. E. (1980). The Analysis of Cross-Classified Categorical Data, 2nd ed. Cambridge, MA: MIT Press.
Fisher, R. A. (1935). The fiducial argument in statistical inference. Annals of Eugenics, 6, 391–398.
Fisher, R. A. (1941). The asymptotic approach to Behren's integral, with further tables for the d test of significance. Annals of Eugenics, 11, 141–172.
Fleiss, J. L. (1981). Statisical Methods for Rates and Proportions, 2nd ed. New York: Wiley.
Fung, K. Y. (1980). Small sample behaviour of some nonparametric multi-sample location tests in the presence of dispersion differences. Statistica Neerlandica, 34, 189–196.
Games, P. A. & Howell, J. (1976). Pairwise multiple comparison procedures with unequal n's and/or variances: A Monte Carlo study. Journal of Educational Statistics, 1, 113–125.
Goodman, L. A. & Kruskal, W. H. (1954). Measures of association for cross-classifications. Journal of the American Statistical Association, 49, 732–736.
Hampel, F. R., Ronchetti, E. M., Rousseeuw, P. J. & Stahel, W. A. (1986). Robust Statistics: The Approach Based on Influence Functions. New York: Wiley.
Harrison, D. & Rubinfeld, D.L. (1978). Hedonic prices and the demand for clean air. Journal of Environmental Economics & Management, 5 , 81–102.
Hayes, A. F. & Cai, L. (2007). Further evaluating the conditional decision rule for comparing two independent means. British Journal of Mathematical and Statistical Psychology, 60, 217–244.
Hayter, A. (1984). A proof of the conjecture that the Tukey-Kramer multiple comparison procedure is conservative. Annals of Statistics, 12, 61–75.

Hayter, A. (1986). The maximum familywise error rate of Fisher's least significant difference test. Journal of the American Statistical Association, 81, 1000–1004.
Heiser, D. A. (2006). Statistical tests, tests of significance, and tests of a hypothesis (Excel). Journal of Modern Applied Statistical Methods, 5, 551–566.
Hettmansperger, T. P. (1984). Statistical Inference Based on Ranks. New York: Wiley.
Hettmansperger, T. P. & McKean, J. W. (1977). A robust alternative based on ranks to least squares in analyzing linear models. Technometrics, 19, 275–284.
Hosmane, B. S. (1986). Improved likelihood ratio tests and Pearson chi-square tests for independence in two dimensional tables. Communications Statistics—Theory and Methods, 15, 1875–1888.
Hosmer, D. W. & Lemeshow, S. (1989). Applied Logistic Regression. New York: Wiley.
Huber, P. J. (1981). Robust Statistics. New York: Wiley.
Iman, R. L., Quade, D. & Alexander, D. A. (1975). Exact probability levels for the Krusakl-Wallis test. Selected Tables in Mathematical Statistics, 3, 329–384.
James, G. S. (1951). The comparison of several groups of observations when the ratios of the population variances are unknown. Biometrika, 38, 324–329.
Jeyaratnam, S. & Othman, A. R. (1985). Test of hypothesis in one-way random effects model with unequal error variances. Journal of Statistical Computation and Simulation, 21, 51–57.
Jones, L. V. & Tukey, J. W. (2000). A sensible formulation of the significance test. Psychological Methods, 5, 411–414.
Kendall, M. G. & Stuart, A. (1973). The Advanced Theory of Statistics, Vol. 2. New York: Hafner.
Keselman, H. J. & Wilcox, R. R. (1999). The "improved" Brown and Forsyth test for mean equality: Some things can't be fixed. Communications in Statistics—Simulation and Computation, 28, 687–698.
Keselman, H. J., Algina, J., Wilcox, R. R. & Kowalchuk, R. K. (2000). Testing repeated measures hypotheses when covariance matrices are heterogeneous: Revisiting the robustness of the Welch-James test again. Educational and Psychological Measurement, 60, 925–938.
Kirk, R. E. (1995). Experimental Design. Monterey, CA: Brooks/Cole.
Kleinbaum, D. G. (1994). Logistic Regression. New York: Springer-Verlag.
Kramer, C. (1956). Extension of multiple range test to group means with unequal number of replications. Biometrics, 12, 307–310.
Li, G. (1985). Robust regression. In D. Hoaglin, F. Mosteller & J. Tukey (Eds.), Exploring Data Tables, Trends, and Shapes, pp. 281–343. New York: Wiley.
Lee, S. & Ahn, C. H. (2003). Modified ANOVA for unequal variances. Communications in Statistics–Simulation and Computation, 32, 987–1004.
Lloyd, C. J. (1999). Statistical Analysis of Categorical Data. New York: Wiley.
Loh, W.-Y. (1987). Does the correlation coefficient really measure the degree of clustering around a line? Journal of Educational Statistics, 12, 235–239.
Mann, H. B. & Whitney, D. R. (1947). On a test of whether one of two random variables is stochastically larger than the other. Annals of Mathematical Statistics, 18, 50–60.
Markowski, C. A. & Markowski, E. P. (1990). Conditions for the effectiveness of a preliminary test of variance. American Statistician, 44, 322–326.
Matuszewski, A. & Sotres, D. (1986). A simple test for the Behrens–Fisher problem. Computational Statistics and Data Analysis, 3, 241–249.
McCullough, B. D. & Wilson, B. (2005). On the accuracy of statistical procedures in Microsoft Excel. Computational Statistics & Data Analysis, 49 1244–1252.
McKean, J. W. & Schrader, R. M. (1984). A comparison of methods for studentizing the sample median. Communications in Statistics— Simulation and Computation, 13, 751–773.
Mehrotra, D. V. (1997). Improving the Brown-Forsythe solution to the generalize d Behrens-fisher problem. Communications in Statistics–Simulation and Computation, 26, 1139–1145.
Micceri, T. (1989). The unicorn, the normal curve, and other improbable creatures. Psychological Bulletin, 105, 156–166.

Miller, R. G. (1976). Least squares regression with censored data. Biometrika, 63, 449–464.

Montgomery, D. C. & Peck, E. A. (1992). Introduction to Linear Regression Analysis. New York: Wiley.

Moser, B. K., Stevens, G. R. & Watts, C. L. (1989). The two-sample *t*-test versus Satterthwaite's approximate F test. Communications in Statistics-Theory and Methods, 18, 3963–3975.

Neuhäuser, M. Lösch, C. & Jöckel, K.-H. (2007). Computational Statistics & Data Analysis, 51, 5055–5060.

Olejnik, S., Li, J., Supattathum, S. & Huberty, C. J. (1997). Multiple testing and statistical power with modified Bonferroni procedures. Journal of Educational and Behavioral Statistics, 22, 389–406.

Pagurova, V. I. (1968). On a comparison of means of two normal samples. Theory of Probability and Its Applications, 13, 527–534.

Pedersen, W. C., Miller, L. C., Putcha-Bhagavatula, A. D. & Yang, Y. (2002). Evolved sex differences in sexual strategies: The long and the short of it. Psychological Science, 13, 157–161.

Powers, D. A. & Xie, Y. (1999). Statistical Methods for Categorical Data Analysis. San Diego, Calif.: Academic Press.

Pratt, J. W. (1968). A normal approximation for binomial, F, beta, and other common, related tail probabilities, I. Journal of the American Statistical Association, 63, 1457–1483.

Ramsey, P. H. (1980). Exact Type I error rates for robustness of Student's *t* test with unequal variances. Journal of Educational Statistics, 5, 337–349.

Reiczigel, J. Zakariás, I. & Rózsa, L. (2005). A bootstrap test of stochastic equality of two populations. American Statistician, 59, 156–161.

Rom, D. M. (1990). A sequentially rejective test procedure based on a modified Bonferroni inequality. Biometrika, 77, 663–666.

Romano, J. P. (1990). On the behavior of randomization tests without a group invariance assumption. Journal of the American Statistical Association 85, 686–692.

Rousseeuw, P. J. & Leroy, A. M. (1987). Robust Regression & Outlier Detection. New York: Wiley.

Rousseeuw, P. J. & van Zomeren, B. C. (1990). Unmasking multivariate outliers and leverage points (with discussion). Journal of the American Statistical Association, 85, 633–639.

Scariano, S. M. & Davenport, J. M. (1986). A four-moment approach and other practical solutions to the Behrens-Fisher problem. Communications in Statistics–Theory and Methods, 15, 1467–1501.

Scheffé, H. (1959). The Analysis of Variance. New York: Wiley.

Schenker, N. & Gentleman, J. F. (2001). On judging the significance of differences by examining the overlap between confidence intervals. American Statistician, 55, 182–186.

Silverman, B. W. (1986). Density Estimation for Statistics and Data Analysis. New York: Chapman and Hall.

Simonoff, J. (2003). Analyzing Categorical Data. New York: Springer-Verlag.

Snedecor, G. W. & Cochran, W. (1967). Statistical Methods. 6th Edition. Ames, Iowa: University Press.

Sockett, E. B., Daneman, D. Clarson & C. Ehrich, R. M. (1987). Factors affecting and patterns of residual insulin secretion during the first year of type I (insulin dependent) diabetes mellitus in children. Diabetes, 30, 453–459.

Staudte, R. G. & Sheather, S. J. (1990). Robust Estimation and Testing. New York: Wiley.

Stigler, S. M. (1986). The History of Statistics: The Measurement of Uncertainty before 1900. Cambridge, Mass: Belknap Press of the Harvard University Press.

Storer, B. E. & Kim, C. (1990). Exact properties of some exact test statistics for comparing two binomial proportions. Journal of the American Statistical Association, 85, 146–155.

Tukey, J. W. & McLaughlin D. H. (1963). Less vulnerable confidence and significance procedures for location based on a single sample: Trimming/Winsorization 1. Sankhya A, 25, 331–352.

Wald, A. (1955). Testing the difference between the means of two normal populations with unknown standard deviations. In T. W. Anderson et al. (Eds.), Selected Papers in Statistics and Probability by Abraham Wald. New York: McGraw-Hill.

Weerahandi, S. (1995). ANOVA under unequal error variances. Biometrics, 51 , 589–599.

Welch, B. L. (1938). The significance of the difference between two means when the population variances are unequal. Biometrika, 29, 350–362.

Welch, B. L. (1951). On the comparison of several mean values: An alternative approach. Biometrika, 38, 330–336.

Wilcox, R. R. (1990). Comparing the means of two independent groups. Biometrical Journal, 32, 771–780.

Wilcox, R. R. (2003). Applying Contemporary Statistical Techniques. San Diego: Academic Press.

Wilcox, R. R. (2005). Introduction to Robust Estimation and Hypothesis Testing. 2nd Edition. San Diego: Academic Press.

Wilcox, R. R. (2006). Comparing medians. Computational Statistics & Data Analysis, 51, 1934–1943.

Wilcox, R. R., Charlin, V. & Thompson, K. L. (1986). New Monte Carlo results on the robustness of the ANOVA F, W, and F^* statistics. Communications in Statistics—Simulation and Computation, 15, 933–944.

Wilcoxon, F. (1945). Individual comparisons by ranking methods. Biometrics, 1, 80–83.

Wu, C. F. J. (1986). Jackknife, bootstrap, and other resampling methods in regression analysis. The Annals of Statistics, 14, 1261–1295.

Wu, P.-C. (2002). Central limit theorem and comparing means, trimmed means one-step M-estimators and modified one-step M-estimators under non-normality. Unpublished doctoral disseration, Dept. of Education, University of Southern California.

Yuen, K. K. (1974). The two sample trimmed t for unequal population variances. Biometrika, 61, 165–170.

Zimmerman, D. W. (2004). A note on preliminary tests of equality of variances. British Journal of Mathematical and Statistical Psychology, 57, 173–182.

Index